Fundamental Constants

Quantity	Symbol	Approximate Value	Current Best Value[†]
Speed of light in vacuum	c	3.00×10^8 m/s	2.99792458×10^8 m/s
Gravitational constant	G	6.67×10^{-11} N·m²/kg²	$6.67259(85) \times 10^{-11}$ N·m²/kg²
Avogadro's number	N_A	6.02×10^{23} mol⁻¹	$6.0221367(36) \times 10^{23}$ mol⁻¹
Gas constant	R	8.315 J/mol·K $= 1.99$ cal/mol·K $= 0.082$ atm·liter/mol·K	$8.314510(70)$ J/mol·K
Boltzmann's constant	k	1.38×10^{-23} J/K	$1.380658(12) \times 10^{-23}$ J/K
Charge on electron	e	1.60×10^{-19} C	$1.60217733(49) \times 10^{-19}$ C
Stefan-Boltzmann constant	σ	5.67×10^{-8} W/m²·K⁴	$5.67051(19) \times 10^{-8}$ W/m²·K⁴
Permittivity of free space	$\epsilon_0 = (1/c^2\mu_0)$	8.85×10^{-12} C²/N·m²	$8.854187817\ldots \times 10^{-12}$ C²/N·m²
Permeability of free space	μ_0	$4\pi \times 10^{-7}$ T·m/A	$1.2566370614\ldots \times 10^{-6}$ T·m/A
Planck's constant	h	6.63×10^{-34} J·s	$6.6260755(40) \times 10^{-34}$ J·s
Electron rest mass	m_e	9.11×10^{-31} kg $= 0.000549$ u $= 0.511$ MeV/c^2	$9.1093897(54) \times 10^{-31}$ kg $= 5.48579903(13) \times 10^{-4}$ u
Proton rest mass	m_p	1.6726×10^{-27} kg $= 1.00728$ u $= 938.3$ MeV/c^2	$1.6726231(10) \times 10^{-27}$ kg $= 1.007276470(12)$ u
Neutron rest mass	m_n	1.6749×10^{-27} kg $= 1.008665$ u $= 939.6$ MeV/c^2	$1.6749286(10) \times 10^{-27}$ kg $= 1.008664904(14)$ u
Atomic mass unit (1 u)		1.6605×10^{-27} kg $= 931.5$ MeV/c^2	$1.6605402(10) \times 10^{-27}$ kg $= 931.49432(28)$ MeV/c^2

[†]Reviewed 1993 by B. N. Taylor, National Institute of Standards and Technology. Numbers in parentheses indicate one standard deviation experimental uncertainties in final digits. Values without parentheses are exact (i.e., defined quantities).

Other Useful Data

Joule equivalent (1 cal)	4.186 J
Absolute zero (0 K)	–273.15°C
Earth: Mass	5.97×10^{24} kg
Radius (mean)	6.38×10^3 km
Moon: Mass	7.35×10^{22} kg
Radius (mean)	1.74×10^3 km
Sun: Mass	1.99×10^{30} kg
Radius (mean)	6.96×10^5 km
Earth-sun distance (mean)	149.6×10^6 km
Earth-moon distance (mean)	384×10^3 km

The Greek Alphabet

Alpha	A	α	Nu	N	ν
Beta	B	β	Xi	Ξ	ξ
Gamma	Γ	γ	Omicron	O	o
Delta	Δ	δ	Pi	Π	π
Epsilon	E	ε	Rho	P	ρ
Zeta	Z	ζ	Sigma	Σ	σ
Eta	H	η	Tau	T	τ
Theta	Θ	θ	Upsilon	Y	υ
Iota	I	ι	Phi	Φ	ϕ, φ
Kappa	K	κ	Chi	X	χ
Lambda	Λ	λ	Psi	Ψ	ψ
Mu	M	μ	Omega	Ω	ω

Values of Some Numbers

$\pi = 3.1415927$	$\sqrt{2} = 1.4142136$	$\ln 2 = 0.6931472$	$\log_{10} e = 0.4342945$
$e = 2.7182818$	$\sqrt{3} = 1.7320508$	$\ln 10 = 2.3025851$	1 rad $= 57.2957795°$

Mathematical Signs and Symbols

$\propto$	is proportional to		$\leq$	is less than or equal to
$=$	is equal to		$\geq$	is greater than or equal to
$\approx$	is approximately equal to		Σ	sum of
$\neq$	is not equal to		$\bar{x}$	average value of x
$>$	is greater than		Δx	change in x
$\gg$	is much greater than		$\Delta x \to 0$	Δx approaches zero
$<$	is less than		$n!$	$n(n-1)(n-2)\ldots(1)$
$\ll$	is much less than			

Unit Conversions (Equivalents)

Length

1 in. = 2.54 cm
1 cm = 0.394 in.
1 ft = 30.5 cm
1 m = 39.37 in. = 3.28 ft
1 mi = 5280 ft = 1.61 km
1 km = 0.621 mi
1 nautical mile (U.S.) = 1.15 mi = 6076 ft = 1.852 km
1 fermi = 1 femtometer (fm) = 10^{-15} m
1 angstrom (Å) = 10^{-10} m
1 light-year (ly) = 9.46×10^{15} m
1 parsec = 3.26 ly = 3.09×10^{16} m

Volume

1 liter (L) = 1000 mL = 1000 cm^3 = 1.0×10^{-3} m^3 = 1.057 quart (U.S.) = 54.6 in.^{3}
1 gallon (U.S.) = 4 qt (U.S.) = 231 in.^{3} = 3.78 L = 0.83 gal (Imperial)
1 m^3 = 35.31 ft^3

Speed

1 mi/h = 1.47 ft/s = 1.609 km/h = 0.447 m/s
1 km/h = 0.278 m/s = 0.621 mi/h
1 ft/s = 0.305 m/s = 0.682 mi/h
1 m/s = 3.28 ft/s = 3.60 km/h
1 knot = 1.151 mi/h = 0.5144 m/s

Angle

1 radian (rad) = 57.30° = 57°18′
1° = 0.01745 rad
1 rev/min (rpm) = 0.1047 rad/s

Time

1 day = 8.64×10^4 s
1 year = 3.156×10^7 s

Mass

1 atomic mass unit (u) = 1.6605×10^{-27} kg
1 kg = 0.0685 slug
[1 kg has a weight of 2.20 lb where $g = 9.81$ m/s^2.]

Force

1 lb = 4.45 N
1 N = 10^5 dyne = 0.225 lb

Energy and Work

1 J = 10^7 ergs = 0.738 ft·lb
1 ft·lb = 1.36 J = 1.29×10^{-3} Btu = 3.24×10^{-4} kcal
1 kcal = 4.18×10^3 J = 3.97 Btu
1 eV = 1.602×10^{-19} J
1 kWh = 3.60×10^6 J = 860 kcal

Power

1 W = 1 J/s = 0.738 ft·lb/s = 3.42 Btu/h
1 hp = 550 ft·lb/s = 746 W

Pressure

1 atm = 1.013 bar = 1.013×10^5 N/m^2
= 14.7 lb/in.^{2} = 760 torr
1 lb/in.^{2} = 6.90×10^3 N/m^2
1 Pa = 1 N/m^2 = 1.45×10^{-4} lb/in.^{2}

SI Derived Units and Their Abbreviations

Quantity	Unit	Abbreviation	In Terms of Base Units[†]
Force	newton	N	kg·m/s^2
Energy and work	joule	J	kg·m^2/s^2
Power	watt	W	kg·m^2/s^3
Pressure	pascal	Pa	kg/(m·s^2)
Frequency	hertz	Hz	s^{-1}
Electric charge	coulomb	C	A·s
Electric potential	volt	V	kg·m^2/(A·s^3)
Electric resistance	ohm	Ω	kg·m^2/(A^2·s^3)
Capacitance	farad	F	A^2·s^4/(kg·m^2)
Magnetic field	tesla	T	kg/(A·s^2)
Magnetic flux	weber	Wb	kg·m^2/(A·s^2)
Inductance	henry	H	kg·m^2/(s^2·A^2)

[†]kg = kilogram (mass), m = meter (length), s = second (time), A = ampere (electric current).

Metric (SI) Multipliers

Prefix	Abbreviation	Value
exa	E	10^{18}
peta	P	10^{15}
tera	T	10^{12}
giga	G	10^9
mega	M	10^6
kilo	k	10^3
hecto	h	10^2
deka	da	10^1
deci	d	10^{-1}
centi	c	10^{-2}
milli	m	10^{-3}
micro	μ	10^{-6}
nano	n	10^{-9}
pico	p	10^{-12}
femto	f	10^{-15}
atto	a	10^{-18}

PHYSICS
for
SCIENTISTS & ENGINEERS

Part 4
PHYSICS
for
SCIENTISTS & ENGINEERS
Third Edition

DOUGLAS C. GIANCOLI

PRENTICE HALL
Upper Saddle River, New Jersey 07458

Editor-in-Chief: Paul F. Corey
Production Editor: Susan Fisher
Executive Editor: Alison Reeves
Development Editor: David Chelton
Director of Marketing: John Tweedale
Senior Marketing Manager: Erik Fahlgren
Assistant Vice President of Production and Manufacturing: David W. Riccardi
Executive Managing Editor: Kathleen Schiaparelli
Manufacturing Manager: Trudy Pisciotti
Art Manager: Gus Vibal
Director of Creative Services: Paul Belfanti
Advertising and Promotions Manager: Elise Schneider
Editor in Chief of Development: Ray Mullaney
Project Manager: Elizabeth Kell
Photo Research: Mary Teresa Giancoli
Photo Research Administrator: Melinda Reo
Copy Editor: Jocelyn Phillips
Editorial Assistant: Marilyn Coco
Cover photo: Onne van der Wal/Young America
Composition: Emilcomp srl / Preparé Inc.

© 2000, 1989, 1984 by Douglas C. Giancoli
Published by Prentice Hall
Upper Saddle River, NJ 07458

Printed in the United States of America

10 9 8 7 6 5 4 3 2 1

ISBN   0-13-029097-1

Prentice-Hall International (UK) Limited, *London*
Prentice-Hall of Australia Pty. Limited, *Sydney*
Prentice-Hall Canada Inc., *Toronto*
Prentice-Hall Hispanoamericana, S.A., *Mexico City*
Prentice-Hall of India Private Limited, *New Delhi*
Prentice-Hall of Japan, Inc., *Tokyo*
Prentice-Hall (*Singapore*) Pte. Ltd.
Editora Prentice-Hall do Brasil, Ltda., *Rio de Janeiro*

CONTENTS

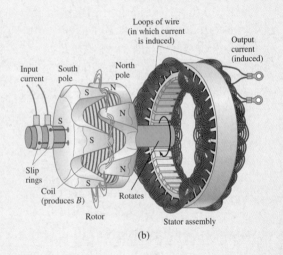

Loops of wire
(in which current
is induced)

Output
current
(induced)

Input
current

South
pole

North
pole

S

N

S

N

S

N

S

Slip
rings

Coil
(produces B)

Rotates

Rotor

Stator assembly

(b)

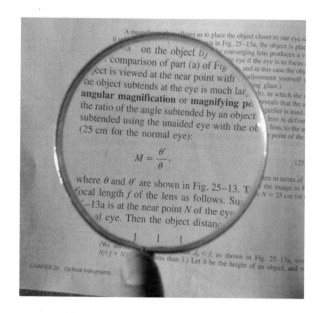

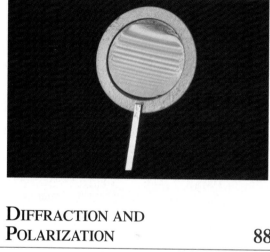

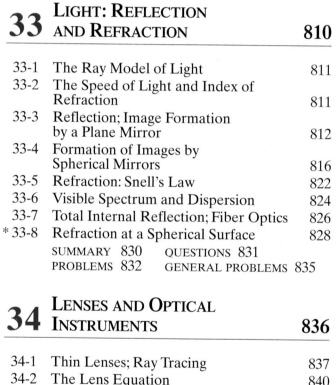

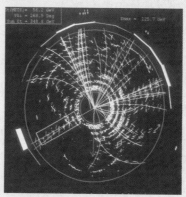

APPENDICES

PREFACE

A Brand New Third Edition

It has been more than ten years since the second edition of this calculus-based introductory physics textbook was published. A lot has changed since then, not only in physics itself, but also in how physics is presented. Research in how students learn has provided textbook authors new opportunities to help students learn physics and learn it well.

This third edition comes in two versions. The standard version covers all of classical physics plus a chapter on special relativity and one on the early quantum theory. The extended version, with modern physics, contains a total of nine detailed chapters on modern physics, ending with astrophysics and cosmology. This book retains the original approach: in-depth physics, concrete and nondogmatic, readable.

This new third edition has many improvements in the physics and its applications. Before discussing those changes in detail, here is a list of some of the overall changes that will catch the eye immediately.

Full color throughout is not just cosmetic, although fine color photographs do help to attract the student readers. More important, full color diagrams allow the physics to be displayed with much greater clarity. We have not stopped at a 4-color process; this book has actually been printed in 5 pure colors (5 passes through the presses) to provide better variety and definition for illustrating vectors and other physics concepts such as rays and fields. I want to emphasize that color is used pedagogically to bring out the physics. For example, different types of vectors are given different colors—see the chart on page xxxi.

Many more diagrams, almost double the number in the previous edition, have all been done or redone carefully using full color; there are many more graphs and many more photographs throughout. See for example in optics where new photographs show lenses and the images they make.

Marginal notes have been added as an aid to students to (i) point out what is truly important, (ii) serve as a sort of outline, and (iii) help students find details about something referred to later that they may not remember so well. Besides such "normal" marginal notes, there are also marginal notes that point out brief *problem solving* hints, and others that point out interesting *applications*.

The great laws of physics are emphasized by giving them a marginal note all in capital letters and enclosed in a rectangle. The most important equations, especially those expressing the great laws, are further emphasized by a tan-colored screen behind them.

Chapter opening photographs have been chosen to illustrate aspects of each chapter. Each was chosen with an eye to writing a caption which could serve as a kind of summary of what is in that chapter, and sometimes offer a challenge. Some chapter-opening photos have vectors or other analysis superimposed on them.

Page layout: complete derivations. Serious attention has been paid to how each page was formatted, especially for page turns. Great effort has been made to keep important derivations and arguments on facing pages. Students then don't have to turn back to check. More important, readers repeatedly see before them, on two facing pages, an important slice of physics.

Two new kinds of Examples: Conceptual Examples and **Estimates.**

New Physics

The whole idea of a new edition is to improve, to bring in new material, and to delete material that is verbose and only makes the book longer or is perhaps too advanced and not so useful. Here is a brief summary of a few of the changes involving the physics iself. These lists are selections, not complete lists.

New discoveries:
- planets revolving around distant stars
- Hubble Space Telescope
- updates in particle physics and cosmology, such as inflation and the age of the universe

New physics topics added:
- new treatment of how to make estimates (Chapter 1), including new Estimating Examples throughout (in Chapter 1, estimating the volume of a lake, and the radius of the Earth)
- symmetry used much more, including for solving problems
- new Tables illustrating the great range of lengths, time intervals, masses, voltages
- gravitation as curvature of space, and black holes (Chapter 6)
- engine efficiency (Chapter 8 as well as Chapter 20)
- rolling with and without slipping, and other useful details of rotational motion (Chapter 10)
- forces in structures including trusses, bridges, arches, and domes (Chapter 12)
- square wave (Chapter 15)
- using the Maxwell distribution (Chapter 18)
- Otto cycle (Chapter 20)
- statistical calculation of entropy change in free expansion (Chapter 20)
- effects of dielectrics on capacitor connected and not (Chapter 24)
- grounding to avoid electric hazards (Chapter 25)
- three phase ac (Chapter 31)
- equal energy in **E** and **B** of EM wave (Chapter 32)
- radiation pressure, EM wave (Chapter 32)
- photos of lenses and mirrors with their images (Chapter 33)
- detailed outlines for ray tracing with mirrors and lenses (Chapters 33, 34)
- lens combinations (Chapter 34)
- new radiation standards (Chapter 43)
- Higgs boson, supersymmetry (Chapter 44)

Modern physics. A number of modern physics topics are discussed in the framework of classical physics. Here are some highlights:
- gravitation as curvature of space, and black holes (Chapter 6)
- planets revolving around distant stars (Chapter 6)
- kinetic energy at relativistic speeds (Chapter 7)
- nuclear collisions (Chapter 9)
- star collapse (Chapter 10)
- galaxy red shift, Doppler (Chapter 16)
- atoms, theory of (Chapters 17, 18, 21)
- atomic theory of thermal expansion (Chapter 17)
- mass of hydrogen atom (Chapter 17)
- atoms and molecules in gases (Chapters 17, 18)
- molecular speeds (Chapter 18)
- equipartition of energy; molar specific heats (Chapter 19)
- star size (Chapter 19)
- molecular dipoles (Chapters 21, 23)
- cathode ray tube (Chapters 23, 27)
- electrons in a wire (Chapter 25)
- superconductivity (Chapter 25)
- discovery and properties of the electron, e/m, oil drop experiment (Chapter 27)
- Hall effect (Chapter 27)

- magnetic moment of electrons (Chapter 27)
- mass spectrometer (Chapter 27)
- velocity selector (Chapter 27)
- electron spin in magnetic materials (Chapter 28)
- light and EM wave emission (Chapter 32)
- spectroscopy (Chapter 36)

Many other examples of modern physics are found as Problems, even in early chapters. Chapters 37 and 38 contain the modern physics topics of Special Relativity, and an introduction to Quantum Theory and Models of the Atom. The longer version of this text, "with Modern Physics," contains an additional seven chapters (for a total of nine) which present a detailed and extremely up-to-date treatment of modern physics: Quantum Mechanics of Atoms (Chapters 38 to 40); Molecules and Condensed Matter (Chapter 41); Nuclear Physics (Chapter 42 and 43); Elementary Particles (Chapter 44); and finally Astrophysics, General Relativity, and Cosmology (Chapter 45).

Revised physics and reorganizations. First of all, a major effort has been made to not throw everything at the students in the first few chapters. The basics have to be learned first; many aspects can come later, when the students are more prepared. Secondly, a great part of this book has been rewritten to make it clearer and more understandable to students. Clearer does not always mean simpler or easier. Sometimes making it "easier" actually makes it harder to understand. Often a little more detail, without being verbose, can make an explanation clearer. Here are a few of the changes, big and small:

- new graphs and diagrams to clarify velocity and acceleration; deceleration carefully treated.
- unit conversion now a new Section in Chapter 1, instead of interrupting kinematics.
- circular motion: Chapter 3 now gives only the basics, with more complicated treatment coming later: non-uniform circular motion in Chapter 5, angular variables in Chapter 10.
- Newton's second law now written throughout as $m\mathbf{a} = \Sigma\mathbf{F}$, to emphasize inclusion of all forces acting on a body.
- Newton's third law follows the second directly, with inertial reference frames placed earlier. New careful discussions to head off confusion when using Newton's third law.
- careful rewriting of chapters on Work and Energy, especially potential energy, conservative and nonconservative forces, and the conservation of energy.
- renewed emphasis that $\Sigma\tau = I\alpha$ is not always valid: only for an axis fixed in an inertial frame or if axis is through the CM (Chapters 10 and 11).
- rolling motion introduced early in Chapter 10, with more details later, including rolling with and without slipping.
- rotating frames of reference, and Coriolis, moved later, to Chapter 11, shortened, optional, but still including why an object does not fall straight down on Earth.
- fluids reduced to a single chapter (13); some topics and details dropped or greatly shortened.
- clearer details on how an object floats (Chapter 13).
- distinction between wave interference in space, and in time (beats) (Chapter 16).
- thermodynamics reduced to four chapters; the old chapters on Heat and on the First Law of Thermodynamics have been combined into one (19), with some topics shortened and a more rational sequence of topics achieved.
- heat transfer now follows the first law of thermodynamics (Chapter 19).
- electric potential carefully rewritten for accuracy (Chapter 23).
- CRT, computer monitors, TV, treated earlier (Chapter 23).
- use of Q_{encl} and I_{encl} for Gauss's and Ampère's laws, with subscripts meaning "enclosed".
- Ohm's law and definition of resistance carefully redone (Chapter 25).
- sources of magnetic field, Chapter 28, reorganized for ease of understanding, with some new material, and deletion of the advanced topic on magnetization vector.
- circuits with L, C, and/or R now introduced via Kirchhoff's loop rule, and clarified in other ways too (Chapters 30, 31).
- streamlined Maxwell's equations, with displacement current downplayed (Chapter 32).
- optics reduced to four chapters; polarization is now placed in the same chapter as diffraction.

New Pedagogy

All of the above mentioned revisions, rewritings, and reorganizations are intended to help students learn physics better. They were done in response to contemporary research in how students learn, as well as to kind and generous input from professors who have read, reviewed, or used the previous editions. This new edition also contains some new elements, especially an increased emphasis on conceptual development:

Conceptual Examples, typically 1 or 2 per chapter, sometimes more, are each a sort of brief Socratic question and answer. It is intended that students will be stimulated by the question to think, or reflect, and come up with a response—before reading the Response given. Here are a few:

- using symmetry (Chapters 1, 44, and elsewhere)
- ball moving upward: misconceptions (Chapter 2)
- reference frames and projectile motion: where does the apple land? (Chapter 3)
- what exerts the force that makes a car move? (Chapter 4)
- Newton's third law clarification: pulling a sled (Chapter 4)
- free-body diagram for a hockey puck (Chapter 4)
- advantage of a pulley (Chapter 4), and of a lever (Chapter 12)
- to push or to pull a sled (Chapter 5)
- which object rolls down a hill faster? (Chapter 10)
- moving the axis of a spinning wheel (Chapter 11)
- tragic collapse (Chapter 12)
- finger at top of a full straw (Chapter 13)
- suction cups on a spacecraft (Chapter 13)
- doubling amplitude of SHM (Chapter 14)
- do holes expand thermally? (Chapter 17)
- simple adiabatic process: stretching a rubber band (Chapter 19)
- charge inside a conductor's cavity (Chapter 22)
- how stretching a wire changes its resistance (Chapter 25)
- series or parallel (Chapter 26)
- bulb brightness (Chapter 26)
- spiral path in magnetic field (Ch. 27)
- practice with Lenz's law (Chapter 29)
- motor overload (Chapter 29)
- emf direction in inductor (Chapter 30)
- photo with reflection—is it upside down? (Chapter 33)
- reversible light rays (Chapter 33)
- how tall must a full-length mirror be? (Chapter 33)
- diffraction spreading (Chapter 36)

Estimating Examples, roughly 10% of all Examples, also a new feature of this edition, are intended to develop the skills for making order-of-magnitude estimates, even when the data are scarce, and even when you might never have guessed that any result was possible at all. See, for example, Section 1–6, Examples 1–5 to 1–8.

Problem Solving, with New and Improved Approaches

Learning how to approach and solve problems is a basic part of any physics course. It is a highly useful skill in itself, but is also important because the process helps bring understanding of the physics. Problem solving in this new edition has a significantly increased emphasis, including some new features.

Problem-solving boxes, about 20 of them, are new to this edition. They are more concentrated in the early chapters, but are found throughout the book. They each outline a step-by-step approach to solving problems in general, and/or specifically for the material being covered. The best students may find these separate "boxes" unnecessary (they can skip them), but many students will find it helpful to be reminded of the general approach and of steps they can take to get started; and, I think, they help to build confidence. The general problem solving box in Section 4–8 is placed there, after students have had some experience wrestling with problems, and so may be strongly motivated to read it with close attention. Section 4–8 can, of course, be covered earlier if desired.

Problem-solving Sections occur in many chapters, and are intended to provide extra drill in areas where solving problems is especially important or detailed.

Examples. This new edition has many more worked-out Examples, and they all now have titles for interest and for easy reference. There are even two new categories of Example: Conceptual, and Estimates, as described above. Regular Examples serve as "practice problems". Many new ones have been added, some of the old ones have been dropped, and many have been reworked to provide greater clarity and detail: more steps are spelled out, more of "why we do it this way", and more discussion of the reasoning and approach. In sum, the idea is "to think aloud with the students", leading them to develop insight. The total number of worked-out Examples is about 30% greater than in the previous edition, for an average of 12 to 15 per chapter. There is a significantly higher concentration of Examples in the early chapters, where drill is especially important for developing skills and a variety of approaches. The level of the worked-out Examples for most topics increases gradually, with the more complicated ones being on a par with the most difficult Problems at the end of each chapter, so that students can see how to approach complex problems. Many of the new Examples, and improvements to old ones, provide relevant applications to engineering, other related fields, and to everyday life.

Problems at the end of each chapter have been greatly increased in quality and quantity. There are over 30% more Problems than in the second edition. Many of the old ones have been replaced, or rewritten to make them clearer, and/or have had their numerical values changed. Each chapter contains a large group of Problems arranged by Section and graded according to difficulty: level I Problems are simple, designed to give students confidence; level II are "normal" Problems, providing more of a challenge and often the combination of two different concepts; level III are the most complex, typically combining different issues, and will challenge even superior students. The arrangement by Section number means only that those Problems depend on material up to and including that Section: earlier material may also be relied upon. The ranking of Problems by difficulty (I, II, III) is intended only as a guide.

General Problems. About 70% of Problems are ranked by level of difficulty (I, II, III) and arranged by Section. New to this edition are General Problems that are unranked and grouped together at the end of each chapter, and account for about 30% of all problems. The average total number of Problems per chapter is about 90. Answers to odd-numbered Problems are given at the back of the book.

Complete Physics Coverage, with Options

This book is intended to give students the opportunity to obtain a thorough background in all areas of basic physics. There is great flexibility in choice of topics so that instructors can choose which topics they cover and which they omit. Sections marked with an asterisk can be considered optional, as discussed more fully on p. xxv. Here I want to emphasize that topics not covered in class can still be read by serious students for their own enrichment, either immediately or later. Here is a partial list of physics topics, not the standard ones, but topics that might not usually be covered, and that represent how thorough this book is in its coverage of basic physics. Section numbers are given in parentheses.

- use of calculus; variable acceleration (2–8)
- nonuniform circular motion (5–4)
- velocity-dependent forces (5–5)
- gravitational versus inertial mass; principle of equivalence (6–8)
- gravitation as curvature of space; black holes (6–9)
- kinetic energy at very high speed (7–5)
- potential energy diagrams (8–9)
- systems of variable mass (9–10)
- rotational plus translational motion (10–11)
- using $\Sigma\tau_{CM} = I_{CM}\alpha_{CM}$ (10–11)
- derivation of $K = K_{CM} + K_{rot}$ (10–11)
- why does a rolling sphere slow down? (10–12)
- angular momentum and torque for a system (11–4)
- derivation of $d\mathbf{L}_{CM}/dt = \Sigma\boldsymbol{\tau}_{CM}$ (11–4)
- rotational imbalance (11–6)
- the spinning top (11–8)
- rotating reference frames; inertial forces (11–9)
- coriolis effect (11–10)
- trusses (12–7)
- flow in tubes: Poiseuille's equation (13–11)
- surface tension and capillarity (13–12)
- physical pendulum; torsion pendulum (14–6)
- damped harmonic motion: finding the solution (14–7)
- forced vibrations; equation of motion and its solution; Q-value (14–8)
- the wave equation (15–5)
- mathematical representation of waves; pressure wave derivation (16–2)
- intensity of sound related to amplitude (16–3)
- interference in space and in time (16–6)
- atomic theory of expansion (17–4)
- thermal stresses (17–5)
- ideal gas temperature scale (17–10)
- calculations using the Maxwell distribution of molecular speeds (18–2)
- real gases (18–3)
- vapor pressure and humidity (18–4)
- van der Waals equation of state (18–5)
- mean free path (18–6)
- diffusion (18–7)
- equipartition of energy (19–8)
- energy availability; heat death (20–8)
- statistical interpretation of entropy and the second law (20–9)
- thermodynamic temperature scale; absolute zero and the third law (20–10)
- electric dipoles (21–11, 23–6)
- experimental basis of Gauss's and Coulomb's laws (22–4)
- general relation between electric potential and electric field (23–2, 23–8)
- electric fields in dielectrics (24–5)
- molecular description of dielectrics (24–6)
- current density and drift velocity (25–8)
- superconductivity (25–9)
- RC circuits (26–4)
- use of voltmeters and ammeters; effects of meter resistance (26–5)
- transducers (26–6)
- magnetic dipole moment (27–5)
- Hall effect (27–8)
- operational definition of the ampere and coulomb (28–3)
- magnetic materials—ferromagnetism (28–7)
- electromagnets and solenoids (28–8)
- hysteresis (28–9)
- paramagnetism and diamagnetism (28–10)
- counter emf and torque; eddy currents (29–5)
- Faraday's law—general form (29–7)
- force due to changing $\mathbf{B}$ is nonconservative (29–7)
- LC circuits and EM oscillations (30–5)
- AC resonance; oscillators (31–6)
- impedance matching (31–7)
- three phase AC (31–8)
- changing electric fields produce magnetic fields (32–1)
- speed of light from Maxwell's equations (32–5)
- radiation pressure (32–8)
- fiber optics (33–7)
- lens combinations (34–3)
- aberrations of lenses and mirrors (34–10)
- coherence (35–4)
- intensity in double-slit pattern (35–5)
- luminous intensity (35–8)
- intensity for single-slit (36–2)
- diffraction for double–slit (36–3)
- limits of resolution, the λ limit (36–4, 36–5)
- resolution of the human eye and useful magnification (36–6)
- spectroscopy (36–8)
- peak widths and resolving power for a diffraction grating (36–9)
- x-rays and x-ray diffraction (36–10)
- scattering of light by the atmosphere (36–12)
- time–dependent Schrödinger equation (39–6)
- wave packets (39–7)
- tunneling through a barrier (39–9)
- free-electron theory of metals (41–6)
- semiconductor electronics (41–9)
- standard model, symmetry, QCD, GUT (44–9, 44–10)
- astrophysics, cosmology (Ch. 45)

New Applications

Relevant applications to everyday life, to engineering, and to other fields such as geology and medicine, provide students with motivation and offer the instructor the opportunity to show the relevance of physics. Applications are a good response to students who ask "Why study physics?" Many new applications have been added in this edition. Here are some highlights:

- airbags (Chapter 2)
- elevator and counterweight (Chapter 4)
- antilock brakes and skidding (Chapter 5)
- geosynchronous satellites (Chapter 6)
- hard drive and bit speed (Chapter 10)
- star collapse (Chapter 10)
- forces within trusses, bridges, arches, domes (Chapter 12)
- the Titanic (Chapter 12)
- Bernoulli's principle: wings, sailboats, TIA, plumbing traps and bypasses (Chapter 13)
- pumps (Chapter 13)
- car springs, shock absorbers, building dampers for earthquakes (Chapter 14)
- loudspeakers (Chapters 14, 16, 27)
- autofocusing cameras (Chapter 16)
- sonar (Chapter 16)
- ultrasound imaging (Chapter 16)
- thermal stresses (Chapter 17)
- R-values, thermal insulation (Ch. 19)
- engines (Chapter 20)
- heat pumps, refrigerators, AC; coefficient of performance (Chapter 20)
- thermal pollution (Chapter 20)
- electric shielding (Chapters 21, 28)
- photocopier (Chapter 21)
- superconducting cables (Chapter 25)
- jump starting a car (Chapter 26)
- aurora borealis (Chapter 27)
- solenoids and electromagnetics (Ch. 28)
- computer memory and digital information (Chapter 29)
- seismograph (Chapter 29)

- tape recording (Chapter 29)
- loudspeaker cross-over network (Ch. 31)
- antennas, for **E** or **B** (Chapter 32)
- TV and radio; AM and FM (Chapter 32)
- eye and corrective lenses (Chapter 34)
- mirages (Chapter 35)
- liquid crystal displays (Chapter 36)
- CAT scans, PET, MRI (Chapter 43)

Some old favorites retained (and improved):

- pressure gauges (Chapter 13)
- musical instruments (Chapter 16)
- humidity (Chapter 18)
- CRT, TV, computer monitors (Ch. 23, 27)
- electric hazards (Chapter 25)
- power in household circuits (Chapter 25)
- ammeters and voltmeters (Chapter 26)
- microphones (Chapters 26, 29)
- transducers (Chapter 26, and elsewhere)
- electric motors (Chapter 27)
- car alternator (Chapter 29)
- electric power transmission (Chapter 29)
- capacitors as filters (Chapter 31)
- impedance matching (Chapter 31)
- fiber optics (Chapter 33)
- cameras, telescopes, microscopes, other optical instruments (Chapter 34)
- lens coatings (Chapter 35)
- spectroscopy (Chapter 36)
- electron microscopes (Chapter 38)
- lasers, holography, CD players (Ch. 40)
- semiconductor electronics (Chapter 41)
- radioactivity (Chapters 42 and 43)

Deletions

Something had to go, or the book would have been too long. Lots of subjects were shortened—the detail simply isn't necessary at this level. Some topics were dropped entirely: polar coordinates; center-of-momentum reference frame; Reynolds number (now a Problem); object moving in a fluid and sedimentation; derivation of Poiseuille's equation; Stoke's equation; waveguide and transmission line analysis; electric polarization and electric displacement vectors; potentiometer (now a Problem); negative pressure; combinations of two harmonic motions; adiabatic character of sound waves; central forces.

Many topics have been shortened, often a lot, such as: velocity-dependent forces; variable acceleration; instantaneous axis; surface tension and capillarity; optics topics such as some aspects of light polarizarion. Many of the brief historical and philosophical issues have been shortened as well.

General Approach

This book offers an in-depth presentation of physics, and retains the basic approach of the earlier editions. Rather than using the common, dry, dogmatic approach of treating topics formally and abstractly first, and only later relating the material to the students' own experience, my approach is to recognize that physics is a description of reality and thus to start each topic with concrete observations and experiences that students can directly relate to. Then we move on to the generalizations and more formal treatment of the topic. Not only does this make the material more interesting and easier to understand, but it is closer to the way physics is actually practiced.

This new edition, even more than previous editions, aims to explain the physics in a readable and interesting manner that is accessible and clear. It aims to teach students by anticipating their needs and difficulties, but without oversimplifying. Physics is all about us. Indeed, it is the goal of this book to help students "see the world through eyes that know physics."

As mentioned above, this book includes of a wide range of Examples and applications from technology, engineering, architecture, earth sciences, the environment, biology, medicine, and daily life. Some applications serve only as examples of physical principles. Others are treated in depth. But applications do not dominate the text—this is, after all, a physics book. They have been carefully chosen and integrated into the text so as not to interfere with the development of the physics but rather to illuminate it. You won't find essay sidebars here. The applications are integrated right into the physics. To make it easy to spot the applications, a new *Physics Applied* marginal note is placed in the margin (except where diagrams in the margin prevent it).

It is assumed that students have started calculus or are taking it concurrently. Calculus is treated gently at first, usually in an optional Section so as not to burden students taking calculus concurrently. For example, using the integral in kinematics, Chapter 2, is an optional Section. But in Chapter 7, on work, the integral is discussed fully for all readers.

Throughout the text, *Système International* (SI) units are used. Other metric and British units are defined for informational purposes. Careful attention is paid to significant figures. When a certain value is given as, say, 3, with its units, it is meant to be 3, not assumed to be 3.0 or 3.00. When we mean 3.00 we write 3.00. It is important for students to be aware of the uncertainty in any measured value, and not to overestimate the precision of a numerical result.

Rather than start this physics book with a chapter on mathematics, I have instead incorporated many mathematical tools, such as vector addition and multiplication, directly in the text where first needed. In addition, the Appendices contain a review of many mathematical topics such as trigonometric identities, integrals, and the binomial (and other) expansions. One advanced topic is also given an Appendix: integrating to get the gravitational force due to a spherical mass distribution.

It is necessary, I feel, to pay careful attention to detail, especially when deriving an important result. I have aimed at including all steps in a derivation, and have tried to make clear which equations are general, and which are not, by explicitly stating the limitations of important equations in brackets next to the equation, such as

$$x = x_0 + v_0 t + \tfrac{1}{2} at^2. \qquad \text{[constant acceleration]}$$

The more detailed introduction to Newton's laws and their use is of crucial pedagogic importance. The many new worked-out Examples include initially fairly simple ones that provide careful step-by-step analysis of how to proceed in solving dynamics problems. Each succeeding Example adds a new element or a new twist that introduces greater complexity. It is hoped that this strategy will enable even less-well-prepared students to acquire the tools for using Newton's laws correctly. If students don't surmount this crucial hurdle, the rest of physics may remain forever beyond their grasp.

Rotational motion is difficult for most students. As an example of attention to detail (although this is not really a "detail"), I have carefully distinguished the position vector (**r**) of a point and the perpendicular distance of that point from an axis, which is called R in this book (see Fig. 10–2). This distinction, which enters particularly in connection with torque, moment of inertia, and angular momentum, is often not made clear—it is a disservice to students to use **r** or r for both without distinguishing. Also, I have made clear that it is not always true that $\Sigma\tau = I\alpha$. It depends on the axis chosen (valid if axis is fixed in an inertial reference frame, or through the CM). To not tell this to students can get them into serious trouble. (See pp. 250, 283, 284.) I have treated rotational motion by starting with the simple instance of rotation about an axis (Chapter 10), including the concepts of angular momentum and rotational kinetic energy. Only in Chapter 11 is the more general case of rotation about a point dealt with, and this slightly more advanced material can be omitted if desired (except for Sections 11–1 and 11–2 on the vector product and the torque vector). The end of Chapter 10 has an optional subsection containing three slightly more advanced Examples, using $\Sigma\tau_{CM} = I_{CM}\alpha_{CM}$: car braking distribution, a falling yo-yo, and a sphere rolling with and without slipping.

Among other special treatments is Chapter 28, Sources of Magnetic Field: here, in one chapter, are discussed the magnetic field due to currents (including Ampère's law and the law of Biot-Savart) as well as magnetic materials, ferromagnetism, paramagnetism, and diamagnetism. This presentation is clearer, briefer, and more of a whole, and all the content is there.

Organization

The general outline of this new edition retains a traditional order of topics: mechanics (Chapters 1 to 12); fluids, vibrations, waves, and sound (Chapter 13 to 16); kinetic theory and thermodynamics (Chapters 17 to 20). In the two-volume version of this text, volume I ends here, after Chapter 20. The text continues with electricity and magnetism (Chapters 21 to 32), light (Chapters 33 to 36), and modern physics (Chapters 37 and 38 in the short version, Chapters 37 to 45 in the extended version "with Modern Physics"). Nearly all topics customarily taught in introductory physics courses are included. A number of topics from modern physics are included with the classical physics chapters as discussed earlier.

The tradition of beginning with mechanics is sensible, I believe, because it was developed first, historically, and because so much else in physics depends on it. Within mechanics, there are various ways to order topics, and this book allows for considerable flexibility. I prefer, for example, to cover statics after dynamics, partly because many students have trouble working with forces without motion. Besides, statics is a special case of dynamics—we study statics so that we can prevent structures from becoming dynamic (falling down)—and that sense of being at the limit of dynamics is intuitively helpful. Nonetheless statics (Chapter 12) can be covered earlier, if desired, before dynamics, after a brief introduction to vector addition. Another option is light, which I have placed after electricity and magnetism and EM waves. But light could be treated immediately after the chapters on waves (Chapters 15 and 16). Special relativity is Chapter 37, but could instead be treated along with mechanics—say, after Chapter 9.

Not every chapter need be given equal weight. Whereas Chapter 4 might require $1\frac{1}{2}$ to 2 weeks of coverage, Chapter 16 or 22 may need only $\frac{1}{2}$ week.

Some instructors may find that this book contains more material than can be covered completely in their courses. But the text offers great flexibility in choice of topics. Sections marked with a star (asterisk) are considered optional. These Sections contain slightly more advanced physics material, or material not usually covered in typical courses, and/or interesting applications. They contain no material needed in later chapters (except perhaps in later optional Sections). This does not imply that all nonstarred Sections must be covered: there still remains considerable flexibility in the choice of material. For a brief course, all optional material could be dropped as well as major parts of Chapters 11, 13, 16, 26, 30, 31, and 36 as well as selected parts of Chapters 9, 12, 19, 20, 32, 34, and the modern physics chapters. Topics not covered in class can be a valuable resource for later study; indeed, this text can serve as a useful reference for students for years because of its wide range of coverage.

Thanks

Some 60 physics professors provided input or direct feedback on every aspect of this textbook. The reviewers and contributors to this third edition are listed below. I owe each a debt of gratitude.

Ralph Alexander, University of Missouri at Rolla

Zaven Altounian, McGill University

Charles R. Bacon, Ferris State University

Bruce Birkett, University of California, Berkeley

Art Braundmeier, Southern Illinois University at Edwardsville

Wayne Carr, Stevens Institute of Technology

Edward Chang, University of Massachusetts, Amherst

Charles Chiu, University of Texas at Austin

Lucien Crimaldi, University of Mississippi

Robert Creel, University of Akron

Alexandra Cowley, Community College of Philadelphia

Timir Datta, University of South Carolina

Gary DeLeo, Lehigh University

John Dinardo, Drexel University

Paul Draper, University of Texas, Arlington

Alex Dzierba, Indiana University

William Fickinger, Case Western University

Jerome Finkelstein, San Jose State University

Donald Foster, Wichita State University

Gregory E. Frances, Montana State University

Lothar Frommhold, University of Texas at Austin

Thomas Furtak, Colorado School of Mines

Edward Gibson, California State University, Sacramento

Christopher Gould, University of Southern California

John Gruber, San Jose State University

Martin den Boer, Hunter College

Greg Hassold, General Motors Institute

Joseph Hemsky, Wright State University

Laurent Hodges, Iowa State University

Mark Holtz, Texas Tech University

James P. Jacobs, University of Montana

James Kettler, Ohio University Eastern Campus

Jean Krisch, University of Michigan

Mark Lindsay, University of Louisville

Eugene Livingston, University of Notre Dame

Bryan Long, Columbia State Community College

Daniel Mavlow, Princeton University

Pete Markowitz, Florida International University

John McCullen, University of Arizona, Tucson

Peter Nemeth, New York University

Hon-Kie Ng, Florida State University

Eugene Patroni, Georgia Institute of Technology

Robert Pelcovits, Brown University

William Pollard, Valdosta State University

Joseph Priest, Miami University

Carl Rotter, West Virginia University

Lawrence Rees, Brigham Young University

Peter Riley, University of Texas at Austin

Roy Rubins, University of Texas at Arlington

Mark Semon, Bates College

Robert Simpson, University of New Hampshire

Mano Singham, Case Western University

Harold Slusher, University of Texas at El Paso

Don Sparks, Los Angeles Pierce Community College

Michael Strauss, University of Oklahoma

Joseph Strecker, Wichita State University

William Sturrus, Youngstown State University

Arthur Swift, University of Massachusetts, Amherst

Leo Takahasi, The Pennsylvania State University

Edward Thomas, Georgia Institute of Technology

Som Tyagi, Drexel University

John Wahr, University of Colorado

Robert Webb, Texas A & M University

James Whitmore, The Pennsylvania State University

W. Steve Quon, Ventura College

I owe special thanks to Irv Miller, not only for many helpful physics discussions, but for having worked out all the Problems and managed the team that also worked out the Problems, each checking the other, and finally for producing the Solutions Manual and all the answers to the odd-numbered Problems at the end of this book. He was ably assisted by Zaven Altounian and Anand Batra.

I am particularly grateful to Robert Pelcovits and Peter Riley, as well as to Paul Draper and James Jacobs, who inspired many of the new Examples, Conceptual Examples, and Problems.

Crucial for rooting out errors, as well as providing excellent suggestions, were the perspicacious Edward Gibson and Michael Strauss, both of whom carefully checked all aspects of the physics in page proof.

Special thanks to Bruce Birkett for input of every kind, from illuminating discussions on pedagogy to a careful checking of details in many sections of this book. I wish also to thank Professors Howard Shugart, Joe Cerny, Roger Falcone and Buford Price for helpful discussions, and for hospitality at the University of California, Berkeley. Many thanks also to Prof. Tito Arecchi at the Istituto Nazionale di Ottica, Florence, Italy, and to the staff of the Institute and Museum for the History of Science, Florence, for their hospitality.

Finally, I wish to thank the superb editorial and production work provided by all those with whom I worked directly at Prentice Hall: Susan Fisher, Marilyn Coco, David Chelton, Kathleen Schiaparelli, Trudy Pisciotti, Gus Vibal, Mary Teresa Giancoli, and Jocelyn Phillips.

The biggest thanks of all goes to Paul Corey, whose constant encouragement and astute ability to get things done, provided the single strongest catalyst.

The final responsibility for all errors lies with me, of course. I welcome comments and corrections.

D.C.G.

AVAILABLE SUPPLEMENTS

For the Student

Student Study Guide and Solutions Manual
Douglas Brandt, Eastern Illinois University. (0-13-021475-2)
Contains chapter objectives, summaries with additional examples, self-study quizzes, key mathematical equations, and complete worked-out solutions to alternate odd problems in the text.

Doing Physics with Spreadsheets: A Workbook
Gordon Aubrecht, T. Kenneth Bolland, and Michael Ziegler, all of The Ohio State University.
(0-13-021474-4)
Designed to introduce students to the use of spreadsheets for solving simple and complex physics problems. Students are either provided with spreadsheets or must construct their own, then use the model to most closely approximate natural behavior. The amount of spreadsheet construction and the complexity of the spreadsheet increases as the student gains experience.

Science on the Internet: A Student's Guide, 1999
Andrew Stull and Carl Adler (0-13-021308-X)
The perfect tool to help students take advantage of the *Physics for Scientists and Engineers, Third Edition* Web page. This useful resource gives clear steps to access Prentice Hall's regularly updated physics resources, along with an overview of general World Wide Web navigation strategies. Available FREE for students when packaged with the text.

Prentice Hall/*New York Times* Themes of the Times — Physics
This unique newspaper supplement brings together a collection of the latest physics-related articles from the pages of *The New York Times*. Updated twice per year and available FREE to students when packaged with the text.

For the Instructor

Instructor's Solutions Manual
Irvin A. Miller, Drexel University.
Print version (0-13-021381-0); Electronic (CD-ROM) version (0-13-021481-7)
Contains detailed worked solutions to every problem in the text. Electronic versions are available in CD-ROM (dual platform for both Windows and Macintosh systems) for instructors with Microsoft Word or Word-compatible software.

Test Item File
Robert Pelcovits, Brown University; David Curott, University of North Alabama; and Edward Oberhofer, University of North Carolina at Charlotte (0-13-021482-5)
Contains over 2200 multiple choice questions, about 25% conceptual in nature. All are referenced to the corresponding Section in the text and ranked by difficulty.

Prentice Hall Custom Test Windows (0-13-021477-9); Macintosh (0-13-021476-0)
Based on the powerful testing technology developed by Engineering Software Associates, Inc. (ESA), Prentice Hall Custom Test includes all questions from the Test Item File and allows instructors to create and tailor exams to their own needs. With the Online Testing Program, exams can also be administered on line and data can then be automatically transferred for evaluation. A comprehensive desk reference guide is included along with online assistance.

Transparency Pack (0-13-021470-1)
Includes approximately 400 full color transparencies of images from the text.

Media Supplements

Physics for Scientists and Engineers Web Site www.prenhall.com/giancoli
A FREE innovative online resource that provides students with a wealth of activities and exercises for each text chapter. Features on the site include:

- Practice Questions, Destinations (links to related sites), NetSearch keywords and algorithmically generated numeric Practice Problems by Carl Adler of East Carolina University.
- Physlet Problems (Java-applet simulations) by Wolfgang Christian of Davidson College.
- Warmups and Puzzles essay questions and Applications from Gregor Novak and Andrew Gavrin at Indiana University-Purdue University, Indianapolis.
- Ranking Task Exercises edited by Tom O'Kuma of Lee College, Curtis Hieggelke of Joliet Junior College and David Maloney of Indiana University-Purdue University, Fort Wayne.

Using Prentice Hall CW '99 technology, the website grades and scores all objective questions, and results can be automatically e-mailed directly to the instructors if so desired. Instructors can also create customized syllabi online and link directly to activities on the Giancoli website.

Presentation Manager CD-ROM
Dual Platform (Windows/Macintosh; 0-13-214479-5)
This CD-ROM enables instructors to build custom sequences of Giancoli text images and Prentice Hall digital media for playback in lecture presentations. The CD-ROM contains all text illustrations, digitized segments from the Prentice Hall *Physics You Can See* videotape as well as additional lab and demonstration videos and animations from the Prentice Hall *Interactive Journey Through Physics* CD-ROM. Easy to navigate with Prentice Hall Presentation Manager software, instructors can preview, sequence, and play back images, as well as perform keyword searches, add lecture notes, and incorporate their own digital resources.

Physics You Can See *Video*
(0-205-12393-7)
Contains eleven two- to five-minute demonstrations of classical physics experiments. It includes segments such as "Coin and Feather" (acceleration due to gravity), "Monkey and Gun" (projectile motion), "Swivel Hips" (force pairs), and "Collapse a Can" (atmospheric pressure).

CAPA: A Computer-Assisted Personalized Approach to Assignments, Quizzes, and Exams
CAPA is an on-line homework system developed at Michigan State University that instructors can use to deliver problem sets with randomized variables for each student. The system gives students immediate feedback on their answers to problems, and records their participation and performance. Prentice Hall has arranged to have half of the even-numbered problems of Giancoli, *Physics for Scientists and Engineers, Third Edition*, coded for use with the CAPA system. For additional information about the CAPA system, please visit the web site at http://www.pa.msu.edu/educ/CAPA/.

WebAssign
WebAssign is a web-based homework delivery, collection, grading, and recording service developed and hosted by North Carolina State University. Prentice Hall will arrange for end-of-chapter problems from Giancoli, *Physics for Scientists and Engineers, Third Edition* to be coded for use with the *WebAssign* system for instructors who wish to take advantage of this service. For more information on the *WebAssign* system and its features, please visit http://webassign.net/info or e-mail webassign@ncsu.edu.

NOTES TO STUDENTS AND INSTRUCTORS ON THE FORMAT

1. Sections marked with a star (*) are considered optional. They can be omitted without interrupting the main flow of topics. No later material depends on them except possibly later starred sections. They may be fun to read though.

2. The customary conventions are used: symbols for quantities (such as m for mass) are italicized, whereas units (such as m for meter) are not italicized. Boldface (**F**) is used for vectors.

3. Few equations are valid in all situations. Where practical, the limitations of important equations are stated in square brackets next to the equation. The equations that represent the great laws of physics are displayed with a tan background, as are a few other equations that are so useful that they are indispensable.

4. The number of significant figures (see Section 1–3) should not be assumed to be greater than given: if a number is stated as (say) 6, with its units, it is meant to be 6 and not 6.0 or 6.00.

5. At the end of each chapter is a set of Questions that students should attempt to answer (to themselves at least). These are followed by Problems which are ranked as level I, II, or III, according to estimated difficulty, with level I Problems being easiest. These Problems are arranged by Section, but Problems for a given Section may depend on earlier material as well. There follows a group of General Problems, which are not arranged by Section nor ranked as to difficulty. Questions and Problems that relate to optional Sections are starred.

6. Being able to solve problems is a crucial part of learning physics, and provides a powerful means for understanding the concepts and principles. This book contains many aids to problem solving: (a) worked-out Examples and their solutions in the text, which are set off with a vertical blue line in the margin, and should be studied as an integral part of the text; (b) special "Problem-solving boxes" placed throughout the text to suggest ways to approach problem solving for a particular topic—but don't get the idea that every topic has its own "techniques," because the basics remain the same; (c) special problem-solving Sections (marked in blue in the Table of Contents); (d) "Problem solving" marginal notes (see point 8 below) which refer to hints for solving problems within the text; (e) some of the worked-out Examples are Estimation Examples, which show how rough or approximate results can be obtained even if the given data are sparse (see Section 1–6); and finally (f) the Problems themselves at the end of each chapter (point 5 above).

7. Conceptual Examples look like ordinary Examples but are conceptual rather than numerical. Each proposes a question or two, which hopefully starts you to think and come up with a response. Give yourself a little time to come up with your own response before reading the Response given.

8. Marginal notes: brief notes in the margin of almost every page are printed in blue and are of four types: (a) ordinary notes (the majority) that serve as a sort of outline of the text and can help you later locate important concepts and equations; (b) notes that refer to the great laws and principles of physics, and these are in capital letters and in a box for emphasis; (c) notes that refer to a problem-solving hint or technique treated in the text, and these say "Problem Solving"; (d) notes that refer to an application of physics, in the text or an Example, and these say "Physics Applied."

9. This book is printed in full color. But not simply to make it more attractive. The color is used above all in the Figures, to give them greater clarity for our analysis, and to provide easier learning of the physical principles involved. The Table on the next page is a summary of how color is used in the Figures, and shows which colors are used for the different kinds of vectors, for field lines, and for other symbols and objects. These colors are used consistently throughout the book.

10. Appendices include useful mathematical formulas (such as derivatives and integrals, trigonometric identities, areas and volumes, expansions), and a table of isotopes with atomic masses and other data. Tables of useful data are located inside the front and back covers.

USE OF COLOR

Vectors

A general vector

   resultant vector (sum) is slightly thicker

   components of any vector are dashed

Displacement (**D**, **r**)

Velocity (**v**)

Acceleration (**a**)

Force (**F**)

   Force on second or

   third object in same figure

Momentum (**p** or m**v**)

Angular momentum (**L**)

Angular velocity (ω)

Torque (τ)

Electric field (**E**)

Magnetic field (**B**)

Electricity and magnetism

Electric field lines

Equipotential lines

Magnetic field lines

Electric charge (+)   + or +

Electric charge (−)   − or −

Electric circuit symbols

Wire

Resistor

Capacitor

Inductor

Battery

Optics

Light rays

Object

Real image
(dashed)

Virtual image
(dashed and paler)

Other

Energy level
(atom, etc.)

Measurement lines |←—1.0 m—→|

Path of a moving
object

Direction of motion
or current

Reflection from still water, as from a glass mirror, can be analyzed using the ray model of light.

Is this picture right side up? How can you tell? What are the clues? See Example 33–2.

In this first chapter on light and optics, we will use the ray model of light to understand the formation of images by mirrors, both plane and curved (spherical), and we will begin examining the refraction of light in transparent media, which will lead us to a study of lenses in the next chapter.

Light:
Reflection and Refraction

The sense of sight is extremely important to us, for it provides us with a large part of our information about the world. How do we see? What is the something called *light* that enters our eyes and causes the sensation of sight? How does light behave so that we can see everything that we do? We saw in Chapter 32 that light can be considered a form of electromagnetic radiation. We now examine the subject of light in detail in the next several chapters.

We see an object in one of two ways: (1) the object may be a source of light, such as a lightbulb, a flame, or a star, in which case we see the light emitted directly from the source; or (2), more commonly, we see an object by light reflected from it. In the latter case, the light may have originated from the sun, artificial lights, or a campfire. An understanding of how bodies *emit* light was not achieved until the 1920s, and this will be discussed in Chapter 38. How light is *reflected* from objects was understood much earlier, and will be discussed in Section 33–3.

33–1 | The Ray Model of Light

A great deal of evidence suggests that *light travels in straight lines* under a wide variety of circumstances. For example, a point source of light like the sun casts distinct shadows, and the beam of a flashlight appears to be a straight line. In fact, we infer the positions of objects in our environment by assuming that light moves from the object to our eyes in straight-line paths. Our orientation to the physical world is based on this assumption.

This reasonable assumption has led to the **ray model** of light. This model assumes that light travels in straight-line paths called light **rays**. Actually, a ray is an idealization; it is meant to represent an extremely narrow beam of light. When we see an object, according to the ray model, light reaches our eyes from each point on the object. Although light rays leave each point in many different directions, normally only a small bundle of these rays can enter an observer's eye, as shown in Fig. 33–1. If the person's head moves to one side, a different bundle of rays will enter the eye from each point.

We saw in Chapter 32 that light can be considered as an electromagnetic wave. Although the ray model of light does not deal with this aspect of light (we discuss the wave nature of light in Chapters 35 and 36), the ray model has been very successful in describing many aspects of light such as reflection, refraction, and the formation of images by mirrors and lenses. Because these explanations involve straight-line rays at various angles, this subject is referred to as **geometric optics**. In ignoring the wave properties of light we must be careful that when the light rays pass by objects or through apertures, these must be large compared to the wavelength of the light (so the wave phenomena of interference and diffraction, as discussed in Chapter 15, can be ignored), and we ignore what happens to the light at the edges of objects until we get to Chapters 35 and 36.

Light rays

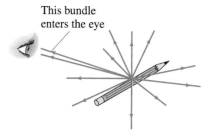

FIGURE 33–1 Light rays come from each single point on an object. A small bundle of rays leaving one point is shown entering a person's eye.

33–2 | The Speed of Light and Index of Refraction

Galileo attempted to measure the speed of light by trying to measure the time required for light to travel a known distance between two hilltops. He stationed an assistant on one hilltop and himself on another and ordered the assistant to lift the cover from a lamp the instant he saw a flash from Galileo's lamp. Galileo measured the time between the flash of his lamp and when he received the light from his assistant's lamp. The time was so short that Galileo concluded it merely represented human reaction time, and that the speed of light must be extremely high.

The first successful determination that the speed of light is finite was made by the Danish astronomer Ole Roemer (1644–1710). Roemer had noted that the carefully measured period of Io, one of Jupiter's moons with an average period of 42.5 h, varied slightly, depending on the relative motion of Earth and Jupiter. When Earth was moving away from Jupiter, the period of the moon was slightly longer, and when Earth was moving toward Jupiter, the period was slightly shorter. He attributed this variation to the extra time needed for light to travel the increasing distance to Earth when Earth is receding, or to the shorter travel time for the decreasing distance when the two planets are approaching one another. Roemer concluded that the speed of light—though great—is finite.

Since then a number of techniques have been used to measure the speed of light. Among the most important were those carried out by the American Albert

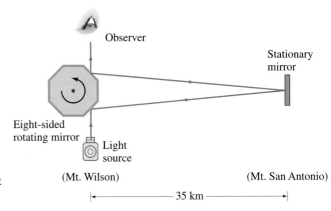

FIGURE 33–2 Michelson's speed-of-light apparatus (not to scale).

Michelson (1852–1931). Michelson used the rotating mirror apparatus diagrammed in Fig. 33–2 for a series of high-precision experiments carried out from 1880 to the 1920s. Light from a source was directed at one face of a rotating eight-sided mirror. The reflected light traveled to a stationary mirror a large distance away and back again as shown. If the rotating mirror was turning at just the right rate the returning beam of light would reflect from one face of the mirror into a small telescope through which the observer looked. If the speed of rotation was only slightly different, the beam would be deflected to one side and would not be seen by the observer. From the required speed of the rotating mirror and the known distance to the stationary mirror, the speed of light could be calculated. In the 1920s, Michelson set up the rotating mirror on the top of Mt. Wilson in southern California and the stationary mirror on Mt. San Antonio, 35 km away. He later measured the speed of light in vacuum using a long evacuated tube.

The accepted value today for the speed of light, c, in vacuum is[‡]

$$c = 2.99792458 \times 10^8 \, \text{m/s}.$$

We usually round this off to

$$3.00 \times 10^8 \, \text{m/s}$$

when extremely precise results are not required. In air, the speed is only slightly less. In other transparent materials such as glass and water, the speed is always less than that in vacuum. For example, in water it travels at about $\frac{3}{4}c$. The ratio of the speed of light in vacuum to the speed v in a given material is called the **index of refraction** n of that material:

$$n = \frac{c}{v}. \tag{33–1}$$

The index of refraction is never less than 1, and its value for various materials is given in Table 33–1. For example, since $n = 2.42$ for diamond, the speed of light in diamond is

$$v = \frac{c}{n} = \frac{(3.00 \times 10^8 \, \text{m/s})}{2.42} = 1.24 \times 10^8 \, \text{m/s}.$$

As we shall see later, n varies somewhat with the wavelength of the light—except in vacuum—so a particular wavelength is specified; in Table 33–1 it is that of yellow light with wavelength $\lambda = 589 \, \text{nm}$.

TABLE 33–1
Indices of refraction[†]

Material	$n = \dfrac{c}{v}$
Vacuum	1.0000
Air (at STP)	1.0003
Water	1.33
Ethyl alcohol	1.36
Glass	
Fused quartz	1.46
Crown glass	1.52
Light flint	1.58
Lucite or Plexiglas	1.51
Sodium chloride	1.53
Diamond	2.42

[†] $\lambda = 589 \, \text{nm}$

Index of refraction

33–3 | Reflection; Image Formation by a Plane Mirror

When light strikes the surface of an object, some of the light is reflected. The rest is either absorbed by the object (and transformed to thermal energy) or, if the object is transparent like glass or water, part of it is transmitted through. For a very shiny object such as a silvered mirror, over 95 percent of the light may be reflected.

[‡] When the meter was redefined in 1983, the speed of light was given this fixed value, and the meter then defined in terms of it (see Section 1–4).

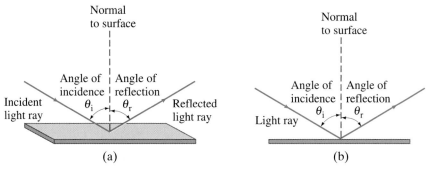

FIGURE 33–3 Law of reflection: (a) Shows an incident ray being reflected at the top of a flat surface; (b) shows a side or "end-on" view, which we will usually use because of its clarity.

When a narrow beam of light strikes a flat surface (Fig. 33–3) we define the **angle of incidence**, θ_i, to be the angle an incident ray makes with the normal to the surface, and the **angle of reflection**, θ_r, to be the angle the reflected ray makes with the normal. It is found that the *incident and reflected rays lie in the same plane with the normal to the surface*, and that

Angle of incidence and
Angle of reflection
(angle with the normal)

the angle of incidence equals the angle of reflection.

Law of reflection

This is the **law of reflection** and is indicated in Fig. 33–3. It was known to the ancient Greeks, and you can confirm it yourself by shining a narrow flashlight beam at a mirror in a darkened room.

When light is incident upon a rough surface, even microscopically rough such as this page, it is reflected in many directions, Fig. 33–4. This is called **diffuse reflection**. The law of reflection still holds, however, at each small section of the surface. Because of diffuse reflection in all directions, an ordinary object can be seen from many different angles. When you move your head to the side, different reflected rays reach your eye from each point on the object (such as this page), Fig. 33–5a. Let us compare diffuse reflection to reflection from a mirror, which is known as *specular* reflection. ("Speculum" is Latin for mirror.) When a narrow beam of light shines on a mirror, the light will not reach your eye unless your eye is positioned at just the right place where the law of reflection is satisfied, as shown in Fig. 33–5b. This is what gives rise to the special image-forming properties of mirrors.

When you look straight in a mirror, you see what appears to be yourself as well as various objects around and behind you, Fig. 33–6. Your face and the other objects look as if they are in front of you, beyond the mirror; but, of course, they are not. What you see in the mirror is an **image** of the objects.

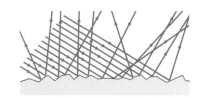

FIGURE 33–4 Diffuse reflection from a rough surface.

Mirrors (specular reflection)

FIGURE 33–6 When you look in a mirror, you see an image of yourself and objects around you. Note that you don't see yourself as others see you, because left and right appear reversed in the image.

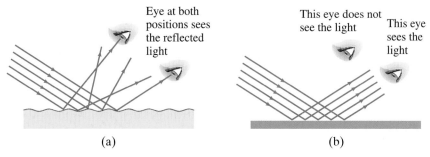

FIGURE 33–5 A beam of light from a flashlight shines on (a) white paper, and (b) a mirror. In part (a), you can see the white light reflected at various points because of diffuse reflection. But in part (b), you see the reflected light only when your eye is placed correctly $(\theta_r = \theta_i)$; this is known as specular reflection. (Galileo, using similar arguments, showed that the Moon must have a rough surface rather than a highly polished surface like a mirror, as some people thought.)

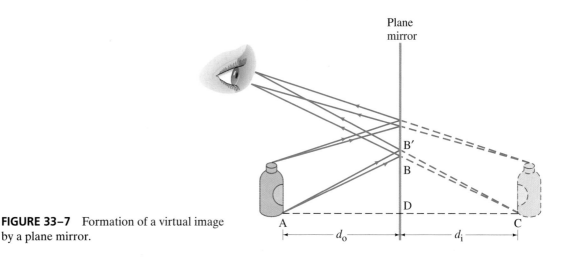

FIGURE 33–7 Formation of a virtual image by a plane mirror.

Figure 33–7 shows how an image is formed by a plane mirror (that is, flat), according to the ray model. We are viewing the mirror, on edge, in the diagram of Fig. 33–7, and the rays are shown reflecting from the front surface. (Good mirrors are generally made by putting a highly reflective metallic coating on one surface of a very flat piece of glass.) Rays from two different points on an object are shown in Fig. 33–7: rays leaving from a point on the top of the bottle, and from a point on the bottom. Rays leave each point on the object going in many directions, but only those that enclose the bundle of rays that reach the eye from the two points are shown. The diverging rays that enter the eye *appear* to come from behind the mirror as shown by the dashed lines. That is, our eyes and brain interpret any rays that enter an eye as having traveled a straight-line path. The point from which each bundle of rays seems to come is one point on the image. For each point on the object, there is a corresponding image point. Let us concentrate on the two rays that leave point A on the object and strike the mirror at points B and B′. The angles ADB and CDB are right angles; and angles ABD and CBD are equal because of the law of reflection. Therefore, the two triangles ABD and CBD are congruent, and the length AD = CD. Thus the image appears as far behind the mirror as the object is in front: the **image distance**, d_i (distance from mirror to image, Fig. 33–7), equals the **object distance**, d_o. From the geometry, we also see that the height of the image is the same as that of the object.

The light rays do not actually pass through the image location itself. It merely *seems* as though the light is coming from the image because our brains interpret any light entering our eyes as having come in a *straight-line path from in front of us*. Because the rays do not actually pass through the image, the image would not appear on paper or film placed at the location of the image. Therefore, it is called a **virtual image**. This is to distinguish it from a **real image** in which the light does pass through the image and which therefore could appear on paper or film placed at the image position. We will see that curved mirrors and lenses can form real images. A movie projector lens, for example, produces a real image that is visible on the screen.

CONCEPTUAL EXAMPLE 33–1 **How tall must a full-length mirror be?**
A woman 1.60 m tall stands in front of a vertical plane mirror. What is the minimum height of the mirror, and how high must its lower edge be above the floor, if she is to see her whole body? Assume her eyes are 10 cm below the top of her head.

RESPONSE The situation is diagrammed in Fig. 33–8. First consider the ray from her foot, AB, which upon reflection becomes BE and enters the eye E. The light from point A (her foot) enters the eye after reflecting at B; so the mirror needs to extend no lower than B. Because the angle of reflection equals the angle of incidence, the height BD is half of the height AE. Because AE = 1.60 m − 0.10 m = 1.50 m, then BD = 0.75 m. Similarly, if the woman is to see the top of her head, the top edge of the mirror only needs to reach point F,

Ray model of how images are formed

Each point of image is where rays intersect

Image distance = object distance

Real and virtual images

➡ **PHYSICS APPLIED**

How tall a mirror is needed to see your whole reflection?

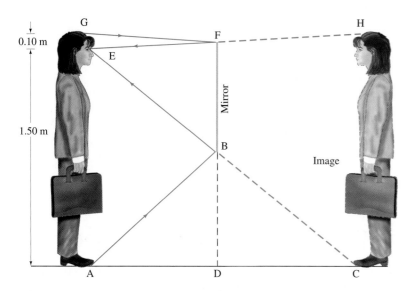

FIGURE 33–8 Seeing oneself in a mirror. Example 33–1.

which is 5 cm below the top of her head (half of GE = 10 cm). Thus, DF = 1.55 m, and the mirror need have a vertical height of only (1.55 m − 0.75 m) = 0.80 m. And the mirror's bottom edge must be 0.75 m above the floor. In general, a mirror need be only half as tall as a person for that person to see all of himself or herself. Does this result depend on the person's distance from the mirror?

CONCEPTUAL EXAMPLE 33–2 | **Is the photo upside down?**   Close examination of the photograph on the first page of this chapter reveals that in the top portion, the image of the Sun is seen clearly, whereas in the lower portion, the image of the Sun is partially blocked by the tree branches. Why isn't the reflection in the water an exact replica of the real scene? Illustrate your answer by drawing a sketch of this situation, showing the Sun, the camera, the branch, and two rays going from the Sun to the camera (one direct and one reflected). Does your sketch tell you if the photograph is right side up?

RESPONSE   We need to draw two diagrams, one assuming the photo on p. 810 is right side up, and another assuming it is upside down. Figure 33–9 is drawn assuming the photo is upside down. In this case, the Sun blocked by the tree would be the direct view, and the full view of the Sun the reflection. Thus, as diagrammed in Fig. 33–9, the ray which reflects off the water and into the camera travels at an angle below the branch, whereas the ray that travels directly to the camera passes through the branches. This works. Try to draw a diagram assuming the photo is right side up (thus assuming that the image of the Sun in the reflection is higher above the horizon than it is as viewed directly). It won't work. The photo on p. 810 is upside down.

Also, what about the people in the photo? Try to draw a diagram showing why they don't appear in the reflection. [*Hint*: Assume they are not sitting on the edge of poolside, but back from the edge a bit.] Then try to draw a diagram of the reverse (i.e., assume the photo is right side up so the people are visible only in the reflection). In general, reflected images are not perfect replicas when different planes (distances) are involved.

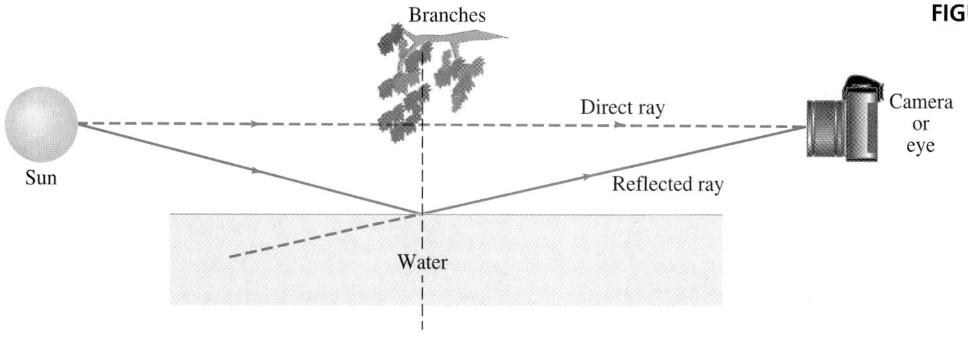

FIGURE 33–9   Example 33–2.

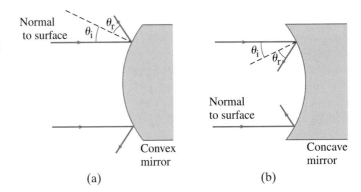

FIGURE 33–10 Mirrors with convex and concave spherical surfaces. Note that $\theta_r = \theta_i$ for each ray.

Normal to surface

θ_i θ_r

Convex mirror

θ_i θ_r

Normal to surface

Concave mirror

(a)

(b)

FIGURE 33–11 (a) A concave cosmetic mirror gives a magnified image. (b) A convex mirror in a store demagnifies and so includes a wide field of view.

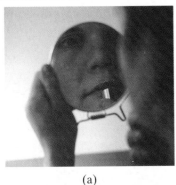

(a)

(b)

33–4 Formation of Images by Spherical Mirrors

Reflecting surfaces do not have to be flat. The most common *curved* mirrors are *spherical*, which means they form a section of a sphere. A spherical mirror is called **convex** if the reflection takes place on the outer surface of the spherical shape so that the center of the mirror surface bulges out toward the viewer (Fig. 33–10a). A mirror is called **concave** if the reflecting surface is on the inner surface of the sphere so that the center of the mirror sinks away from the viewer (like a "cave"), Fig. 33–10b. Concave mirrors are used as shaving or cosmetic mirrors (Fig. 33–11a), and convex mirrors are sometimes used on cars and trucks (rearview mirrors) and in shops (to watch for thieves), because they take in a wide field of view (Fig. 33–11b).

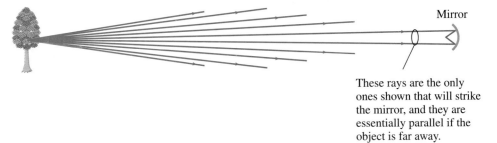

Mirror

These rays are the only ones shown that will strike the mirror, and they are essentially parallel if the object is far away.

FIGURE 33–12 If the object's distance is large compared to the size of the mirror (or lens), the rays are nearly parallel. They are parallel for an object at infinity (∞).

To see how spherical mirrors form images, we first consider an object that is very far from a concave mirror. For a distant object, as shown in Fig. 33–12, the rays from each point on the object that reach the mirror will be nearly parallel. *For an object infinitely far away* (the Sun and stars approach this), *the rays would be precisely parallel.* Now consider such parallel rays falling on a concave mirror as in Fig. 33–13. The law of reflection holds for each of these rays at the point each strikes the mirror. As can be seen, they are not all brought to a single point. In

FIGURE 33–13 Parallel rays striking a concave spherical mirror do not focus at precisely a single point.

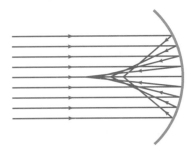

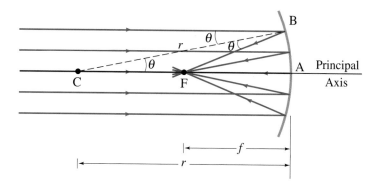

FIGURE 33–14 Rays parallel to the principal axis of a spherical mirror come to a focus at F, called the focal point, as long as the mirror is small in width as compared to its radius of curvature, r, so that the rays are "paraxial"—that is, make only small angles with the axis.

order to form a sharp image, the rays must come to a point. Thus a spherical mirror will not make as sharp an image as a plane mirror will. However, as we show below, if the mirror is small compared to its radius of curvature, so that the reflected rays make only a *small angle* upon reflection, then the rays will cross each other at very nearly a single point, or **focus**, as shown in Fig. 33–14. In the case shown, the rays are parallel to the **principal axis**, which is defined as the straight line perpendicular to the curved surface at its center (line CA in the diagram). The point F, where rays parallel to the principal axis come to a focus, is called the **focal point** of the mirror. The distance between F and the center of the mirror, length FA, is called the **focal length**, f, of the mirror. Another way of defining the focal point is to say that it is the *image point for an object infinitely far away* along the principal axis. The image of the Sun, for example, would be at F.

Small angle approximation

Focal point
Focal length

Now we will show, for a mirror whose reflecting surface is small compared to the radius of curvature, that the rays very nearly meet at a common point, F, and we will also calculate the focal length f. In this approximation, we consider only rays that make a small angle with the principal axis; such rays are called **paraxial rays**, and their angles are exaggerated in Fig. 33–14 only so the labeling will be clear. First we consider a ray that strikes the mirror at B in Fig. 33–14. The point C is the center of curvature of the mirror (the center of the sphere of which the mirror is a part). So the dashed line CB is equal to r, the radius of curvature, and CB is normal to the mirror's surface at B. The incoming ray that hits the mirror at B makes an angle θ with this normal, and hence the reflected ray, BF, also makes an angle θ with the normal. Note also from the geometry that angle BCF is also θ as shown. The triangle CBF is isosceles because two of its angles are equal. Thus length CF = BF. We assume the mirror has a width or diameter that is small compared to its radius of curvature, so the angles are small, and the length FB is nearly equal to length FA. In this approximation, FA = FC. But FA = f, the focal length, and CA = 2 FA = r. Thus the focal length is half the radius of curvature:

$$f = \frac{r}{2}.\qquad\qquad\text{(33–2)}$$

Focal length of mirror

This argument assumed only that the angle θ was small, so the same result applies for all other paraxial rays. Thus all paraxial rays pass through the same point F in the paraxial approximation.

Since it is only approximately true that the rays come to a perfect focus at F, the more curved the mirror, the worse the approximation (Fig. 33–13) and the more blurred the image. This "defect" of spherical mirrors is called **sperical aberration**; we will discuss it more with regard to lenses in Chapter 34. A *parabolic* reflector, on the other hand, will reflect the rays to a perfect focus. However, because parabolic shapes are much harder to make and thus much more expensive, spherical mirrors are used for most purposes. (Many astronomical telescopes use parabolic reflectors.) We consider here only spherical mirrors and we will assume that they are small compared to their radius of curvature so that the image is sharp and Eq. 33–2 holds.

Parabolic mirror

We saw that for an object at infinity the image is located at the focal point of a concave spherical mirror, where $f = r/2$. But where does the image lie for an object not at infinity? First consider the object shown as an arrow in Fig. 33–15, which is placed at point O between F and C. Let us determine where the image will be for a given point O′ on the object. To do this we can draw several rays and make sure these reflect from the mirror such that the angle of reflection equals the angle of incidence. Determining the image position is simplified if we deal with three particularly simple rays. These are the rays labeled 1, 2, and 3 in Fig. 33–15 and we draw them leaving object point O′ as follows:

Ray 1 is drawn parallel to the axis; therefore after reflection it must pass along a line through F (as we saw in Fig. 33–14, and drawn here in Fig. 33–15a).

Ray 2 leaves O′ and is made to pass through F; therefore it must reflect so it is parallel to the axis (Fig. 33–15b).

Ray 3 is chosen to be perpendicular to the mirror, and so is drawn so that it passes through C, the center of curvature; it is along a radius of the spherical surface, and because it is perpendicular to the mirror it will be reflected back on itself (Fig. 33–15c).

The point at which these rays cross is the image point I′. All other rays from the same object point will pass through this image point. To find the image point for any object point, only these three particular rays need to be used. Actually, only two of these rays are needed, but the third serves as a check.

We have shown the image point in Fig. 33–15 only for a single point on the object. Other points on the object are imaged nearby, so a complete image of the object is formed, as shown by the dashed arrow in Fig. 33–15c. Because the light actually passes through the image itself, this is a real image that will appear on a piece of paper or film placed there. This can be compared to the virtual image formed by a plane mirror (the light does not actually pass through that image, Fig. 33–7).

The image in Fig. 33–15 can be seen by the eye when the eye is placed to the left of the image so that some of the rays diverging from each point on the image (as point I′) can enter the eye as shown in Fig. 33–15c. (See also Figs. 33–1 and 33–7.)

➡ RAY DIAGRAM

Finding the image position for a curved mirror

Real image

FIGURE 33–15 Rays leave point O′ on the object (an arrow). Shown are the three most useful rays for determining where the image I′ is formed. [Note that our mirror's height is not small compared to f, so our diagram will not give the precise position of the image.]

(a) Ray 1 goes out from O′ parallel to the axis and reflects through F.

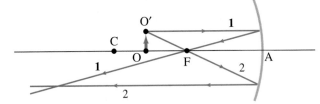

(b) Ray 2 goes through F and then reflects back parallel to the axis.

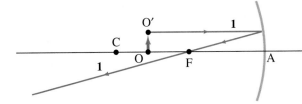

(c) Ray 3 heads out perpendicular to mirror and then reflects back on itself and goes through C (center of curvature)

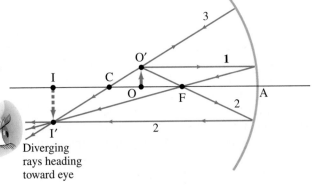

Diverging rays heading toward eye

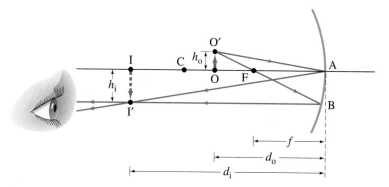

FIGURE 33–16 Diagram for deriving the mirror equation. For the derivation, we assume the mirror size is small compared to its radius of curvature.

Image points can be determined, roughly, by drawing the three rays as described above, but high accuracy is hard to obtain. For one thing it is difficult to draw, maintaining small angles for the "paraxial" rays as we assumed. However, it is possible to derive an equation that gives the image distance if the object distance and radius of curvature of the mirror are known. To do this, we refer to Fig. 33–16. The distance of the object from the center of the mirror, called the **object distance**, is labeled d_o. The **image distance** is labeled d_i. The height of the object OO′ is called h_o and the height of the image, I′I, is h_i. Two rays are shown, O′FBI′ (same as ray 2 in the previous figure) and O′AI′. The ray O′AI′ obeys the law of reflection, of course, so the two right triangles O′AO and I′AI are similar. Therefore, we have

$$\frac{h_o}{h_i} = \frac{d_o}{d_i}.$$

For the other ray, O′FBI′, the triangles O′FO and AFB are also similar since the length $AB = h_i$ (in our approximation of a mirror that is small compared to its radius) and $FA = f$, the focal length of the mirror. Therefore,

$$\frac{h_o}{h_i} = \frac{OF}{FA} = \frac{d_o - f}{f}.$$

The left sides of the two preceding expressions are the same, so we can equate the right sides:

$$\frac{d_o}{d_i} = \frac{d_o - f}{f}.$$

We now divide both sides by d_o and rearrange to obtain

$$\frac{1}{d_o} + \frac{1}{d_i} = \frac{1}{f}. \tag{33–3}$$

Mirror equation

This is the equation we were seeking. It is called the **mirror equation** and relates the object and image distances to the focal length f (where $f = r/2$).

The **lateral magnification**, m, of a mirror is defined as the height of the image divided by the height of the object. From our first set of similar triangles above, we can write:

$$m = \frac{h_i}{h_o} = -\frac{d_i}{d_o}. \tag{33–4}$$

Lateral magnification for curved mirror

The minus sign in Eq. 33–4 is inserted as a convention. Indeed, we must be careful about the signs of all quantities in Eqs. 33–3 and 33–4. Sign conventions are chosen so as to give the correct locations and orientations of images, as predicted by ray diagrams. The sign conventions we use are: the image height h_i is positive if the image is upright, and negative if inverted, relative to the object (assuming h_o is taken as positive); d_i and d_o are positive if image and object are on the reflecting side of the mirror (as in Fig. 33–16), but if either image or object is behind the mirror, the corresponding distance is negative (an example can be seen in Fig. 33–17, Example 33–5). Thus the magnification (Eq. 33–4) is positive for an upright image and negative for an inverted image. We will summarize sign conventions more fully after discussing convex mirrors later in this Section.

Sign conventions

EXAMPLE 33–3 **Image in a concave mirror.** A 1.50-cm-high diamond ring is placed 20.0 cm from a concave mirror whose radius of curvature is 30.0 cm. Determine (a) the position of the image, and (b) its size.

SOLUTION The focal length $f = r/2 = 15.0$ cm. The ray diagram is basically like that shown in Fig. 33–15 and Fig. 33–16, since the object is between F and C. Referring to either Fig. 33–15 or Fig. 33–16, we have CA $= r = 30.0$ cm, FA $= f = 15.0$ cm, and OA $= d_o = 20.0$ cm. (a) From Eq. 33–3

$$\frac{1}{d_i} = \frac{1}{f} - \frac{1}{d_o} = \frac{1}{15.0 \text{ cm}} - \frac{1}{20.0 \text{ cm}} = 0.0167 \text{ cm}^{-1}.$$

➡ **PROBLEM SOLVING**

Remember to take the reciprocal

So $d_i = 1/0.0167 \text{ cm}^{-1} = 60.0$ cm. The image is 60.0 cm from the mirror on the same side as the object.
(b) From Eq. 33–4, the lateral magnification is

$$m = -\frac{60.0 \text{ cm}}{20.0 \text{ cm}} = -3.00.$$

Therefore the image height is

$$h_i = m h_o = (-3.00)(1.5 \text{ cm}) = -4.5 \text{ cm}.$$

The minus sign reminds us that the image is inverted, as in Figs. 33–15 and 33–16.

CONCEPTUAL EXAMPLE 33–4 **Reversible rays.** If the object in Example 33–3 is placed instead where the image is (see Fig. 33–16), where will the new image be?

RESPONSE The mirror equation is symmetric in d_o and d_i. Thus the new image will be where the old object was. Indeed, in Fig. 33–16 we need only reverse the direction of the rays to get our new situation.

EXAMPLE 33–5 **Object closer to concave mirror.** A 1.00-cm-high object is placed 10.0 cm from a concave mirror whose radius of curvature is 30.0 cm. (a) Draw a ray diagram to locate (approximately) the position of the image. (b) Determine the position of the image and the magnification analytically.

SOLUTION (a) Since $f = r/2 = 15.0$ cm, the object is between the mirror and the focal point. We draw the three rays as described earlier (Fig. 33–15) and this is shown in Fig. 33–17. Ray 1 leaves the tip of our object heading toward the mirror parallel to the axis, and reflects through F. Ray 2 cannot head toward F because it would not strike the mirror; so ray 2 must head as if it started at F (dashed line) and heads to the mirror and then is reflected parallel to the principal axis. Ray 3 is perpendicular to the mirror, as before. The rays reflected from the mirror diverge and so never meet at a point. They appear, however, to be coming from a point behind the mirror. This point is the image which is thus behind the mirror and *virtual*. (Why?)
(b) We use Eq. 33–3 to find d_i when $d_o = 10.0$ cm:

$$\frac{1}{d_i} = \frac{1}{15.0 \text{ cm}} - \frac{1}{10.0 \text{ cm}} = \frac{2-3}{30.0 \text{ cm}} = -\frac{1}{30.0 \text{ cm}}.$$

Therefore, $d_i = -30.0$ cm. The minus sign means the image is behind the mirror. The lateral magnification is $m = -d_i/d_o = -(-30.0 \text{ cm})/(10.0 \text{ cm}) = +3.00$. So the image is 3.00 times larger than the object; the plus sign indicates that the image is upright (which is consistent with the ray diagram, Fig. 33–17).
[Note that the image distance cannot be obtained accurately by measuring on Fig. 33–17, because our diagram violates the paraxial ray assumption (we had to, to make all rays clearly visible).]

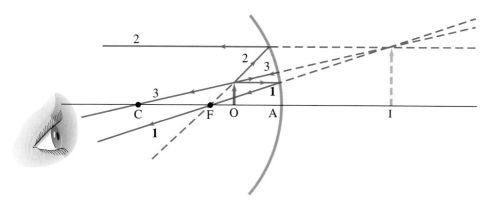

FIGURE 33–17 Object placed within the focal point F. The image is *behind* the mirror and is *virtual*, Example 33–5. [Note that the vertical scale (height of object = 1.0 cm) is different from the horizontal (OA = 10.0 cm) for ease of drawing, and this will affect the precision of the drawing.]

It is useful to compare Figs. 33–15 and 33–17. We can see that if the object is within the focal point, as in Fig. 33–17, the image is virtual, upright, and magnified. This is how a shaving or cosmetic mirror is used—you must place your head within the focal point if you are to see yourself right side up (Fig. 33–11a). If the object is *beyond* the focal point, as in Fig. 33–15, the image is real and inverted (upside down—and hard to use!). Whether the magnification is greater or less than 1.0 in the latter case depends on the position of the object relative to the center of curvature, point C.

The mirror equation also holds for a plane mirror: the focal length is $f = r/2 = \infty$, and Eq. 33–3 gives $d_i = -d_o$.

The analysis used for concave mirrors can be applied to **convex** mirrors. Even the mirror equation (Eq. 33–3) holds for a convex mirror, although the quantities involved must be carefully defined. Figure 33–18a shows parallel rays falling on a convex mirror. Again spherical aberration will be present, but we assume the mirror's size is small compared to its radius of curvature. The reflected rays diverge, but seem to come from point F behind the mirror. This is the **focal point**, and its distance from the center of the mirror is the **focal length**, f. It is easy to show that again $f = r/2$. We see that an object at infinity produces a virtual image in a convex mirror. Indeed, no matter where the object is placed on the reflecting side of a convex mirror, the image will be virtual and erect, as indicated in Fig. 33–18b. To find the image we draw rays 1 and 3 according to the rules used before on the concave mirror, as shown in Fig. 33–18b. Note that although rays 1 and 3 don't actually pass through points F and C, the line along which they are drawn do (shown dashed).

➡ **PHYSICS APPLIED**

Magnifying mirror

Convex mirrors—analysis

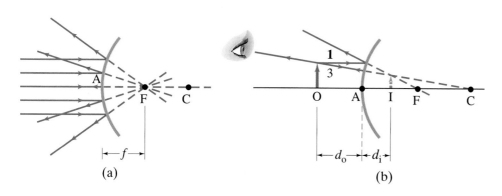

FIGURE 33–18 Convex mirror: (a) the focal point is at F, behind the mirror; (b) the image I of the object at O is virtual, upright, and smaller than the object. [Not to scale for Example 33–6.]

(a)

(b)

The mirror equation, Eq. 33–3, holds for convex mirrors but the focal length f must be considered negative, as must the radius of curvature. The proof is left as a Problem. It is also left as a Problem to show that Eq. 33–4 for the magnification is also valid.

➡ **PHYSICS APPLIED**

Convex rearview mirror

EXAMPLE 33–6 **Convex rearview mirror.** A convex rearview car mirror has a radius of curvature of 16 m. Determine the location of the image and its magnification for an object 10.0 m from the mirror.

SOLUTION The ray diagram will be like Fig. 33–18, but the large object distance $(d_o = 10.0\,\text{m})$ makes a precise drawing difficult. We have a convex mirror, so r is negative by convention. Specifically, $r = -16\,\text{m}$, so $f = -8.0\,\text{m}$. The mirror equation gives

$$\frac{1}{d_i} = \frac{1}{f} - \frac{1}{d_o} = \frac{1}{-8.0\,\text{m}} - \frac{1}{10.0\,\text{m}} = -\frac{18}{80\,\text{m}}.$$

So $d_i = -80\,\text{m}/18 = -4.4\,\text{m}$, or 4.4 m behind the mirror. The lateral magnification is $m = -d_i/d_o = -(-4.4\,\text{m})/(10.0\,\text{m}) = 0.44$.

So the upright image is a little less than half the size of the object. Rearview mirrors sometimes come with warnings that objects look farther away (appear smaller) than they really are; judging distances in a spherical mirror requires skill.

PROBLEM SOLVING Spherical Mirrors

1. Always draw a ray diagram even though you are going to make an analytic calculation—the diagram serves as a check, even if not precise. Draw at least two, and preferably three, of the easy-to-draw rays as described in Fig. 33–15. Generally draw the rays starting from a point on the object to the left of the mirror and moving to the right.

2. Use Eqs. 33–3 and 33–4; it is crucially important to follow the sign conventions.

3. **Sign Conventions**
   (a) When the object, image, or focal point is on the reflecting side of the mirror (on the left in all our drawings), the corresponding distance is considered positive. If any of these points is behind the mirror (on the right) the corresponding distance is negative.[†]
   (b) The image height h_i is positive if the image is upright, and negative if inverted, relative to the object (that is, if h_o is taken as positive).

[†]Object distances are positive for material objects, but can be negative in systems with more than one mirror or lens—see Section 34–3.

33–5 Refraction: Snell's Law

When light passes from one medium into another with a different index of refraction (Section 33–2), part of the incident light is reflected at the boundary. The remainder passes into the new medium. If a ray of light is incident at an angle to the surface (other than perpendicular), the ray is bent as it enters the new medium. This bending is called **refraction**. Figure 33–19a shows a ray passing from air into water. The angle θ_1 is the angle the incident ray makes with the perpendicular to the surface and is called the **angle of incidence**. The angle θ_2 is the **angle of refraction**, the angle the refracted ray makes with the perpendicular. Notice that the ray bends toward the normal when entering the water. This is always the case

Angle of refraction

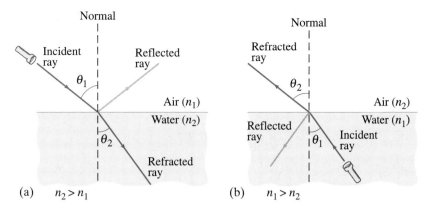

FIGURE 33–19 Refraction.
(a) Light refracted when passing from air (n_1) into water (n_2): $n_2 > n_1$.
(b) Light refracted when passing from water (n_1) into air (n_2): $n_1 > n_2$.

(a) $n_2 > n_1$

(b) $n_1 > n_2$

when the ray enters a medium where the speed of light is *less*. If light travels from one medium into a second where its speed is *greater*, the ray bends away from the normal; this is shown in Fig. 33–19b for a ray traveling from water to air.

Refraction is responsible for a number of common optical illusions. For example, a person standing in waist-deep water appears to have shortened legs. As shown in Fig. 33–20, the rays leaving the person's foot are bent at the surface. The observer's brain assumes the rays to have traveled a straight-line path, and so the feet appear to be higher than they really are. Similarly, when you put a pencil in water, it appears to be bent (Fig. 33–21).

The angle of refraction depends on the speed of light in the two media and on the incident angle. An analytical relation between θ_1 and θ_2 was arrived at experimentally about 1621 by Willebrord Snell (1591–1626). It is known as **Snell's law** and is written:

$$n_1 \sin \theta_1 = n_2 \sin \theta_2. \tag{33–5}$$

θ_1 is the angle of incidence and θ_2 is the angle of refraction; n_1 and n_2 are the respective indices of refraction in the materials. See Fig. 33–19. The incident and refracted rays lie in the same plane, which also includes the perpendicular to the surface. Snell's law is the basic **law of refraction**. Snell's law can be derived from the wave theory of light (Chapter 35), and in fact we did derive it in Section 15–10 where Eq. 15–19 is just a combination of Eqs. 33–5 and 33–1.

It is clear from Snell's law that if $n_2 > n_1$, then $\theta_2 < \theta_1$. That is, if light enters a medium where n is greater (and its speed less), then the ray is bent toward the normal. And if $n_2 < n_1$, then $\theta_2 > \theta_1$, so the ray bends away from the normal. This is what we saw in Fig. 33–19.

➡ PHYSICS APPLIED

Optical illusions

*Snell's law
(law of refraction)*

FIGURE 33–20 Ray diagram showing why a person's legs look shorter when standing in waist-deep water: the path of light traveling from the bather's foot to the observer's eye bends at the water's surface, and our brain interprets the light as having traveled in a straight line, from higher up (dashed line).

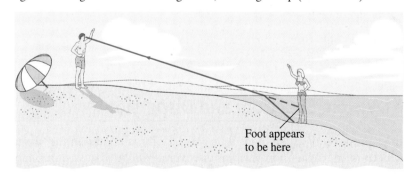

Foot appears to be here

FIGURE 33–21 A pencil in water looks bent even when it isn't.

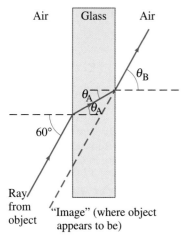

Air | Glass | Air

θ_B

θ_A

θ_A

60°

Ray from object

"Image" (where object appears to be)

FIGURE 33–22 Light passing through a piece of glass (Example 33–7).

FIGURE 33–23 Example 33–8.

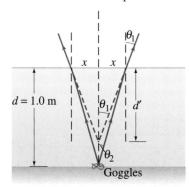

θ_1

x | x

$d = 1.0$ m

θ_1

d'

θ_2

Goggles

EXAMPLE 33–7 **Refraction through flat glass.** Light strikes a flat piece of uniformly thick glass at an incident angle of 60°, as shown in Fig. 33–22. If the index of refraction of the glass is 1.50, (a) what is the angle of refraction θ_A in the glass; (b) what is the angle θ_B at which the ray emerges from the glass?

SOLUTION (a) We assume the incident ray is in air, so $n_1 = 1.00$ and $n_2 = 1.50$. Then, from Eq. 33–5 we have

$$\sin \theta_A = \frac{1.00}{1.50} \sin 60° = 0.577,$$

so $\theta_A = 35.2°$.

(b) Since the faces of the glass are parallel, the incident angle at the second surface is just θ_A, so $\sin \theta_A = 0.577$. This time $n_1 = 1.50$ and $n_2 = 1.00$. Thus, $\theta_B \ (= \theta_2)$ is

$$\sin \theta_B = \frac{1.50}{1.00} \sin \theta_A = 0.866,$$

and $\theta_B = 60.0°$. The direction of the beam is thus unchanged by passing through a flat piece of glass of uniform thickness. It should be clear that this works for any angle of incidence. The ray is displaced slightly to one side, however. You can observe this by looking through a piece of glass (near its edge) at some object and then moving your head to the side so that you see the object directly.

EXAMPLE 33–8 **Apparent depth of a pool.** A swimmer has dropped her goggles in the shallow end of a pool, marked as 1.0 m deep. But the goggles don't look that deep. Why? How deep do the goggles appear to be when you look straight down into the water?

SOLUTION The ray diagram of Fig. 33–23 shows why the water seems less deep than it actually is. Rays traveling upward from the goggles on the bottom are refracted *away* from the normal as they exit the water. The rays appear to be diverging from a point higher in the water. To calculate the apparent depth d', given a real depth $d = 1.0$ m, we use Snell's law with $n_1 = 1$ for air and $n_2 = 1.33$ for water: $\sin \theta_1 = n_2 \sin \theta_2$. We are considering only small angles, so $\sin \theta \approx \tan \theta \approx \theta$, with θ in radians. So Snell's law becomes

$$\theta_1 \approx n_2 \theta_2.$$

From Fig. 33–23, we see that

$$\theta_1 \approx \tan \theta_1 = \frac{x}{d'} \quad \text{and} \quad \theta_2 \approx \tan \theta_2 = \frac{x}{d}.$$

Putting these into Snell's law, $\theta_1 \approx n_2 \theta_2$, we get

$$\frac{x}{d'} \approx n_2 \frac{x}{d} \quad \text{or} \quad d' \approx \frac{d}{n_2} = \frac{1.0 \text{ m}}{1.33} = 0.75 \text{ m}.$$

The pool seems only three-fourths as deep as it actually is.

33–6 Visible Spectrum and Dispersion

An obvious property of visible light is its color. Color is related to the wavelengths or frequencies of the light. (How this was discovered will be discussed in Chapter 35.) Visible light—that to which our eyes are sensitive—falls in the wavelength range of about 400 nm to 750 nm.[†] This is known as the **visible spectrum**, and within it lie

[†]Sometimes the angstrom (Å) unit is used when referring to light: $1 \text{ Å} = 1 \times 10^{-10}$ m. Then visible light falls in the wavelength range of 4000 Å to 7500 Å.

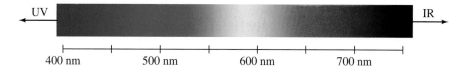

UV   IR

| 400 nm | 500 nm | 600 nm | 700 nm |

FIGURE 33–24 The spectrum of visible light, showing the range of wavelengths for the various colors.

the different colors from violet to red, as shown in Fig. 33–24. Light with wavelength shorter than 400 nm is called **ultraviolet** (UV), and light with wavelength greater than 750 nm is called **infrared** (IR).[‡] Although human eyes are not sensitive to UV or IR, some types of photographic film do respond to them.

A prism separates white light into a rainbow of colors, as shown in Fig. 33–25. This happens because the index of refraction of a material depends on the wavelength, as shown for several materials in Fig. 33–26. White light is a mixture of all visible wavelengths, and when incident on a prism, as in Fig. 33–27, the different wavelengths are bent to varying degrees. Because the index of refraction is greater for the shorter wavelengths, violet light is bent the most and red the least as indicated. This spreading of white light into the full spectrum is called **dispersion**.

FIGURE 33–25 White light passing through a prism is broken down into its constituent colors.

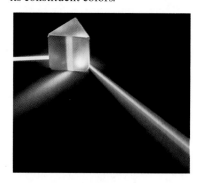

FIGURE 33–26 Index of refraction as a function of wavelength for various transparent solids.

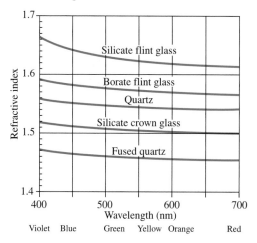

FIGURE 33–27 White light dispersed by a prism into the visible spectrum.

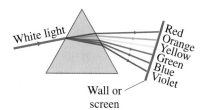

Rainbows are a spectacular example of dispersion—by drops of water. You can see rainbows when you look at falling water droplets with the Sun at your back. Figure 33–28 shows how red and violet rays are bent by spherical water droplets and are reflected off the back surface. Red is bent the least and so reaches the observer's eyes from droplets higher in the sky, as shown in the diagram. Thus the top of the rainbow is red.

➡ **PHYSICS APPLIED**

Rainbows

[‡]The complete electromagnetic spectrum is illustrated in Fig. 32–12.

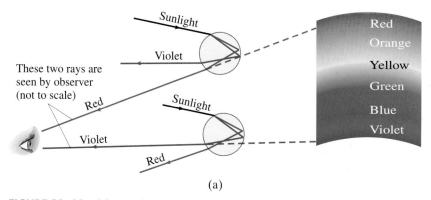

FIGURE 33–28 (a) Ray diagram explaining how a rainbow (b) is formed.

SECTION 33–6   **825**

33–7 | Total Internal Reflection; Fiber Optics

When light passes from one material into a second material where the index of refraction is less (say, from water into air), the light bends away from the normal, as for rays I and J in Fig. 33–29. At a particular incident angle, the angle of refraction will be 90°, and the refracted ray would skim the surface (ray K) in this case. The incident angle at which this occurs is called the **critical angle**, θ_C. From Snell's law, θ_C is given by

Critical angle

$$\sin \theta_C = \frac{n_2}{n_1} \sin 90° = \frac{n_2}{n_1}.$$

(33–6)

For any incident angle less than θ_C there will be a refracted ray, although part of the light will also be reflected at the boundary. However, for incident angles greater than θ_C, Snell's law would tell us that $\sin \theta_2$ is greater than 1.00. Yet the sine of an angle can never be greater than 1.00. In this case there is no refracted ray at all, and *all of the light is reflected*, as for ray L in Fig. 33–29. This effect is called **total internal reflection**. But note that total internal reflection can occur only when light strikes a boundary where the medium beyond has a lower index of refraction.

Total internal reflection

FIGURE 33–29 Since $n_2 < n_1$, light rays are totally internally reflected if $\theta > \theta_C$ as for ray L. If $\theta < \theta_C$, as for rays I and J, only a part of the light is reflected, and the rest is refracted.

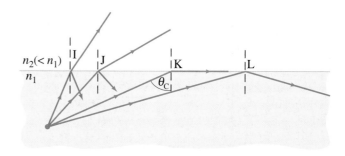

CONCEPTUAL EXAMPLE 33–9 | **View up from under water.** Describe what a person would see who looked up at the world from beneath the perfectly smooth surface of a lake or swimming pool.

RESPONSE For an air–water interface, the critical angle is given by

$$\sin \theta_C = \frac{1.00}{1.33} = 0.750.$$

Therefore, $\theta_C = 49°$. Thus the person would see the outside world compressed into a circle whose edge makes a 49° angle with the vertical. Beyond this angle, the person would see reflections from the sides and bottom of the pool or lake (Fig. 33–30).

FIGURE 33–30 (a) Light rays, and (b) view looking upward from beneath the water (the surface of the water must be very smooth).

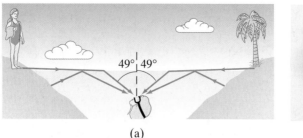

(a)

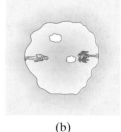

(b)

Diamonds

Diamonds achieve their brilliance from a combination of dispersion (previous Section) and total internal reflection. Because diamonds have a very high index of refraction of about 2.4, the critical angle for total internal reflection is

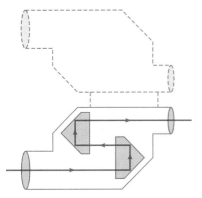

FIGURE 33–31 Total internal reflection of light by prisms in binoculars.

FIGURE 33–32 Light reflected totally at the interior surface of a glass or transparent plastic fiber.

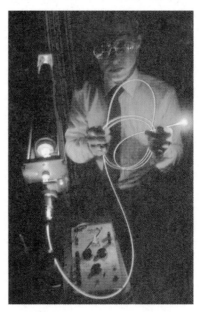

FIGURE 33–33 Total internal reflection within the tiny fibers of this light pipe makes it possible to transmit light in complex paths with minimal loss.

FIGURE 33–34 (a) How a fiber-optic image is made. (b) Example of a fiber-optic device inserted through the nose, and the image seen.

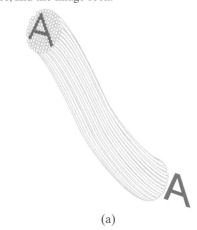

(a)

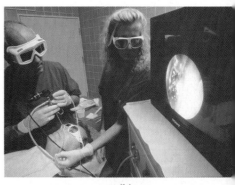

(b)

only 25°. Incident light therefore strikes many of the internal surfaces before it strikes one at less than 25° and emerges. After many such reflections, the light has traveled far enough that the colors have become sufficiently separated to be seen individually and brilliantly by the eye after leaving the crystal.

Many optical instruments, such as binoculars, use total internal reflection within a prism to reflect light. The advantage is that very nearly 100 percent of the light is reflected, whereas even the best mirrors reflect somewhat less than 100 percent. Thus the image is brighter. For glass with $n = 1.50$, $\theta_C = 41.8°$. Therefore, 45° prisms will reflect all the light internally, if oriented as shown in the binoculars of Fig. 33–31.

Total internal reflection is the principle behind **fiber optics**. Glass and plastic fibers as thin as a few micrometers in diameter can now be made. A bundle of such tiny fibers is called a **light pipe** or cable, and light can be transmitted along it with almost no loss because of total internal reflection. Figure 33–32 shows how light traveling down a thin fiber makes only glancing collisions with the walls so that total internal reflection occurs. Even if the light pipe is bent into a complicated shape, the critical angle still won't be exceeded, so light is transmitted practically undiminished to the other end (see Fig. 33–33). Very small losses do occur, mainly by reflection at the ends and absorption within the fiber.

Important applications of fiber-optic cables are in telecommunications and medicine. They are being used to transmit telephone calls, video signals, and computer data. The signal is a modulated light beam (a light beam whose intensity can be varied) and is transmitted at a much higher rate and with less loss and less interference than an electrical signal in a copper wire. Fibers have been developed that can support over one hundred separate wavelengths, each modulated to carry up to 10 gigabits (10^{10} bits) of information per second. That amounts to a terabit (10^{12} bits) per second for the full one hundred wavelengths. The sophisticated use of fiber optics to transmit a clear picture, is particularly useful in medicine, Fig. 33–34. For example, a patient's lungs can be examined by inserting a light pipe known as a bronchoscope through the mouth and down the bronchial tube. Light is sent down an outer set of fibers to illuminate the lungs. The reflected light returns up a central core set of fibers. Light directly in front of each fiber travels up that fiber. At the opposite end, a viewer sees a series of bright and dark spots, much like a TV screen—that is, a picture of what lies at the opposite end. Lenses are used at each end: at the object end to bring the rays in parallel, and at the viewing end as a telescope. The image may be viewed directly or on a TV monitor or film. The fibers must be optically insulated from one another, usually by a thin coating of material whose refractive index is less than that of the fiber. The fibers must be arranged precisely parallel to one another if the picture is to be clear. The more fibers there are, and the smaller they are, the more detailed the picture. Such instruments, including bronchoscopes, colonoscopes (for viewing the colon) and endoscopes (stomach or other organs) are extremely useful for examining hard-to-reach places.

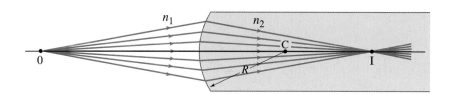

FIGURE 33–35 Rays from a point O on an object will be focused at a single image point I by a spherical boundary between two transparent materials $(n_2 > n_1)$, as long as the rays make small angles with the axis.

*$\boxed{\text{33–8}}$ Refraction at a Spherical Surface

We now examine the refraction of rays at the spherical surface of a transparent material. Such a surface could be one face of a lens or the cornea of the eye. To be general, let us consider an object which is located in a medium whose index of refraction is n_1, and rays from each point on the object can enter a medium whose index of refraction is n_2. The radius of curvature of the spherical boundary is R, and its center of curvature is at point C, Fig. 33–35. We now show that all rays leaving a point O on the object will be focused at a single point I, the image point, if we consider only paraxial rays, rays that make a small angle with the axis. To do so, we consider a single ray that leaves point O as shown in Fig. 33–36. From Snell's law, Eq. 33–5, we have

$$n_1 \sin \theta_1 = n_2 \sin \theta_2.$$

We are assuming that angles $\theta_1, \theta_2, \alpha, \beta,$ and γ are small, so $\sin \theta \approx \theta$ (in radians), and Snell's law becomes, approximately,

$$n_1 \theta_1 = n_2 \theta_2.$$

Also, $\beta + \phi = 180°$ and $\theta_2 + \gamma + \phi = 180°$, so

$$\beta = \gamma + \theta_2.$$

Similarly for triangle OPC,

$$\theta_1 = \alpha + \beta.$$

These three relations can be combined to yield

$$n_1 \alpha + n_2 \gamma = (n_2 - n_1)\beta.$$

Since we are considering only the case of small angles, we can write, approximately,

$$\alpha = \frac{h}{d_o}, \qquad \beta = \frac{h}{R}, \qquad \gamma = \frac{h}{d_i},$$

where d_o and d_i are the object and image distances and h is the height as shown in Fig. 33–36. We substitute these into the previous equation, divide through by h, and obtain

$$\frac{n_1}{d_o} + \frac{n_2}{d_i} = \frac{n_2 - n_1}{R}. \qquad \text{(33–7)}$$

For a given object distance d_o, this equation tells us d_i, the image distance, does not depend on the angle of a ray. Hence all paraxial rays meet at the same point I. This

FIGURE 33–36 Diagram for proving that all paraxial rays from O focus at the same point $I(n_2 > n_1)$.

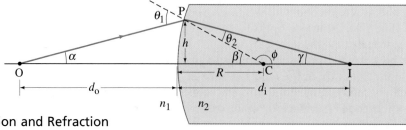

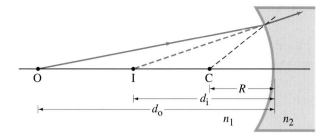

FIGURE 33–37 Rays from O refracted by a concave surface form a virtual image $(n_2 > n_1)$.

is true, of course, only for rays that make small angles with the axis and with each other. This is equivalent to assuming that the width of the refracting spherical surface is small compared to its radius of curvature. If this assumption is not true, the rays will not converge to a point; there will be spherical aberration, just as for a mirror (see Fig. 33–13), and the image will be blurry. (Spherical aberration will be discussed further in Section 34–10.)

We derived Eq. 33–7 using Fig. 33–36 for which the spherical surface is convex (as viewed by the incoming ray). It is also valid for a concave surface—as can be seen using Fig. 33–37—if we make the following conventions:

1. If the surface is convex (so the center of curvature C is on the side of the surface opposite to that from which the light comes), R is positive; if the surface is concave (C on the same side from which the light comes) R is negative.

2. The image distance, d_i, follows the same convention: positive if on the opposite side from where the light comes, negative if on the same side.

3. The object distance is positive if on the same side from which the light comes (this is the normal case, although when several surfaces bend the light it may not be so), otherwise it is negative.

For the case shown in Fig. 33–37 with a concave surface, both R and d_i are negative when used in Eq. 33–7. Note, in this case, that the image is virtual.

EXAMPLE 33–10 **Apparent depth II.** A person looks vertically down into a 1.0-m-deep pool. How deep does the pool appear to be?

SOLUTION Example 33–8 solved this problem using Snell's law. Here we use Eq. 33–7. A ray diagram is shown in Fig. 33–38. Point O represents a point on the pool's bottom. The rays diverge and appear to come from point I, the image. We have $d_o = 1.0$ m and, for a flat surface $R = \infty$. Then Eq. 33–7 becomes

$$\frac{1.33}{1.0\,\text{m}} + \frac{1.00}{d_i} = \frac{(1.00 - 1.33)}{\infty} = 0$$

Hence $d_i = -(1.0\,\text{m})/(1.33) = -0.75\,\text{m}$. So the pool appears to be only three-fourths as deep as it actually is, the same result we got in Example 33–8. The minus sign tells us the image point I is on the same side of the surface as O, and the image is virtual. At angles other than vertical, this conclusion must be modified.

FIGURE 33–38 Example 33–10.

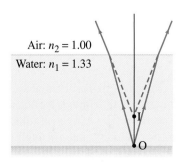

Air: $n_2 = 1.00$
Water: $n_1 = 1.33$

EXAMPLE 33–11 **A spherical "lens."** A point source of light is placed at a distance of 25.0 cm from the center of a glass sphere ($n = 1.5$) of radius 10.0 cm, Fig. 33–39. Find the image of the source.

SOLUTION As shown in Fig. 33–39, there are two refractions, and we treat them successively, one at a time. The light rays from the source first refract from the convex glass surface facing the source. We analyze this refraction ignoring the back side of the sphere, treating it as if it were in the shape shown in Fig. 33–35. Using Eq. 33–7, assuming paraxial rays, with $n_1 = 1.0$, $n_2 = 1.5$, $R = 10.0$ cm and $d_o = 25.0$ cm $- 10.0$ cm $= 15.0$ cm, we solve for the image distance as formed at surface 1, d_{i1}:

$$\frac{1}{d_{i1}} = \frac{1}{1.5}\left(\frac{1.5 - 1.0}{10.0\ \text{cm}} - \frac{1.0}{15.0\ \text{cm}}\right) = -\frac{1}{90.0\ \text{cm}}.$$

Thus, the image of the first refraction is located 90.0 cm *to the left* of the front surface. This image now serves as the object for the refraction occuring at the back surface (surface 2) of the sphere. This surface is concave so $R = -10.0$ cm, and we consider a ray close to the axis. Then the object distance is $d_{o2} = 90.0$ cm $+ 2(10.0$ cm$) = 110.0$ cm, and Eq. 33–7 yields, with $n_1 = 1.5$, $n_2 = 1.0$,

$$\frac{1}{d_{i2}} = \frac{1}{1.0}\left(\frac{1.0 - 1.5}{-10.0\ \text{cm}} - \frac{1.5}{110.0\ \text{cm}}\right) = \frac{4.0}{110.0\ \text{cm}}$$

so $d_{i2} = 28$ cm. Thus, the final image is located a distance 28 cm from the back side of the sphere.

FIGURE 33–39 Example 33–11.

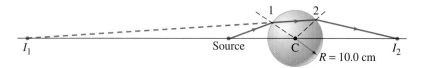

Summary

Light appears to travel along straight-line paths, called **rays**, at a speed v that depends on the **index of refraction**, n, of the material; that is

$$v = \frac{c}{n},$$

where c is the speed of light in vacuum.

When light reflects from a flat surface, *the angle of reflection equals the angle of incidence*. This **law of reflection** explains why mirrors can form **images**.

In a plane mirror, the image is virtual, upright, the same size as the object, and is as far behind the mirror as the object is in front.

A spherical mirror can be concave or convex. A concave spherical mirror focuses parallel rays of light (light from a very distant object) to a point called the **focal point**. The distance of this point from the mirror is the **focal length** f of the mirror and

$$f = \frac{r}{2}$$

where r is the radius of curvature of the mirror.

Parallel rays falling on a convex mirror reflect from the mirror as if they diverged from a common point behind the mirror. The distance of this point from the mirror is the focal length and is considered negative for a convex mirror.

For a given object, the position and size of the image formed by a mirror can be found by ray tracing. Algebraically, the relation between image and object distances, d_i and d_o, and the focal length f, is given by the **mirror equation**:

$$\frac{1}{d_o} + \frac{1}{d_i} = \frac{1}{f}.$$

The ratio of image height to object height, which equals the magnification m, is

$$m = \frac{h_i}{h_o} = -\frac{d_i}{d_o}.$$

If the rays that converge to form an image actually pass through the image, so the image would appear on film or a screen placed there, the image is said to be a **real image**. If the rays do not actually pass through the image, the image is a **virtual image**.

When light passes from one transparent medium into another, the rays bend or refract. The **law of refraction (Snell's law)** states that

$$n_1 \sin \theta_1 = n_2 \sin \theta_2,$$

where n_1 and θ_1 are the index of refraction and angle with the normal to the surface for the incident ray, and n_2 and θ_2 are for the refracted ray.

The wavelength of light determines its color; the **visible spectrum** extends from 400 nm (violet) to about 750 nm (red).

Glass prisms (and other transparent materials) can break white light down into its constituent colors because the index of refraction varies with wavelength, a phenomenon known as **dispersion**.

When light rays reach the boundary of a material where the index of refraction decreases, the rays will be **totally internally reflected** if the incident angle, θ_1, is such that Snell's law would predict $\sin \theta_2 > 1$; this occurs if θ_1 exceeds the critical angle θ_C given by

$$\sin \theta_C = \frac{n_2}{n_1}.$$

Questions

1. What would be the appearance of the Moon if it had (a) a rough surface; (b) a polished mirrorlike surface?
2. Archimedes is said to have burned the whole Roman fleet in the harbor of Syracuse by focusing the rays of the Sun with a huge spherical mirror. Is this reasonable?
3. If a concave mirror produces a real image, is the image necessarily inverted?
4. When you use a concave mirror, you cannot see an inverted image of yourself unless you place your head beyond the center of curvature C. Yet you can see an inverted image of another object placed between C and F, as in Fig. 33–15. Explain. [*Hint*: You can see a real image only if your eye is behind the image, so that the image can be formed.]
5. What is the focal length of a plane mirror? What is the magnification of a plane mirror?
6. Does the mirror equation, Eq. 33–3, hold for a plane mirror? Explain.
7. How might you determine the speed of light in a solid, rectangular, transparent object?
8. When you look at the Moon's reflection from a ripply sea, it appears elongated (Fig. 33–40). Explain.

9. What is the angle of refraction when a light ray meets the boundary between two materials perpendicularly?
10. When you look down into a swimming pool or a lake, you are likely to underestimate its depth. How does the apparent depth vary with the viewing angle? (Use ray diagrams.)
11. Draw a ray diagram to show why a stick looks bent when part of it is under water (Fig. 33–21).
12. When a wide beam of parallel light enters water at an angle, the beam broadens. Explain.
13. Your eye looks into an aquarium and views a fish inside. One ray of light that emerges from the tank from the fish is shown in Fig. 33–41. Also shown is the apparent position of the fish as seen by the eye ball. In the drawing, indicate the approximate position of the actual fish. Briefly justify your answer.

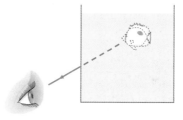

FIGURE 33–41
Question 13.

14. How can you "see" a round drop of water on a table even though the water is transparent and colorless?
15. Can a light ray traveling in air be totally reflected when it strikes a smooth water surface if the incident angle is right?
16. When you look up at an object in air from beneath the surface in a swimming pool, does the object appear to be the same size as when you see it directly in air? Explain.
17. What type of mirror is shown in Fig. 33–42?

FIGURE 33–42
Question 17.

FIGURE 33–40
Question 8.

Problems

Section 33–2

1. (I) What is the speed of light in (a) ethyl alcohol, (b) lucite?

2. (I) The speed of light in ice is 2.29×10^8 m/s. What is the index of refraction of ice?

3. (I) How long does it take light to reach us from the Sun, 1.50×10^8 km away?

4. (I) Our nearest star (other than the Sun) is 4.2 light years away. That is, it takes 4.2 years for the light to reach Earth. How far away is it in meters?

5. (II) Light is emitted from an ordinary light bulb filament in wave-train bursts of about 10^{-8} s in duration. What is the length in space of such wave trains?

6. (II) The speed of light in a certain substance is 92 percent of its value in water. What is the index of refraction of that substance?

7. (II) What is the minimum angular speed at which Michelson's eight-sided mirror would have had to rotate in order that light would be reflected into an observer's eye by succeeding mirror faces (Fig. 33–2)?

Section 33–3

8. (I) Suppose that you want to take a photograph of yourself as you look at your image in a mirror 1.8 m away. For what distance should the camera lens be focused?

9. (II) Stand up two plane mirrors so they form a right angle as in Fig. 33–43. When you look into this double mirror, you see yourself as others see you, instead of reversed as in a single mirror. Make a ray diagram to show how this occurs.

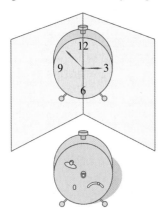

FIGURE 33–43
Problems 9 and 14.

10. (II) A person whose eyes are 1.54 m above the floor stands 2.30 m in front of a vertical plane mirror whose bottom edge is 40 cm above the floor, Fig. 33–44. What is the horizontal distance x to the base of the wall supporting the mirror of the nearest point on the floor that can be seen reflected in the mirror?

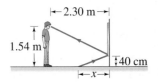

FIGURE 33–44
Problem 10.

11. (II) Two plane mirrors meet at a 135° angle, Fig. 33–45. If light rays strike one mirror at 40° as shown, at what angle ϕ do they leave the second mirror?

FIGURE 33–45
Problem 11.

12. (II) Suppose you are 80 cm from a plane mirror. What area of the mirror is used to reflect the rays entering one eye from a point on the tip of your nose if your pupil diameter is 5.0 mm?

13. (II) Show that if two plane mirrors meet at an angle ϕ, a single ray reflected successively from both mirrors is deflected through an angle of 2ϕ independent of the incident angle. Assume $\phi < 90°$ and that only two reflections, one from each mirror, take place.

14. (III) Suppose a third mirror is placed beneath the two shown in Fig. 33–43, so that all three are perpendicular to each other. (a) Show that for such a "corner reflector," any incident ray will return in its original direction after three reflections. (b) What happens if it makes only two reflections?

Section 33–4

15. (I) A solar cooker, really a concave mirror pointed at the Sun, focuses the Sun's rays 18.2 cm in front of the mirror. What is the radius of the spherical surface from which the mirror was made?

16. (I) How far from a concave mirror (radius 22.0 cm) must an object be placed if its image is to be at infinity?

17. (II) Show with ray diagrams that the magnification of a concave mirror is less than 1 if the object is beyond the center of curvature C, and is greater than 1 if it is within this point.

18. (II) If you look at yourself in a shiny Christmas tree ball with a diameter of 9.0 cm when your face is 25.0 cm away from it, where is your image? Is it real or virtual? Is it upright or inverted?

19. (II) A mirror at an amusement park shows an upright image of any person who stands 1.5 m in front of it. If the image is three times the person's height, what is the radius of curvature of the mirror?

20. (II) A dentist wants a small mirror that, when 2.00 cm from a tooth, will produce a 5.0× upright image. What kind of mirror must be used and what must its radius of curvature be?

21. (II) Some rearview mirrors produce images of cars to your rear that are smaller than they would be if the mirror were flat. Are the mirrors concave or convex? What is a mirror's radius of curvature if cars 20.0 m away appear 0.33 their normal size?

22. (II) A luminous object 3.0 mm high is placed 20 cm from a convex mirror of radius of curvature 20 cm. (a) Show by ray tracing that the image is virtual, and estimate the image distance. (b) Show that to compute this (negative) image distance from Eq. 33–3, it is sufficient to let the focal length be −10 cm. (c) Compute the image size, using Eq. 33–4.

23. (II) (a) Where should an object be placed in front of a concave mirror so that it produces an image at the same location as the object? (b) Is the image real or virtual? (c) Is the image inverted or erect? (d) What is the magnification of the image?

24. (II) The image of a distant tree is virtual and very small when viewed in a curved mirror. The image appears to be 14.0 cm behind the mirror. What kind of mirror is it, and what is its radius of curvature?

25. (II) Use the mirror equation to show that the magnitude of the magnification of a concave mirror is less than 1 if the object is beyond the center of curvature C($d_o > r$), and is greater than 1 if the object is within C($d_o < r$).

26. (II) Show, using a ray diagram, that the magnification m of a convex mirror is $m = -d_i/d_o$, just as for a concave mirror. [Hint: Consider a ray from the top of the object that reflects at the center of the mirror.]

27. (II) Use ray diagrams to show that the mirror equation, Eq. 33–3, is valid for a convex mirror as long as f is considered negative.

28. (II) The magnification of a convex mirror is +0.55× for objects 3.2 m from the mirror. What is the focal length of this mirror?

29. (II) A 4.5-cm tall object is placed 28 cm in front of a spherical mirror. It is desired to produce a virtual image that is erect and 3.5 cm tall. (a) What type of mirror should be used? (b) Where is the image located? (c) What is the focal length of the mirror? (d) What is the radius of curvature of the mirror?

30. (II) A shaving/cosmetic mirror is designed to magnify your face by a factor of 1.3 when your face is placed 20.0 cm in front of it. (a) What type of mirror is it? (b) Describe the type of image that it makes of your face. (c) Calculate the required radius of curvature for the mirror.

31. (III) A short thin object (like a short length of wire) of length l is placed on, and parallel to, the principal axis of a spherical mirror. Show that its image has length $l' = m^2 l$ so the *longitudinal magnification* is equal to $-m^2$ where m is the lateral magnification (Eq. 33–4). Why the minus sign? [Hint: Find the image positions for both ends of the rod, and assume l is very small.]

Section 33–5

32. (I) A flashlight beam strikes the surface of a pane of glass ($n = 1.50$) at a 63° angle to the normal. What is the angle of refraction?

33. (I) A diver shines a flashlight upward from beneath the water at a 32.5° angle to the vertical. At what angle does the light leave the water?

34. (I) A light beam coming from an underwater spotlight exits the water at an angle of 76.0°. At what angle of incidence did it hit the air–water interface from below the surface?

35. (I) Rays of the Sun are seen to make a 43.0° angle to the vertical beneath the water. At what angle above the horizon is the Sun?

36. (II) Light is incident on an equilateral glass prism at a 45.0° angle to one face, Fig. 33–46. Calculate the angle at which light emerges from the opposite face. Assume that $n = 1.50$.

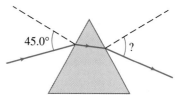

FIGURE 33–46
Problems 36 and 52.

37. (II) In searching the bottom of a pool at night, a watchman shines a narrow beam of light from his flashlight, 1.3 m above the water level, onto the surface of the water at a point 2.7 m from his foot at the edge of the pool (Fig. 33–47). Where does the spot of light hit the bottom of the pool, measured from the wall beneath his foot, if the pool is 2.1 m deep?

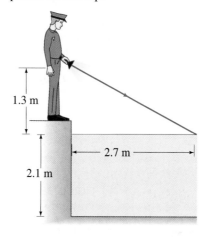

FIGURE 33–47
Problem 37.

38. (II) A beam of light in air strikes a slab of crown glass ($n = 1.52$) and is partially reflected and partially refracted. Find the angle of incidence if the angle of reflection is twice the angle of refraction.

39. (II) If the medium to the left of the glass in Fig. 33–22 is different than that to the right (so there are three different materials of different index of refraction), show that the angle of refraction into the third medium (on the right of the glass) is the same as if the light passed from the first medium to the third directly (as if the glass were of zero thickness).

40. (II) An aquarium filled with water has flat glass sides whose index of refraction is 1.58. A beam of light from outside the aquarium strikes the glass at a 43.5° angle to the perpendicular (Fig. 33–48). What is the angle of this light ray when it enters (a) the glass, and then (b) the water? (c) What would be the refracted angle if the ray entered the water directly?

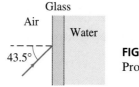

FIGURE 33–48
Problem 40.

41. (II) Prove in general that for a light beam incident on a uniform layer of transparent material, as in Fig. 33–22, the direction of the emerging beam is parallel to the incident beam, independent of the incident angle θ. Assume the medium on the two sides is the same.

42. (III) A light ray is incident on a flat piece of glass as in Fig. 33–22. Show that if the incident angle θ is small, the ray is displaced a distance $d = t\theta(n - 1)/n$, where t is the thickness of the glass and θ is in radians.

Section 33–6

43. (I) By what percent, approximately, does the speed of red light (700 nm) exceed that of violet light (400 nm) in silicate flint glass? (See Fig. 33–26).

44. (I) By what percent is the speed of blue light (450 nm) less than the speed of red light (700 nm), in silicate flint glass (see Fig. 33–26).

45. (II) A light beam strikes a piece of glass at a 60.00° incident angle. The beam contains two wavelengths, 450.0 nm and 700.0 nm, for which the index of refraction of the glass is 1.4820 and 1.4742, respectively. What is the angle between the two refracted beams?

46. (II) A parallel beam of light containing two wavelengths, $\lambda_1 = 450$ nm and $\lambda_2 = 650$ nm, enters the silicate flint glass of an equilateral prism as shown in Fig. 33–49. At what angle does each beam leave the prism (give angle with normal to the face)?

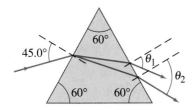

FIGURE 33–49
Problem 46.

Section 33–7

47. (I) What is the critical angle for the interface between water and lucite? To be internally reflected, the light must start in which material?

48. (I) The critical angle for a certain liquid-air surface is 51.3°. What is the index of refraction of the liquid?

49. (II) A beam of light is emitted in a pool of water from a depth of 82.0 cm. Where must it strike the air–water interface, relative to the spot directly above it, in order that the light does *not* exit the water?

50. (II) A ray of light enters a light fiber at an angle of 15.0° with the long axis of the fiber, as in Fig. 33–50. Calculate the distance the light ray travels between successive reflections off the sides of the fiber. Assume that the fiber has an index of refraction of 1.55 and is 1.40×10^{-4} m in diameter.

FIGURE 33–50 Problem 50.

51. (II) A beam of light is emitted 8.0 cm beneath the surface of a liquid and strikes the surface 7.0 cm from the point directly above the source. If total internal reflection occurs, what can you say about the index of refraction of the liquid?

52. (III) Suppose a ray strikes the left face of the prism in Fig. 33–46 at 45° as shown, but is totally internally reflected at the opposite side. If the apex angle (at the top) is $\theta = 75°$, what can you say about the index of refraction of the prism?

53. (III) A beam of light enters the end of an optic fiber as shown in Fig. 33–51. Show that we can guarantee total internal reflection at the side surface of the material (at point a), if the index of refraction is greater than about 1.42. In other words, regardless of the angle α, the light beam reflects back into the material at point a.

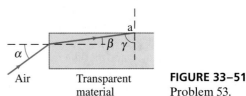

FIGURE 33–51
Problem 53.

54. (III) (a) What is the minimum index of refraction for a glass or plastic prism to be used in binoculars (Fig. 33–31) so that total internal reflection occurs at 45°? (b) Will binoculars work if its prisms (assume $n = 1.50$) are immersed in water? (c) What minimum n is needed if the prisms are immersed in water?

* Section 33–8

* 55. (II) A 12.0-cm-thick plane piece of glass ($n = 1.50$) lies on the surface of a 12.0-cm-deep pool of water. How far below the top of the glass does the bottom of the pool seem, as viewed from directly above?

* 56. (II) A fish is swimming in water inside a thin spherical glass bowl of uniform thickness. Assuming the radius of curvature of the bowl is 25.0 cm, locate the image of the fish if the fish is located: (a) at the center of the bowl; (b) 20.0 cm from the side of the bowl between the observer and the center of the bowl. [*Hint*: For (a) what is the angle of incidence of the light as it strikes the water–glass interface? Given this, where is the image? Finally, verify this by computation, noticing that R must be negative.]

* 57. (III) Show that Eq. 33–7 is valid for both convex and concave spherical surfaces and for differently located objects and images as long as the conventions discussed in Section 33–8 are adhered to. Show this by using diagrams similar to Fig. 33–36 for all possible cases. Assume both $n_2 > n_1$ and then $n_2 < n_1$.

* 58. (III) A coin lies at the bottom of a 1.00 m deep pool. If a viewer sees it at a 45° angle, where is the image of the coin, relative to the coin? [*Hint*: The image is found by tracing back to the intersection of two rays.]

General Problems

59. Two plane mirrors are facing each other 2.0 m apart as in Fig. 33–52. You stand 1.5 m away from one of these mirrors and look into it. You will see multiple images of yourself. (*a*) How far away from you are the first three images of yourself in the mirror in front of you? (*b*) Which way are these first three images facing, toward you or away from you?

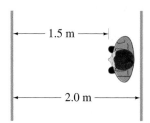

FIGURE 33–52
Problem 59.

60. We wish to determine the depth of a swimming pool filled with water by measuring the width ($x = 5.50$ m) and then noting that the bottom edge of the pool is just visible at an angle of 14.0° above the horizontal as shown in Fig. 33–53. Calculate the depth of the pool.

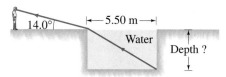

FIGURE 33–53   Problem 60.

61. A 1.70-m-tall person stands 3.80 m from a convex mirror and notices that he looks precisely half as tall as he does in a plane mirror placed at the same distance. What is the radius of curvature of the convex mirror? (Assume that $\sin\theta \approx \theta$.)

62. The critical angle of a certain piece of plastic in air is $\theta_C = 37.3°$. What is the critical angle of the same plastic if it is immersed in water?

63. Each student in a physics lab is assigned to find the location where a bright object may be placed in order that a concave mirror with radius of curvature $r = 40$ cm, will produce an image three times the size of the object. Two students complete the assignment at different times using identical equipment, but when they compare notes later, they discover that their answers for the object distance are not the same. Explain why they do not necessarily need to repeat the lab, and justify your response with a calculation.

64. A kaleidoscope makes symmetric patterns with two plane mirrors having a 60° angle between them as shown in Fig. 33–54. Draw the location of the images (some of them images of images) of the object placed between the mirrors.

FIGURE 33–54
Problem 64.

65. If the apex angle of a prism is $\phi = 72°$ (see Fig. 33–55), what is the minimum incident angle for a ray if it is to emerge from the opposite side (i.e., not be totally internally reflected), given $n = 1.56$?

66. When light passes through a prism, the angle that the refracted ray makes relative to the incident ray is called the deviation angle δ, Fig. 33–55. Show that this angle is a minimum when the ray passes through the prism symmetrically, perpendicular to the bisector of the apex angle ϕ, and show that the minimum deviation angle, δ_m, is related to the prism's index of refraction n by

$$n = \frac{\sin\frac{1}{2}(\phi + \delta_m)}{\sin\phi/2}.$$

67. *Fermat's principle* states that "light travels between two points along that path which requires the least time, as compared to other nearby paths." From Fermat's principle derive (*a*) the law of reflection ($\theta_i = \theta_r$) and (*b*) the law of refraction (Snell's law). [*Hint*: Choose two appropriate points so that a ray between them can undergo reflection or refraction. Draw a rough path for a ray between these points, and write down an expression of the time required for light to travel the arbitrary path chosen. Then take the derivative to find the minimum.]

68. The end faces of a cylindrical glass rod ($n = 1.54$) are perpendicular to the sides. Show that a light ray entering an end face at any angle will be totally internally reflected inside the rod when it strikes the sides. Assume the rod is in air. What if it were in water?

69. Suppose Fig. 33–35 shows a cylindrical rod whose end has a radius of curvature $R = 2.0$ cm, and the rod is immersed in water with index of refraction of 1.33. The rod has index of refraction 1.50. Find the location of the image of an object 2.0 mm high located 20 cm away from the rod.

70. An optical fiber is a long transparent cylinder of diameter d and index of refraction n. If this fiber is bent sharply, some light hitting the side of the cylinder may escape rather than reflect back into the fiber (Fig. 33–56). What is the smallest radius of curvature at a short bent section for which total internal reflection will be assured for light initially travelling parallel to the axis of the fiber?

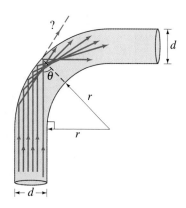

FIGURE 33–56   Problem 70.

Of the many optical devices we discuss in this chapter, the magnifying glass is the simplest. Here it is magnifying a page that describes how it works according to the ray model.

In this chapter we examine thin lenses in detail, seeing how to determine image position as a function of object position and the focal length of the lens, based on the ray model. We will then examine various optical devices from cameras and eyeglasses to telescopes and microscopes.

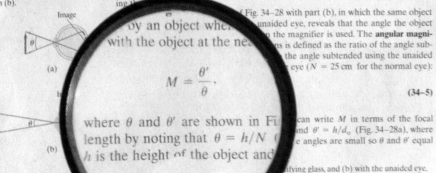

FIGURE 34-26 Photo of a magnifying glass and the image it makes.

FIGURE 34-27 When the same object is viewed at a shorter distance, the image on the retina is greater, so the object appears larger and more detail can be seen. The angle θ that the object subtends in (a) is greater than in (b).

34-7 | Magnifying Glass

Much of the remainder of this chapter will deal with optical devices that are used to produce magnified images of objects. We first discuss the **simple magnifier**, or **magnifying glass**, which is simply a converging lens, Fig. 34-26.

How large an object appears, and how much detail we can see on it, depends on the size of the image it makes on the retina. This, in turn, depends on the angle subtended by the object at the eye. For example, a penny held 30 cm from the eye looks twice as tall as one held 60 cm away because the angle it subtends is twice as great (Fig. 34-27). When we want to examine detail on an object, we bring it up close to our eyes so that it subtends a greater angle. However, our eyes can accommodate only up to a point (the near point), and we will assume a standard distance of 25 cm as the near point in what follows.

A magnifying glass allows us to place the object closer to our eye so that it subtends a greater angle. As shown in Fig. 34-28a, the object is placed at the focal point or just within it. Then the converging lens produces a virtual image, which must be at least 25 cm from the eye if the eye is to focus on it. If the eye is relaxed, the image will be at infinity, and in this case the object is exactly at the focal point. (You make this slight adjustment yourself when you "focus" on the object by moving ... of Fig. 34-28 with part (b), in which the same object ... unaided eye, reveals that the angle the object ... the magnifier is used. The **angular magni-** ... is defined as the ratio of the angle sub- ... the angle subtended using the unaided eye ($N = 25$ cm for the normal eye):

$$M = \frac{\theta'}{\theta},$$

(34-5)

... can write M in terms of the focal ... and $\theta' = h/d_o$ (Fig. 34-28a), where ... e angles are small so θ and θ' equal

where θ and θ' are shown in Fi... length by noting that $\theta = h/N$ ... h is the height of the object and ... ifying glass, and (b) with the unaided eye.

CHAPTER 34

Lenses and Optical Instruments

The laws of reflection and refraction, particularly the latter, are the basis for explaining the operation of many optical instruments. In this chapter we discuss and analyze simple lenses using the model of ray optics discussed in the previous chapter. We then analyze a number of optical instruments, from the magnifying glass and the human eye to telescopes and microscopes. Figure 34-1 shows a variety of lenses.

FIGURE 34-1 (a) Converging and (b) diverging lenses, shown in cross section. (c) Photo of a converging lens (on the left) and a diverging lens. (d) Converging lenses (above), and diverging lenses, lying flat, and raised off the paper to form images.

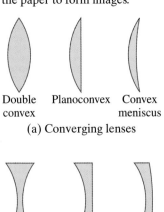

Double convex   Planoconvex   Convex meniscus

(a) Converging lenses

Double concave   Planoconcave   Concave meniscus

(b) Diverging lenses

(c)

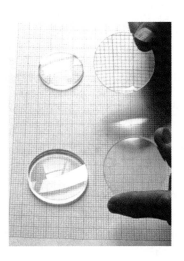

(d)

34-1 | Thin Lenses; Ray Tracing

The most important simple optical device is no doubt the thin lens. The development of optical devices using lenses dates to the sixteenth and seventeenth centuries, although the earliest record of eyeglasses dates from the late thirteenth century.[†] Today we find lenses in eyeglasses, cameras, magnifying glasses, telescopes, binoculars, microscopes, and many specialized instruments. A thin lens is usually circular in cross section, and its two faces are portions of a sphere. (Although cylindrical surfaces are also possible, we will concentrate on spherical.) The two faces can be concave, convex, or plane; several types are shown in Fig. 34-1, in cross section (previous page). The importance of lenses is that they form images of objects, as shown in Fig. 34-2.

FIGURE 34-2 Converging lens (in holder) forms an image (large "F" on screen at right) of a bright object (glowing "F" at the left).

Consider the rays parallel to the axis of the double convex lens which is shown in cross section in Fig. 34-3a. We assume the lens is made of glass or transparent plastic, so its index of refraction is greater than that of the air outside. The **axis** of a lens is a straight line passing through the very center of the lens and perpendicular to its two surfaces (Fig. 34-3). From Snell's law, we can see that each ray in Fig. 34-3a is bent toward the axis at both lens surfaces (note the dashed lines indicating the normals to each surface for the top ray). If rays parallel to the axis fall on a thin lens, they will be focused to a point called the **focal point**, **F**. This will not be precisely true for a lens with spherical surfaces. But it will be very nearly true—that is, parallel rays will be focused to a tiny region that is nearly a point—if the diameter of the lens is small compared to the radii of curvature of the two lens surfaces. This criterion is satisfied by a **thin lens**, one that is very thin compared to its diameter, and we consider only thin lenses here.

[†]Rounded gemstones used as magnifiers probably date from much earlier.

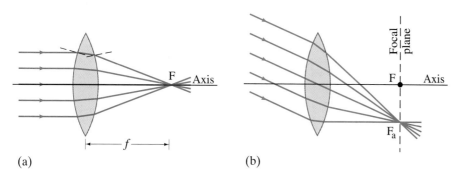

(a)　　　　　　　　　(b)

FIGURE 34-3 Parallel rays are brought to a focus by a converging thin lens.

The rays from a point on a distant object are essentially parallel—see Fig. 33–12. Therefore we can also say that *the focal point is the image point for an object at infinity on the principal axis.* Thus, the focal point of a lens can be found by locating the point where the Sun's rays (or those of some other distant object) are brought to a sharp image, Fig. 34–4. The distance of the focal point from the center of the lens is called the **focal length**, *f*. A lens can be turned around so that light can pass through it from the opposite side. The focal length is the same on both sides, as we shall see later, even if the curvatures of the two lens surfaces are different. If parallel rays fall on a lens at an angle, as in Fig. 34–3b, they focus at a point F_a. The plane in which all points such as F and F_a fall is called the **focal plane** of the lens.

Any lens[†] that is thicker in the center than at the edges will make parallel rays converge to a point, and is called a **converging lens** (see Fig. 34–1a). Lenses that are thinner in the center than at the edges (Fig. 34–1b) are called **diverging lenses** because they make parallel light diverge, as shown in Fig. 34–5. The focal point, F, of a diverging lens is defined as that point from which refracted rays, originating from parallel incident rays, seem to emerge as shown in the Figure. And the distance from F to the lens is called the **focal length**, just as for a converging lens.

Optometrists and ophthalmologists, instead of using the focal length, use the reciprocal of the focal length to specify the strength of eyeglass (or contact) lenses. This is called the **power**, *P*, of a lens:

$$P = \frac{1}{f}. \tag{34–1}$$

The unit for lens power[‡] is the diopter (D), which is an inverse meter: $1\,D = 1\,m^{-1}$. For example, a 20-cm-focal-length lens has a power $P = 1/0.20\,m = 5.0\,D$. We will mainly use the focal length here, but we will refer again to the power of a lens when we discuss eyeglass lenses in Section 34–6.

The most important parameter of a lens is its focal length *f*. For a converging lens, *f* is easily measured by finding the image point for the Sun or other distant objects. Once *f* is known, the image position can be found for any object. To find the image point by drawing rays would be difficult if we had to determine all the refractive angles. Instead, we can do it very simply by making use of certain facts we already know, such as that a ray parallel to the axis of the lens passes (after refraction) through the focal point. In fact, to find an image point, we need consider only the three rays indicated in Fig. 34–6, which shows an arrow as the object and a converging lens forming an image to the right. These rays, emanating from a single point on the object, are drawn as if the lens were infinitely thin, and we show only a single sharp bend within the lens instead of the refractions at each surface. These three rays are drawn as follows:

Ray 1 is drawn parallel to the axis; therefore it is refracted by the lens so that it passes along a line through the focal point F, Fig. 34–6a. (See also Fig. 34–3a.)

Ray 2 is drawn on a line passing through the other focal point F′ (front side of lens in Fig. 34–6) and emerges from the lens parallel to the axis, Fig. 34–6b.

Ray 3 is directed toward the very center of the lens, where the two surfaces are essentially parallel to each other; this ray therefore emerges from the lens at the same angle as it entered; as we saw in Example 33–8, the ray would be displaced slightly to one side, but since we assume the lens is thin, we draw ray 3 straight through as shown.

Actually, any two of these rays will suffice to locate the image point, which is the point where they intersect. Drawing the third can serve as a check.

[†]We are assuming the lens has an index of refraction greater than that of the surrounding material, such as a glass or plastic lens in air, which is the usual situation.

[‡]Note that lens power has nothing to do with power as the rate of doing work or transforming energy (Section 8–8).

Focal length of lens

FIGURE 34–4 Image of the Sun burning a hole, almost, on a piece of paper.

Power of lens

FIGURE 34–5 Diverging lens.

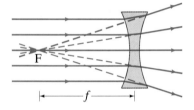

➡ **RAY DIAGRAM**

Finding the image position formed by a thin lens

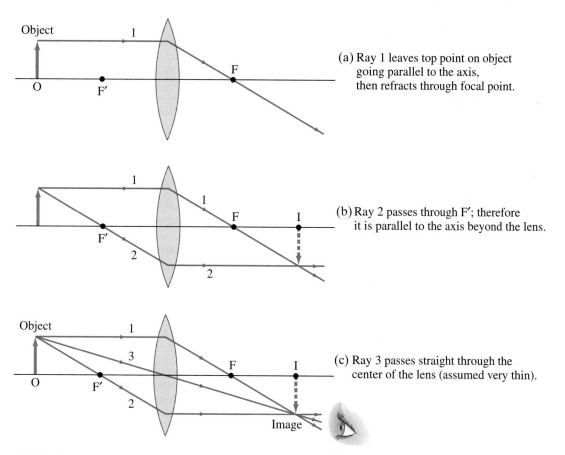

(a) Ray 1 leaves top point on object going parallel to the axis, then refracts through focal point.

(b) Ray 2 passes through F′; therefore it is parallel to the axis beyond the lens.

(c) Ray 3 passes straight through the center of the lens (assumed very thin).

FIGURE 34–6 Finding the image by ray tracing for a converging lens. Rays leave each point on the object. Shown are the three most useful rays, leaving the tip of the object, for determining where the image of that point is formed.

In this way we can find the image point for one point of the object (the top of the arrow in Fig. 34–6). The image points for all other points on the object can be found similarly to determine the complete image of the object. Because the rays actually pass through the image for the case shown in Fig. 34–6, it is a **real image** (see page 814). The image could be detected by film, or actually seen on a white surface placed at the position of the image (Fig. 34–2).

The image can also be seen directly by the eye when the eye is placed behind the image, as shown in Fig. 34–6c, so that some of the rays diverging from each point on the image enter the eye.[†] See Fig. 34–7.

[†] Why, in order to see the image, the rays must be diverging from each point on the image will be discussed in Section 34–6, but is essentially because we see real objects when diverging rays from each point enter the eye as shown in Fig. 33–1.

(a)

(b)

FIGURE 34–7 (a) A converging lens can form a real image (here of a distant building) on a screen. (b) That real image is also directly visible to the eye. Figure 34–1d shows images seen by the eye made by both diverging and converging lenses.

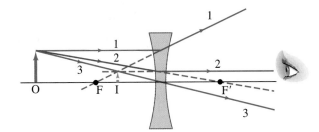

FIGURE 34–8 Finding the image by ray tracing for a diverging lens.

→ **RAY TRACING**

For a diverging lens

By drawing the same three rays we can determine the image position for a diverging lens, as shown in Fig. 34–8. Note that ray 1 is drawn parallel to the axis, but does not pass through the focal point F′ behind the lens. Instead it seems to come from the focal point F in front of the lens (dashed line). Ray 2 is directed toward F′ and is refracted parallel by the lens. Ray 3 passes directly through the center of the lens. The three refracted rays seem to emerge from a point on the left of the lens. This is the image, I. Because the rays do not pass through the image, it is a **virtual image**. Note that the eye does not distinguish between real and virtual images—both are visible.

34–2 The Lens Equation

We now derive an equation that relates the image distance to the object distance and the focal length of the lens. This will make the determination of image position quicker and more accurate than doing ray tracing. Let d_o be the object distance, the distance of the object from the center of the lens, and d_i be the image distance, the distance of the image from the center of the lens; and let h_o and h_i refer to the heights of the object and image. Consider the two rays shown in

FIGURE 34–9 Deriving the lens equation for a converging lens.

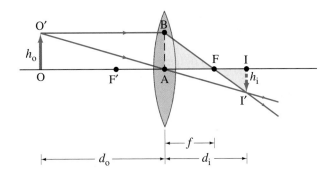

Fig. 34–9 for a converging lens (assumed to be very thin). The triangles FI′I and FBA (highlighted in yellow in Fig. 34–9) are similar because angle AFB equals angle IFI′; so

$$\frac{h_i}{h_o} = \frac{d_i - f}{f},$$

since length AB = h_o. Triangles OAO′ and IAI′ are similar. Therefore,

$$\frac{h_i}{h_o} = \frac{d_i}{d_o}.$$

We equate the right sides of these two equations, divide by d_i, and rearrange to obtain

LENS EQUATION

$$\frac{1}{d_o} + \frac{1}{d_i} = \frac{1}{f}.$$

(34–2)

This is called the **lens equation**. It relates the image distance d_i to the object

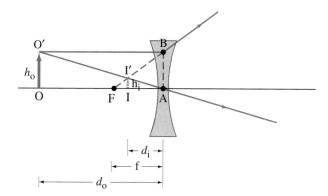

FIGURE 34–10 Deriving the lens equation for a diverging lens.

distance d_o and the focal length f. It is the most useful equation in geometric optics. (Interestingly, it is exactly the same as the mirror equation, Eq. 33–3). Note that if the object is at infinity, then $1/d_o = 0$, so $d_i = f$. Thus the focal length is the image distance for an object at infinity, as mentioned earlier.

We can derive the lens equation for a diverging lens using Fig. 34–10. Triangles IAI′ and OAO′ are similar; and triangles IFI′ and AFB are similar. Thus (noting that length AB = h_o)

$$\frac{h_i}{h_o} = \frac{d_i}{d_o} \quad \text{and} \quad \frac{h_i}{h_o} = \frac{f - d_i}{f}.$$

When these are equated and simplified, we obtain

$$\frac{1}{d_o} - \frac{1}{d_i} = -\frac{1}{f}.$$

This equation becomes the same as Eq. 34–2 if we make f and d_i negative. That is, we take f to be *negative for a diverging lens*, and d_i negative when the image is on the same side of the lens as the light comes from. Thus Eq. 34–2 will be valid for both converging and diverging lenses, and for *all* situations, if we use the following **sign conventions**:

1. The focal length is positive for converging lenses and negative for diverging lenses.
2. The object distance is positive if it is on the side of the lens from which the light is coming (this is usually the case, although when lenses are used in combination, it might not be so); otherwise, it is negative.
3. The image distance is positive if it is on the opposite side of the lens from where the light is coming; if it is on the same side, d_i is negative. Equivalently, the image distance is positive for a real image and negative for a virtual image.
4. The height of the image, h_i, is positive if the image is upright, and negative if the image is inverted relative to the object. (h_o is always taken as positive.)

➡ **PROBLEM SOLVING**

Sign conventions for lenses

Magnification

The **lateral magnification**, m, of a lens is defined as the ratio of the image height to object height, $m = h_i/h_o$. From Figs. 34–9 and 34–10 and the conventions just stated, we have

$$m = \frac{h_i}{h_o} = -\frac{d_i}{d_o}. \tag{34–3}$$

Lateral magnification of a lens

For an upright image the magnification is positive, and for an inverted image m is negative.

From convention 1 above, it follows that the power (Eq. 34–1) of a converging lens, in diopters, is positive, whereas the power of a diverging lens is negative. A converging lens is sometimes referred to as a **positive lens**, and a diverging lens as a **negative lens**.

1. As always, read and reread the problem.
2. Draw a ray diagram, precise if possible, but even a rough one can serve as confirmation of analytic results. Draw at least two, and preferably three, of the easy-to-draw rays described in Figs. 34–6 and 34–8.
3. For analytic solutions, solve for unknowns in the lens equation (Eq. 34–2) and the magnification (Eq. 34–3). The lens equation (Eq. 34–2) involves reciprocals— avoid the obvious error of forgetting to take the reciprocal.
4. Follow the **Sign Conventions** above.
5. Check that your analytic answers are consistent with your ray diagram.

EXAMPLE 34–1 **Image formed by converging lens.** What is (*a*) the position, and (*b*) the size, of the image of a large 7.6-cm-high flower placed 1.00 m from a +50.0-mm-focal-length camera lens?

SOLUTION Figure 34–11 is a rough ray diagram, showing only rays 1 and 3 for a single point on the flower. We see that the image ought to be a little behind the focal point, F, to the right of the lens. (*a*) We find the image position analytically using the lens equation, Eq. 34–2. The camera lens is converging, with $f = +5.00$ cm, and $d_o = 100$ cm, and so the lens equation gives

$$\frac{1}{d_i} = \frac{1}{f} - \frac{1}{d_o} = \frac{1}{5.00 \text{ cm}} - \frac{1}{100 \text{ cm}} = \frac{20.0 - 1.0}{100 \text{ cm}}.$$

Then

$$d_i = \frac{100 \text{ cm}}{19.0} = 5.26 \text{ cm},$$

or 52.6 mm behind the lens. Notice that the image is 2.6 mm farther from the lens than the image for an object at infinity. Indeed, when focusing a camera lens, the closer the object is to the camera, the farther the lens must be from the film. (*b*) The magnification is

$$m = -\frac{d_i}{d_o} = -\frac{5.26 \text{ cm}}{100 \text{ cm}} = -0.0526;$$

so

$$h_i = mh_o = (-0.0526)(7.6 \text{ cm}) = -0.40 \text{ cm}.$$

The image is 4.0 mm high and is inverted ($m < 0$), as in Fig. 34–9, and shown in our sketch, Fig. 34–11.

FIGURE 34–11
Example 34–1.
(Not to scale.)

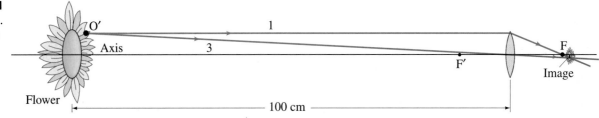

EXAMPLE 34–2 **Object close to converging lens.** An object is placed 10 cm from a 15-cm-focal-length converging lens. Determine the image position and size (*a*) analytically, and (*b*) using a ray diagram.

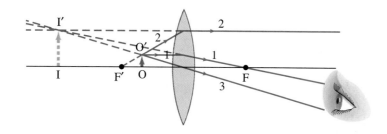

SOLUTION (a) Given $f = 15$ cm and $d_o = 10$ cm, then

$$\frac{1}{d_i} = \frac{1}{15 \text{ cm}} - \frac{1}{10 \text{ cm}} = -\frac{1}{30 \text{ cm}},$$

and $d_i = -30$ cm. (Remember to take the reciprocal!) Because d_i is negative, the image must be virtual and on the same side of the lens as the object. The magnification

$$m = -\frac{d_i}{d_o} = -\frac{-30 \text{ cm}}{10 \text{ cm}} = 3.0.$$

➡ **PROBLEM SOLVING**

Don't forget to take the reciprocal

The image is three times as large as the object and is upright. This lens is being used as a simple magnifying glass, which we discuss in more detail in Section 34–7. (b) The ray diagram is shown in Fig. 34–12, and confirms the result in part (a). For point O' on the top of the object, ray 1 is easy to draw. But ray 2 may take some thought: if we draw it heading toward F', it is going the wrong way—so we have to draw it as if coming from F' (and so dashed), striking the lens, and then going out parallel to the principal axis. We project it backward with a dashed line, as we must do also for ray 1, in order to find where they meet. Ray 3 is easy to draw, through the lens center, and it meets the other two at the image point, I'.

From this last Example and Fig. 34–12, we readily see that, whenever an object is placed between a converging lens and its focal point, the image is virtual.

EXAMPLE 34–3 **Diverging lens.** Where must a small insect be placed if a 25-cm-focal-length diverging lens is to form a virtual image 20 cm in front of the lens?

SOLUTION The ray diagram is basically that of Fig. 34–10 because our lens here is diverging and our image is in front of the lens within the focal distance. (It would be a valuable exercise to draw the ray diagram to scale, precisely, now.) Since $f = -25$ cm and $d_i = -20$ cm, then Eq. 34–2 gives

$$\frac{1}{d_o} = \frac{1}{f} - \frac{1}{d_i} = -\frac{1}{25 \text{ cm}} + \frac{1}{20 \text{ cm}} = \frac{-4 + 5}{100 \text{ cm}} = \frac{1}{100 \text{ cm}}.$$

So the object must be 100 cm in front of the lens.

34–3 | Combinations of Lenses

We now consider Examples illustrating how to deal with lenses used in combination. In general, when light passes through more than one lens, we find the image formed by the first lens as if it were alone. This image becomes the *object* for the second lens, and we find the image then formed by this second lens, which is the final image if there are only two lenses. The total magnification will be the product of the separate magnifications of each lens as we shall see.

Multiple lenses: image formed by first lens is object for second lens

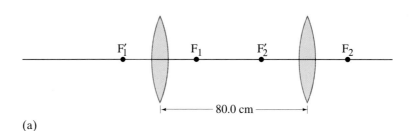

(a)

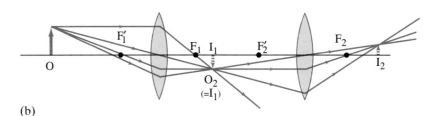

FIGURE 34–13 Example 34–4.   (b)

EXAMPLE 34–4 A two-lens system. Two converging lenses, with focal lengths $f_1 = 20.0$ cm and $f_2 = 25.0$ cm, are placed 80.0 cm apart, as shown in Fig. 34–13a. An object is placed 60.0 cm in front of the first lens as shown in Fig. 34–13b. Determine (a) the position, and (b) the magnification, of the final image formed by the combination of the two lenses.

SOLUTION   (a) The object is a distance $d_{o1} = +60.0$ cm from the first lens, and this lens forms an image whose position can be calculated using the lens equation:

$$\frac{1}{d_{i1}} = \frac{1}{f_1} - \frac{1}{d_{o1}} = \frac{1}{20.0 \text{ cm}} - \frac{1}{60.0 \text{ cm}} = \frac{3 - 1}{60.0 \text{ cm}} = \frac{1}{30.0 \text{ cm}}.$$

Careful!
Note that object distance
for second lens is **not**
equal to the image
distance for first lens

So the first image I_1 is at $d_{i1} = 30.0$ cm behind the first lens. This image becomes the object for the second lens. It is a distance $d_{o2} = 80.0$ cm $-$ 30.0 cm $= 50.0$ cm in front of lens 2, as shown in Fig. 34–13b. The image formed by the second lens, again using the lens equation, is at a distance d_{i2} from the second lens:

$$\frac{1}{d_{i2}} = \frac{1}{f_2} - \frac{1}{d_{o2}} = \frac{1}{25.0 \text{ cm}} - \frac{1}{50.0 \text{ cm}} = \frac{4 - 2}{100.0 \text{ cm}} = \frac{2}{100.0 \text{ cm}}.$$

Hence $d_{i2} = 50.0$ cm behind lens 2. This is the final image—see Fig. 34–13b.
(b) The first lens has a magnification (Eq. 34–3)

$$m_1 = -\frac{d_{i1}}{d_{o1}} = -\frac{30.0 \text{ cm}}{60.0 \text{ cm}} = -0.500.$$

Thus, the first image is inverted and is half as high as the object: again Eq. 34–3,

$$h_{i1} = m_1 h_{o1} = -0.500\, h_{o1}.$$

The second lens takes this image as object and changes its height by a factor

$$m_2 = -\frac{d_{i2}}{d_{o2}} = -\frac{50.0 \text{ cm}}{50.0 \text{ cm}} = -1.000.$$

The final image height is $\left(\text{remember } h_{o2} \text{ is the same as } h_{i1}\right)$:

$$h_{i2} = m_2 h_{o2} = m_2 h_{i1} = m_2 m_1 h_{o1}.$$

Total magnification is
$m = m_1 m_2$

We see from this equation that the total magnification is the product of m_1 and m_2, which here equals $(-1.000)(-0.500) = +0.500$, or $\frac{1}{2}$ the original height and upright.

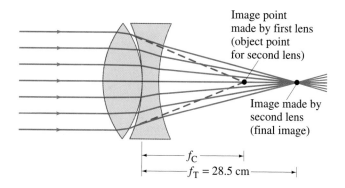

Image point
made by first lens
(object point
for second lens)

Image made by
second lens
(final image)

f_C

$f_T = 28.5$ cm

FIGURE 34–14 Determining the
focal length of a diverging lens.
Example 34–5.

EXAMPLE 34–5 **Measuring _f_ for a diverging lens.** To measure the focal
length of a diverging lens, a converging lens is placed in contact with it, as shown
in Fig. 34–14. The Sun's rays are focused by this combination at a point 28.5 cm,
behind the lenses as shown. If the converging lens has a focal length f_C of 16.0 cm,
what is the focal length f of the diverging lens? Assume both lenses are thin and
the space between them is negligible.

SOLUTION Rays from the Sun are focused 28.5 cm behind the combination, so
the focal length of the total combination is $f_T = 28.5$ cm. If the diverging lens
were absent, the converging lens would form the image at its focal point—that is,
at a distance $f_C = 16.0$ cm behind it (dashed lines in Fig. 34–14). When the diverging lens is placed next to the converging lens, we treat the image formed by the
first lens as the _object_ for the second lens. Since this object lies to the right of the
diverging lens, this is a situation where d_o is negative (see the sign conventions,
page 841). Thus, for the diverging lens, the object is virtual and $d_o = -16.0$ cm.
The diverging lens forms the image of this virtual object at a distance $d_i = 28.5$ cm
away (this was given). Thus,

$$\frac{1}{f_D} = \frac{1}{d_o} + \frac{1}{d_i} = \frac{1}{-16.0 \text{ cm}} + \frac{1}{28.5 \text{ cm}} = -0.0274 \text{ cm}^{-1}.$$

We take the reciprocal to find $f_D = -1/(0.0274 \text{ cm}^{-1}) = -36.5$ cm. Note that the
converging lens must be "stronger" than the diverging lens—that is, it must have
a focal length whose magnitude is less than that of the diverging lens—if this
technique is to work.

34–4 Lensmaker's Equation

In this Section, we will prove that parallel rays are brought to a focus at a _single_
point for a thin lens. At the same time, we will also derive an equation that relates
the focal length of a lens to the radii of curvature of its two surfaces, which is
known as the lensmaker's equation.

In Fig. 34–15, a ray parallel to the axis of a lens is refracted at the front surface of the lens at point A_1 and is refracted at the back surface at point A_2. This
ray then passes through point F, which we call the focal point for this ray. Point A_1
is a height h_1 above the axis, and point A_2 is height h_2 above the axis. C_1 and C_2 are
the centers of curvature of the two lens surfaces; so the length $C_1 A_1 = R_1$, the
radius of curvature of the front surface, and $C_2 A_2 = R_2$ is the radius of the second surface. The thickness of the lens has been grossly exaggerated so the various
angles would be clear. But we will assume that the lens is actually very thin and
that angles between the rays and the axis are small. In this approximation, $h_1 \approx h_2$,
and the sines and tangents of all the angles will be equal to the angles themselves
in radians. For example, $\sin \theta_1 \approx \tan \theta_1 \approx \theta_1$ (radians).

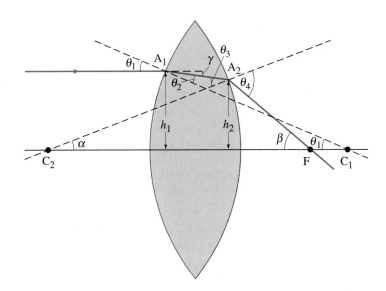

FIGURE 34–15 Diagram of a ray passing through a lens for derivation of the lensmaker's equation.

To this approximation, then, Snell's law tells us that

$$\theta_1 = n\theta_2$$

$$\theta_4 = n\theta_3$$

where n is the index of refraction of the glass, and we assume that the lens is surrounded by air ($n = 1$). Notice also in Fig. 34–15 that

$$\theta_1 \approx \sin\theta_1 = \frac{h_1}{R_1}$$

$$\alpha \approx \frac{h_2}{R_2}$$

$$\beta \approx \frac{h_2}{f}.$$

The last follows because the distance from F to the lens (assumed very thin) is f. From the diagram, the angle γ is

$$\gamma = \theta_1 - \theta_2.$$

A careful examination of Fig. 34–15 shows also that

$$\alpha = \theta_3 - \gamma.$$

This can be seen by drawing a horizontal line to the left from point A_2, which divides the angle θ_3 into two parts. The upper part equals γ and the lower part equals α. (The opposite angles between an oblique line and two parallel lines are equal.) Thus, $\theta_3 = \gamma + \alpha$. Finally, by drawing a horizontal line to the right from point A_2, we divide θ_4 into two parts. The upper part is α and the lower is β. Thus

$$\theta_4 = \alpha + \beta.$$

We now combine all these equations:

$$\alpha = \theta_3 - \gamma = \frac{\theta_4}{n} - (\theta_1 - \theta_2) = \frac{\alpha}{n} + \frac{\beta}{n} - \theta_1 + \theta_2,$$

or

$$\frac{h_2}{R_2} = \frac{h_2}{nR_2} + \frac{h_2}{nf} - \frac{h_1}{R_1} + \frac{h_1}{nR_1}.$$

Because the lens is thin, $h_1 \approx h_2$ and all h's can be canceled from all the numerators.

We then multiply through by n and rearrange to find that

$$\frac{1}{f} = (n-1)\left(\frac{1}{R_1} + \frac{1}{R_2}\right).$$

(34–4)   *Lensmaker's equation*

This is called the **lensmaker's equation**. It relates the focal length of a lens to the radii of curvature of its two surfaces and its index of refraction. Notice that f does not depend on h_1 or h_2. Thus the position of the point F does not depend on where the ray strikes the lens. Hence, all rays parallel to the axis of a thin lens will pass through the same point F, which we wished to prove.

In our derivation, both surfaces are convex and R_1 and R_2 are considered positive.[†] Equation 34–4 also works for lenses with one or both surfaces concave; but for a concave surface, the radius must be considered *negative*.

Notice in Eq. 34–4 that the equation is symmetrical in R_1 and R_2. Thus, if a lens is turned around so that light impinges on the other surface, the focal length is the same even if the two lens surfaces are different.

EXAMPLE 34–6 **Calculating f for a converging lens.** A convex meniscus lens (Figs. 34–1a and 34–16) is made from glass with $n = 1.50$. The radius of curvature of the convex surface is 22.4 cm and that of the concave surface is 46.2 cm. (*a*) What is the focal length? (*b*) Where will it focus an object 2.00 m away?

SOLUTION (*a*) $R_1 = 22.4$ cm and $R_2 = -46.2$ cm; the latter is negative because it refers to the concave surface. Then

$$\frac{1}{f} = (1.50 - 1.00)\left(\frac{1}{22.4\text{ cm}} - \frac{1}{46.2\text{ cm}}\right)$$
$$= 0.0115\text{ cm}^{-1}.$$

So

$$f = \frac{1}{0.0115\text{ cm}^{-1}} = 87\text{ cm}$$

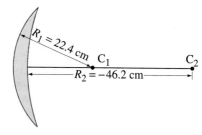

FIGURE 34–16 Example 34–6.

and the lens is converging. Notice that if we turn the lens around so that $R_1 = -46.2$ cm and $R_2 = +22.4$ cm, we get the same result.
(*b*) From the lens equation, with $f = 0.87$ m and $d_o = 2.00$ m, we have

$$\frac{1}{d_i} = \frac{1}{f} - \frac{1}{d_o} = \frac{1}{0.87\text{ m}} - \frac{1}{2.00\text{ m}}$$
$$= 0.65\text{ m}^{-1},$$

so $d_i = 1/0.65\text{ m}^{-1} = 1.53$ m.

EXAMPLE 34–7 **Calculating f for a diverging lens.** A Lucite planoconcave lens (see Fig. 34–1b) has one flat surface and the other has $R = -18.4$ cm. What is the focal length?

SOLUTION From Table 33–1, n for Lucite is 1.51. A plane surface has infinite radius of curvature; if we call this R_1, then $1/R_1 = 0$. Therefore,

$$\frac{1}{f} = (1.51 - 1.00)\left(-\frac{1}{18.4\text{ cm}}\right).$$

So $f = (-18.4\text{ cm})/0.51 = -36$ cm, and the lens is diverging.

[†]Some books use a different convention—for example, R_1 and R_2 are considered positive if their centers of curvature are to the right of the lens, in which case a minus sign appears in their equivalent of Eq. 34–4.

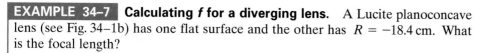

In the remainder of this chapter, we discuss briefly some of the more common optical instruments, including the camera, our eyes and corrective lenses, the magnifying glass, telescopes, and microscopes. We start with the camera.

The basic elements of a **camera** are a lens, a light-tight box, a shutter to let light pass through the lens only briefly, and a sensitized plate or piece of film (Fig. 34–17). When the shutter is opened, light from external objects in the field of view is focused by the lens as an image on the film. The film contains light-sensitive chemicals that undergo change when light strikes them. In the development process, chemical reactions cause the changed areas to turn opaque so that the image is recorded on the film.[†] You can see the image yourself if you can remove the back of the camera and then view through a piece of tissue or wax paper (on which the image can form) placed at the position of the film with the shutter open.

There are three main adjustments on good-quality cameras: shutter speed, *f*-stop, and focusing (see Fig. 34–18), and we now discuss them.

Shutter speed. This refers to how long the shutter is open and the film exposed. It may vary from a second or more ("time exposures") to $\frac{1}{1000}$ s or less. To avoid blurring from camera movement, speeds faster than $\frac{1}{100}$ s are normally used. If the object is moving, faster shutter speeds are needed to "stop" the action. A shutter can be "behind the lens," as in Fig. 34–17, or a "focal plane" shutter which is a moveable curtain just in front of the film.

f-stop. The amount of light reaching the film must be carefully controlled to avoid **underexposure** (too little light for any but the brightest objects to show up) or **overexposure** (too much light, so that all bright objects look the same, with a consequent lack of contrast and a "washed-out" appearance). To control the exposure, a "stop" or iris diaphragm, whose opening is of variable diameter, is placed behind the lens (Fig. 34–17). The size of the opening is varied to compensate for bright or dark days, the sensitivity of the film[‡] used, and for different shutter speeds. The size of the opening is specified by the ***f*-stop**, defined as

$$f\text{-stop} = \frac{f}{D},$$

where f is the focal length of the lens and D is the diameter of the opening. For example, when a 50-mm-focal-length lens has an opening $D = 25\,\text{mm}$, we say it is set at $f/2$. When the lens is set at $f/8$, the opening is only $6\frac{1}{4}$ mm $\left(50/6\frac{1}{4} = 8\right)$. The faster the shutter speed, or the darker the day, the greater the opening must be to get a proper exposure. This corresponds to a smaller *f*-stop number. The smaller the *f*-stop number, the more light passes through the lens to the film. The smallest *f*-number of a lens (largest opening) is referred to as the *speed* of the lens. It is common to find $f/2.0$ lenses today, and some even faster. The advantage of a fast lens is that it allows pictures to be taken under poor lighting conditions. Good quality lenses consist of several elements to reduce the defects present in simple thin lenses (Section 34–10). Standard *f*-stop markings on good lenses are 1.0, 1.4, 2.0, 2.8, 4.0, 5.6, 8, 11, 16, 22, and 32. Each of these stops corresponds to a diameter reduction by a factor of about $\sqrt{2} = 1.4$. Because the amount of light reaching the film is proportional to the *area* of the opening, and therefore proportional to the diameter squared, each standard *f*-stop corresponds to a factor of 2 in light intensity reaching the film.

[†]This is called a *negative*, because the black areas correspond to bright objects and vice versa. The same process occurs during printing to produce a black-and-white "positive" picture from the negative. Color film makes use of three dyes corresponding to the primary colors.

[‡]Different films have different sensitivities to light, referred to as the "film speed," and specified as an "ASA" number; a "faster" film is more sensitive and needs less light to produce a good image.

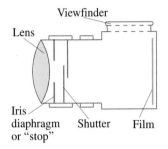

Viewfinder

Lens

Iris diaphragm or "stop"   Shutter   Film

FIGURE 34–17 A simple camera.

FIGURE 34–18 On this camera, the *f*-stops and the focusing ring are on the camera lens. Shutter speeds are selected on the small wheel on top of the camera body.

(a)   (b)

FIGURE 34–19 Photos with camera focused (a) on nearby object with distant object blurry, and (b) on more distant object with nearby object blurry.

Focusing. Focusing is the operation of placing the lens at the correct position relative to the film for the sharpest image. The image distance is a minimum for objects at infinity (the symbol ∞ is used for infinity) and is equal to the focal length. For closer objects, the image distance is greater than the focal length, as can be seen from the lens equation, $1/f = 1/d_o + 1/d_i$. To focus on nearby objects, the lens must therefore be moved away from the film, and this is usually done by turning a ring on the lens.

     If the lens is focused on a nearby object, a sharp image of it will be formed, but distant objects may be blurry (Fig. 34–19). The rays from a point on the distant object will be out of focus—they will form a circle on the film as shown (exaggerated) in Fig. 34–20. The distant object will thus produce an image consisting of overlapping circles and will be blurred. These circles are called **circles of confusion**. To include near and distant objects in the same photo, you can try setting the lens focus at an intermediate position. For a given distance setting, there is a range of distances over which the circles of confusion will be small enough that the images will be reasonably sharp. This is called the **depth of field**. For a particular choice of circle of confusion diameter as upper limit (typically taken to be 0.03 mm for 35-mm cameras), the depth of field varies with the lens opening. If the lens opening is smaller, only rays through the central part of the lens are accepted, and these form smaller circles of confusion for a given object distance. Hence, at smaller lens openings, a greater range of object distances will fit within the circle of confusion criterion, so that the depth of field is greater.

Focusing

Depth of field

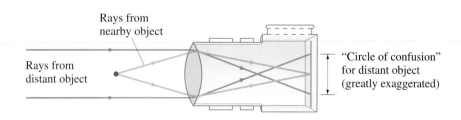

Rays from nearby object

Rays from distant object

"Circle of confusion" for distant object (greatly exaggerated)

FIGURE 34–20 When the lens is positioned to focus on a nearby object, points on a distant object produce circles and are therefore blurred. (The effect is shown greatly exaggerated.)

Camera lenses are categorized into normal, telephoto, and wide angle, according to focal length and film size. A **normal lens** is one that covers the film with a field of view that corresponds approximately to that of normal vision. A normal lens for 35-mm film has a focal length in the vicinity of 50 mm.[†] A **telephoto lens**, as its name implies, acts like a telescope to magnify images. They have longer focal lengths than a normal lens. As we saw in Section 34–2 (Eq. 34–3), the height of the image for a given object distance is proportional to the image distance, and the image distance will be greater for a lens with longer focal length. For distant objects, the image height is very nearly proportional to the focal length (can you prove this?). Thus a 200-mm telephoto lens for use with a 35-mm camera gives a 4× magnification over the normal 50-mm lens. A **wide-angle lens** has a shorter focal length than normal: a wider field of view is included and objects appear smaller. A **zoom lens** is one whose focal length can be changed so that you seem to zoom up to, or away from, the subject as you change the focal length.

Telephoto and wide-angle lenses

CONCEPTUAL EXAMPLE 34–8 **Shutter speed.** To improve the depth of field, you stop down your camera lens by two *f*-stops (say, from *f*/4 to *f*/8). What should you do to the shutter speed to maintain the same exposure?

RESPONSE The amount of light admitted by the lens is proportional to the area of the lens opening. Reducing the lens opening by one *f*-stop reduces the diameter by a factor of $\sqrt{2}$, and the area by a factor of 2. Stopping down by two *f*-stops reduces the area of the lens opening by a factor of 4. To maintain the same exposure, the shutter must be open 4 times as long. So if the shutter speed had been $\frac{1}{250}$ s, you have to increase it to $\frac{1}{60}$ s.

34–6 The Human Eye; Corrective Lenses

The eye

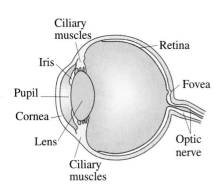

Ciliary muscles
Iris
Pupil
Cornea
Lens
Ciliary muscles
Retina
Fovea
Optic nerve

FIGURE 34–21 Diagram of a human eye.

The human eye resembles a camera in its basic structure (Fig. 34–21), but is far more sophisticated. The eye is an enclosed volume into which light passes through a lens. A diaphragm, called the **iris** (the colored part of your eye), adjusts automatically to control the amount of light entering the eye. The hole in the iris through which light passes (the **pupil**) is black because no light is reflected from it (it's a hole), and very little light is reflected back out from the interior of the eye. The **retina**, which plays the role of the film in a camera, is on the curved rear surface. It consists of a complex array of nerves and receptors known as *rods* and *cones* which act to change light energy into electrical signals that travel along the nerves. The reconstruction of the image from all these tiny receptors is done mainly in the brain, although some analysis is apparently done in the complex interconnected nerve network at the retina itself. At the center of the retina is a small area called the **fovea**, about 0.25 mm in diameter, where the cones are very closely packed and the sharpest image and best color discrimination are found.

Unlike a camera, the eye contains no shutter. The equivalent operation is carried out by the nervous system, which analyzes the signals to form images at the rate of about 30 per second. This can be compared to motion picture or television cameras, which operate by taking a series of still pictures at a rate of 24 (movies) or 30 (U.S. television) per second. Their rapid projection on the screen gives the appearance of motion.

The lens of the eye does little of the bending of the light rays. Most of the refraction is done at the front surface of the **cornea** (index of refraction = 1.376), which also acts as a protective covering. The lens acts as a fine adjustment for

Focusing the eye

[†]Note that a "35-mm camera" uses film that is 35 mm wide; that 35 mm is not to be confused with a focal length.

focusing at different distances. This is accomplished by the ciliary muscles (Fig. 34–21), which change the curvature of the lens so that its focal length is changed. To focus on a distant object, the muscles are relaxed and the lens is thin, Fig. 34–22a, and parallel rays focus at the focal point (on the retina). To focus on a nearby object, the muscles contract, causing the center of the lens to be thicker, Fig. 34–22b, thus shortening the focal length so that images of nearby objects can be focused on the retina, behind the focal point. This focusing adjustment is called **accommodation**.

The closest distance at which the eye can focus clearly is called the **near point** of the eye. For young adults it is typically 25 cm, although younger children can often focus on objects as close as 10 cm. As people grow older, the ability to accommodate is reduced and the near point increases. A given person's **far point** is the farthest distance at which an object can be seen clearly. For some purposes it is useful to speak of a **normal eye** (a sort of average over the population), which is defined as one having a near point of 25 cm and a far point of infinity. To check your own near point, place this book close to your eye and slowly move it away until the type is sharp.

The "normal" eye is more of an ideal than a commonplace. A large part of the population have eyes that do not accommodate within the normal range of 25 cm to infinity, or have some other defect. Two common defects are nearsightedness and farsightedness. Both can be corrected to a large extent with lenses—either eyeglasses or contact lenses.

Nearsightedness, or *myopia*, refers to an eye that can focus only on nearby objects. The far point is not infinity but some shorter distance, so distant objects are not seen clearly. It is usually caused by an eyeball that is too long, although sometimes it is the curvature of the cornea that is too great. In either case, images of distant objects are focused in front of the retina. A diverging lens, because it causes parallel rays to diverge, allows the rays to be focused at the retina (Fig. 34–23a) and thus corrects this defect.

Farsightedness, or *hyperopia*, refers to an eye that cannot focus on nearby objects. Although distant objects are usually seen clearly, the near point is somewhat greater than the "normal" 25 cm, which makes reading difficult. This defect is caused by an eyeball that is too short or (less often) by a cornea that is not sufficiently curved. It is corrected by a converging lens, Fig. 34–23b. Similar to hyperopia is *presbyopia*, which refers to the lessening ability of the eye to accommodate as one ages, and the near point moves out. Converging lenses also compensate for this.

Astigmatism is usually caused by an out-of-round cornea or lens so that point objects are focused as short lines, which blurs the image. Astigmatism is corrected with the use of a cylindrical lens which, for eyes that are nearsighted or farsighted, is superimposed on the spherical surface, so that the radius of curvature of the correcting lens is different in different planes.

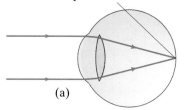

Focal point of lens and cornea

(a)

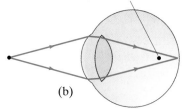

Focal point of lens and cornea

(b)

FIGURE 34–22 Accommodation by a normal eye: (a) lens relaxed, focused at infinity; (b) lens thickened, focused on a nearby object.

➡ P H Y S I C S   A P P L I E D

Corrective lenses

Nearsightedness

Farsightedness

Astigmatism

FIGURE 34–23 Correcting eye defects with lenses: (a) a nearsighted eye, which cannot focus clearly on distant objects, can be corrected by use of a diverging lens; (b) a farsighted eye, which cannot focus clearly on nearby objects, can be corrected by use of a converging lens.

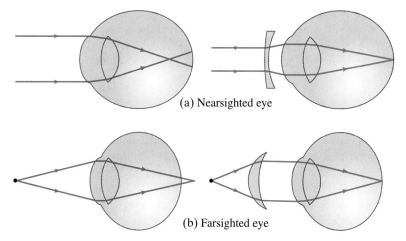

(a) Nearsighted eye

(b) Farsighted eye

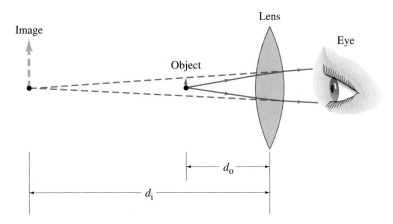

FIGURE 34–24 Lens of reading glasses (Example 34–9).

EXAMPLE 34–9 **Farsighted eye.** A particular farsighted person has a near point of 100 cm. Reading glasses must have what lens power so that this person can read a newspaper at a distance of 25 cm? Assume the lens is very close to the eye.

SOLUTION When the object is placed 25 cm from the lens, we want the image to be 100 cm away on the *same* side of the lens, and so it will be virtual, Fig. 34–24. Thus, $d_o = 25$ cm, $d_i = -100$ cm, and the lens equation gives

$$\frac{1}{f} = \frac{1}{25 \text{ cm}} + \frac{1}{-100 \text{ cm}} = \frac{4 - 1}{100 \text{ cm}} = \frac{1}{33 \text{ cm}}.$$

So $f = 33$ cm $= 0.33$ m. The power P of the lens is $P = 1/f = +3.0$ D. The plus sign indicates that it is a converging lens.

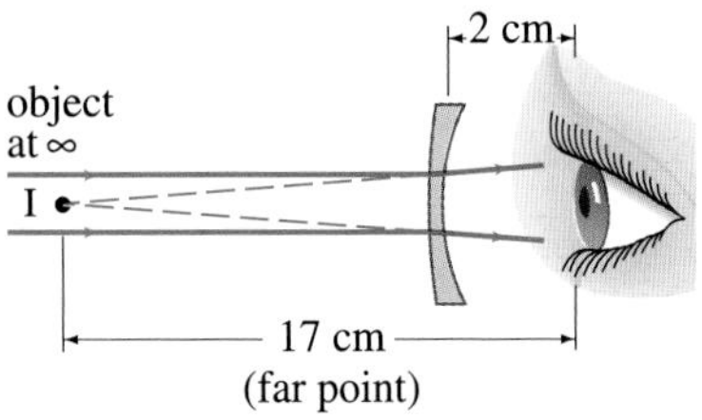

EXAMPLE 34–10 **Nearsighted eye.** A nearsighted eye has near and far points of 12 cm and 17 cm, respectively. (*a*) What lens power is needed for this person to see distant objects clearly, and (*b*) what then will be the near point? Assume that the lens is 2.0 cm from the eye (typical for eyeglasses).

SOLUTION (*a*) First we determine the power of the lens needed to focus objects at infinity, when the eye is relaxed. For a distant object $(d_o = \infty)$, as shown in Fig. 34–25, the lens must put the image 17 cm from the eye (its far point), which is 15 cm in front of the lens; hence $d_i = -15$ cm. We use the lens equation to solve for the focal length of the needed lens:

$$\frac{1}{f} = -\frac{1}{15 \text{ cm}} + \frac{1}{\infty} = -\frac{1}{15 \text{ cm}}.$$

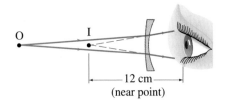

object at ∞

I •

17 cm (far point)

FIGURE 34–25 Example 34–10a.

FIGURE 34–26 Example 34–10b.

O •

I

12 cm (near point)

So $f = -15$ cm $= -0.15$ m or $P = 1/f = -6.7$ D. The minus sign indicates that it must be a diverging lens.

(*b*) To determine the near point when wearing the glasses, we note that a sharp image will be 12 cm from the eye (its near point, see Fig. 34–26), which is 10 cm from the lens; so $d_i = -0.10$ m and the lens equation gives

$$\frac{1}{d_o} = \frac{1}{f} - \frac{1}{d_i} = -\frac{1}{0.15 \text{ m}} + \frac{1}{0.10 \text{ m}} = \frac{1}{0.30 \text{ m}}.$$

So $d_o = 30$ cm, which means the near point when the person is wearing glasses is 30 cm in front of the lens.

Contact lenses

Contact lenses could be used to correct the eye in Example 34–10. Since contacts are placed directly on the cornea, we would not subtract out the 2.0 cm for the image distances. That is, for distant objects $d_i = -17$ cm, so $P = 1/f = -5.9$ D (diopters). The new near point would be 41 cm. Thus we see that a contact lens and an eyeglass lens will require slightly different powers, or focal lengths, for the same eye because of their different placements relative to the eye.

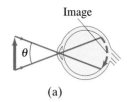

34–7 Magnifying Glass

Much of the remainder of this chapter will deal with optical devices that are used to produce magnified images of objects. We first discuss the **simple magnifier**, or **magnifying glass**, which is simply a converging lens (see chapter opening photo).

How large an object appears, and how much detail we can see on it, depends on the size of the image it makes on the retina. This, in turn, depends on the angle subtended by the object at the eye. For example, a penny held 30 cm from the eye looks twice as tall as one held 60 cm away because the angle it subtends is twice as great (Fig. 34–27). When we want to examine detail on an object, we bring it up close to our eyes so that it subtends a greater angle. However, our eyes can accommodate only up to a point (the near point), and we will assume a standard distance of 25 cm as the near point in what follows.

A magnifying glass allows us to place the object closer to our eye so that it subtends a greater angle. As shown in Fig. 34–28a, the object is placed at the focal point or just within it. Then the converging lens produces a virtual image, which must be at least 25 cm from the eye if the eye is to focus on it. If the eye is relaxed, the image will be at infinity, and in this case the object is exactly at the focal point. (You make this slight adjustment yourself when you "focus" on the object by moving the magnifying glass.)

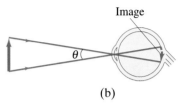

FIGURE 34–27 When the same object is viewed at a shorter distance, the image on the retina is greater, so the object appears larger and more detail can be seen. The angle θ that the object subtends in (a) is greater than in (b).

FIGURE 34–28 Leaf viewed (a) through a magnifying glass, and (b) with the unaided eye, with the eye focused at its near point.

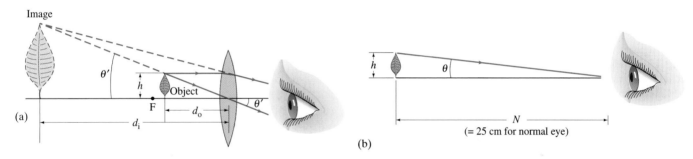

A comparison of part (a) of Fig. 34–28 with part (b), in which the same object is viewed at the near point with the unaided eye, reveals that the angle the object subtends at the eye is much larger when the magnifier is used. The **angular magnification** or **magnifying power**, M, of the lens is defined as the ratio of the angle subtended by an object when using the lens, to the angle subtended using the unaided eye, with the object at the near point N of the eye ($N = 25$ cm for the normal eye):

$$M = \frac{\theta'}{\theta}, \tag{34–5}$$

where θ and θ' are shown in Fig. 34–28. We can write M in terms of the focal length by noting that $\theta = h/N$ (Fig. 34–28b) and $\theta' = h/d_o$ (Fig. 34–28a), where h is the height of the object and we assume the angles are small so θ and θ' equal their sines and tangents. If the eye is relaxed (for least eye strain), the image will be at infinity and the object will be precisely at the focal point; see Fig. 34–29. Then $d_o = f$ and $\theta' = h/f$. Thus

$$M = \frac{\theta'}{\theta} = \frac{h/f}{h/N} = \frac{N}{f}. \quad \begin{bmatrix} \text{eye focused at } \infty; \\ N = 25 \text{ cm for normal eye} \end{bmatrix} \tag{34–6a}$$

We see that the shorter the focal length of the lens, the greater the magnification.

FIGURE 34–29 With the eye relaxed, the object is placed at the focal point, and the image is at infinity. Compare to Fig. 34–28a where the image is at the eye's near point.

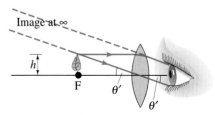

Magnification of a simple magnifier

[However, simple single-lens magnifiers are limited to about 2 or 3× because of distortion due to spherical aberration (Section 34–10).] The magnification of a given lens can be increased a bit by moving the lens and adjusting your eye so it focuses on the image at the eye's near point. In this case, $d_i = -N$ (see Fig. 34–28a) if your eye is very near the magnifier. Then the object distance d_o is given by

$$\frac{1}{d_o} = \frac{1}{f} - \frac{1}{d_i} = \frac{1}{f} + \frac{1}{N}.$$

We see from this equation that $d_o < f$, as shown in Fig. 34–28a, since $N/(f + N)$ must be less than 1. With $\theta' = h/d_o$ the magnification is

$$M = \frac{\theta'}{\theta} = \frac{h/d_o}{h/N} = \frac{N}{d_o} = N\left(\frac{1}{f} + \frac{1}{N}\right)$$

or

$$M = \frac{N}{f} + 1. \qquad \left[\begin{array}{c}\text{eye focused at near point, } N; \\ N = 25\,\text{cm for normal eye}\end{array}\right] \quad \textbf{(34–6b)}$$

We see that the magnification is slightly greater when the eye is focused at its near point, rather than relaxed.

EXAMPLE 34–11 ESTIMATE A jeweler's loupe. An 8-cm-focal-length converging lens is used as a "jeweler's loupe," which is a magnifying glass. Estimate (a) the magnification when the eye is relaxed, and (b) the magnification if the eye is focused at its near point $N = 25$ cm.

SOLUTION (a) With the relaxed eye focused at infinity, $M = N/f = 25\,\text{cm}/8\,\text{cm} \approx 3\times$. (b) The magnification when the eye is focused at its near point ($N = 25$ cm), and the lens is near the eye, is:

$$M = 1 + \frac{N}{f} = 1 + \frac{25}{8} \approx 4\times.$$

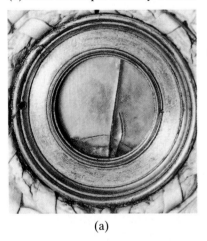

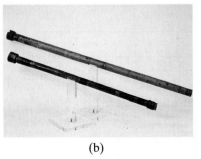

FIGURE 34–30 (a) Objective lens (mounted now in an ivory frame) from the telescope with which Galileo made his world-shaking discoveries, including the moons of Jupiter. (b) Later telescopes made by Galileo.

(a)

(b)

*$\boxed{\text{34–8}}$ Telescopes

A telescope is used to magnify objects that are very far away. In most cases, the object can be considered to be at infinity.

Galileo, although he did not invent it,[†] developed the telescope into a usable and important instrument. He was the first to examine the heavens with the telescope (Fig. 34–30), and he made world-shaking discoveries (the moons of Jupiter, the phases of Venus, sunspots, the structure of the Moon's surface, that the Milky Way is made up of a huge number of individual stars, among others).

Several types of **astronomical telescope** exist. The common **refracting** type, sometimes called **Keplerian**, contains two converging lenses located at opposite ends of a long tube, as diagrammed in Fig. 34–31. The lens closest to the object is called the **objective lens** and forms a real image I_1 of the distant object in the plane of its focal point F_o (or near it if the object is not at infinity). Although this image, I_1, is smaller than the original object, it subtends a greater angle and is very close to the second lens, called the **eyepiece**, which acts as a magnifier. That is, the eye-

---

[†] Galileo built his first telescope in 1609 after having heard of such an instrument existing in Holland. The first telescopes magnified only 3 to 4 times, but Galileo soon made a 30-power instrument. The firs Dutch telescope seems to date from about 1604, but there is a reference suggesting it may have been copied from an Italian telescope built as early as 1590. Kepler gave a ray description (1611) of the Keplerian telescope, which is named for him because he first described it, although he did not build it.

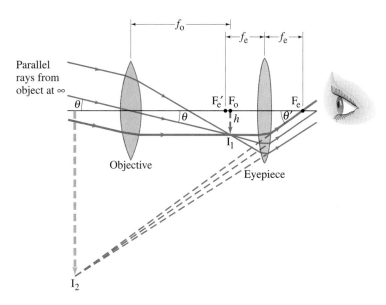

FIGURE 34–31 Astronomical telescope (refracting). Parallel light from one point on a distant object $(d_o = \infty)$ is brought to a focus by the objective lens in its focal plane. This image (I_1) is magnified by the eyepiece to form the final image I_2. Only two of the rays shown are standard rays described in Fig. 34–6.

piece magnifies the image produced by the objective to produce a second, greatly magnified image, I_2, which is virtual, and inverted. If the viewing eye is relaxed, the eyepiece is adjusted so the image I_2 is at infinity. Then the real image I_1 is at the focal point F'_e of the eyepiece, and the distance between the lenses is $f_o + f_e$ for an object at infinity.

To find the total magnification of this telescope, we note that the angle an object subtends as viewed by the unaided eye is just the angle θ subtended at the telescope objective. From Fig. 34–31 we can see that $\theta \approx h/f_o$, where h is the height of the image I_1 and we assume θ is small so that $\tan\theta \approx \theta$. Note, too, that the thickest of the three rays drawn in Fig. 34–31 is parallel to the axis before it strikes the eyepiece and therefore passes through the eyepiece focal point F_e. Thus, $\theta' \approx h/f_e$ and the total magnifying power (angular magnification) of this telescope is

$$M = \frac{\theta'}{\theta} = -\frac{f_o}{f_e}, \qquad (34\text{–}7)$$

Telescope magnification

where we have inserted a minus sign to indicate that the image is inverted. To achieve a large magnification, the objective lens should have a long focal length and the eyepiece a short focal length.

FIGURE 34–32 This large refracting telescope was built in 1897 and is housed at Yerkes Observatory in Wisconsin. The objective lens is 102 cm (40 inches) in diameter, and the telescope tube is about 19 m long.

EXAMPLE 34–12 **Telescope magnification.** The largest refracting telescope in the world is located at the Yerkes Observatory in Wisconsin, Fig. 34–32. It is referred to as a "40-inch" telescope, meaning that the diameter of the objective is 40 inches, or 102 cm. The objective has a focal length of 19 m, and the eyepiece has a focal length of 10 cm. (*a*) Calculate the total magnifying power of this telescope. (*b*) Estimate the length of the telescope.

SOLUTION (*a*) From Eq. 34–7 we find,

$$M = -\frac{f_o}{f_e} = -\frac{19\,\text{m}}{0.10\,\text{m}} = -190\times.$$

(*b*) For a relaxed eye, the image I_1 is at the focal point of both the eyepiece and the objective lenses. The distance between the two lenses is thus $f_o + f_e \approx 19\,\text{m}$, which is essentially the length of the telescope.

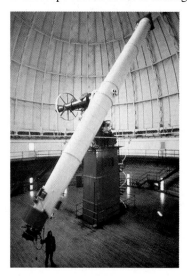

For an astronomical telescope to produce bright images of distant stars, the objective lens must be large to allow in as much light as possible. Indeed, the diameter of the objective (and hence its "light-gathering power") is an important parameter for an astronomical telescope, which is why the largest ones are specified by giving the objective diameter (such as the 200-inch Hale telescope on Palomar Mountain). The construction and grinding of large lenses is very difficult. Therefore,

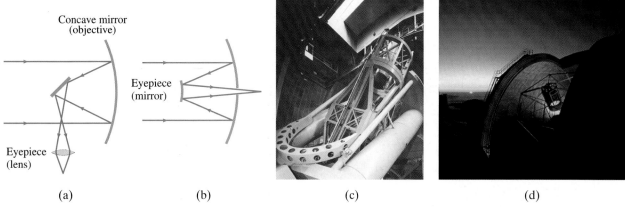

(a)   (b)   (c)   (d)

Concave mirror
(objective)

Eyepiece
(mirror)

Eyepiece
(lens)

FIGURE 34–33 A concave mirror can be used as the objective of an astronomical telescope. Either a lens (a) or a mirror (b) can be used as the eyepiece. Arrangement (a) is called the Newtonian focus and (b) the Cassegrainian focus. Other arrangements are also possible. (c) The 200-inch (mirror diameter) Hale telescope on Palomar Mountain in California. (d) The 10-meter Keck telescope on Mauna Kea, Hawaii. The Keck combines thirty-six 1.8 meter six-sided mirrors into the equivalent of a very large (10-m diameter) single reflector.

Reflecting telescopes

the largest telescopes are **reflecting telescopes** that use a curved mirror as the objective, Fig. 34–33, since a mirror has only one surface to be ground and can be supported along its entire surface† (a large lens, supported at its edges, would sag under its own weight). Normally, the eyepiece lens or mirror (see Fig. 34–33) is removed so that the real image formed by the objective mirror can be recorded directly on film.

A **terrestrial telescope** (for use in viewing objects on Earth), unlike its astronomical counterpart, must provide an upright image. Two designs are shown in Fig. 34–34. The **Galilean** type shown in part (a), which Galileo used for his great astronomical discoveries, has a diverging lens as eyepiece which intercepts the converging rays from the objective lens before they reach a focus, and acts to form a virtual upright image. This design is often used in opera glasses. The tube is reasonably short, but the field of view is small. The second type, shown in Fig. 34–34b, is often called a **spyglass** and makes use of a third lens ("field lens") that acts to make the image upright as shown. A spyglass must be quite long. The most practical design today is the **prism binocular** which was shown in Fig. 33–31. The objective and eyepiece are converging lenses. The prisms reflect the rays by total internal reflection and shorten the physical size of the device, and they also act to produce an upright image. One prism reinverts the image in the vertical plane, the other in the horizontal plane.

FIGURE 34–34 Terrestrial telescopes that produce an upright image: (a) Galilean; (b) spyglass, or field-lens, type.

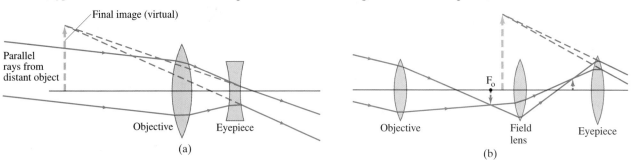

Final image (virtual)

Parallel rays from distant object

Objective   Eyepiece

(a)

Objective F_o Field lens   Eyepiece

(b)

* | 34–9 | Compound Microscope

The compound **microscope**, like the telescope, has both objective and eyepiece (or ocular) lenses, Fig. 34–35. The design is different from that for a telescope because a microscope is used to view objects that are very close, so the object distance is

---

† Another advantage of mirrors is that they exhibit no chromatic aberration because the light doesn't pass through them; and they can be ground in a parabolic shape to correct for spherical aberration (Section 34–10). The reflecting telescope was first proposed by Newton.

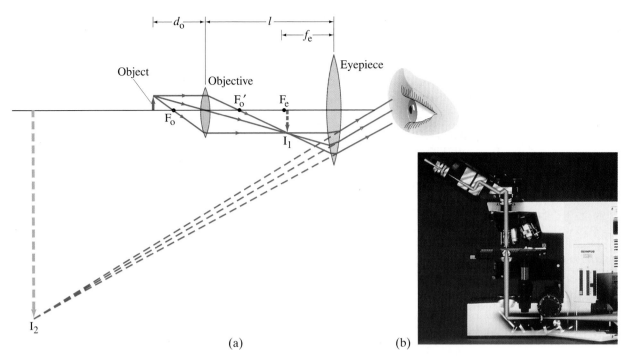

FIGURE 34–35 Compound microscope: (a) ray diagram, (b) photograph (illumination comes from the lower right, then up through the slide holding the object).

very small. The object is placed just beyond the objective's focal point as shown in Fig. 34–35a. The image I_1 formed by the objective lens is real, quite far from the lens, and much enlarged. This image is magnified by the eyepiece into a very large virtual image, I_2, which is seen by the eye and is inverted.

The overall magnification of a microscope is the product of the magnifications produced by the two lenses. The image I_1 formed by the objective is a factor m_o greater than the object itself. From Fig. 34–35a and Eq. 34–3 for the lateral magnification of a simple lens, we have

$$m_o = \frac{h_i}{h_o} = \frac{d_i}{d_o} = \frac{l - f_e}{d_o}, \tag{34–8}$$

where d_o and d_i are the object and image distances for the objective lens, l is the distance between the lenses (equal to the length of the barrel), and we ignored the minus sign in Eq. 34–3 which only tells us that the image is inverted. We set $d_i = l - f_e$, which is true only if the eye is relaxed, so that the image I_1 is at the eyepiece focal point F_e. The eyepiece acts like a simple magnifier. If we assume that the eye is relaxed, the eyepiece angular magnification M_e is (from Eq. 34–6a)

$$M_e = \frac{N}{f_e}, \tag{34–9}$$

where the near point $N = 25$ cm for the normal eye. Since the eyepiece enlarges the image formed by the objective, the overall angular magnification M is the product of the lateral magnification of the objective lens, m_o, times the angular magnification, M_e, of the eyepiece lens (Eqs. 34–8 and 34–9):

$$M = M_e m_o = \left(\frac{N}{f_e}\right)\left(\frac{l - f_e}{d_o}\right) \tag{34–10a}$$

$$\approx \frac{Nl}{f_e f_o}. \tag{34–10b}$$

Magnification

of

microscope

The approximation, Eq. 34–10b, is accurate when f_e and f_o are small compared to l, so $l - f_e \approx l$ and $d_o \approx f_o$ (Fig. 34–35a). This is a good approximation for large magnifications, since these are obtained when f_o and f_e are very small (they are in the denominator of Eq. 34–10b). In order to make lenses of very short focal length, which can be done best for the objective, compound lenses involving several elements must be used to avoid serious aberrations, as discussed in the next Section.

EXAMPLE 34–13 **Microscope.** A compound microscope consists of a 10× eyepiece and a 50× objective 17.0 cm apart. Determine (a) the overall magnification, (b) the focal length of each lens, and (c) the position of the object when the final image is in focus with the eye relaxed. Assume a normal eye, so $N = 25$ cm.

SOLUTION (a) The overall magnification is $(10\times)(50\times) = 500\times$.
(b) The eyepiece focal length is (Eq. 34–9) $f_e = N/M_e = 25 \text{ cm}/10 = 2.5$ cm. For the objective lens, it is easier to next find d_o (part (c)) before we find f_o because we can use Eq. 34–8. Solving for d_o, we find

$$d_o = \frac{l - f_e}{m_o} = \frac{(17.0 \text{ cm} - 2.5 \text{ cm})}{50} = 0.29 \text{ cm}.$$

Then, from the lens equation with $d_i = l - f_e = 14.5$ cm (see Fig. 34–35a),

$$\frac{1}{f_o} = \frac{1}{d_o} + \frac{1}{d_i} = \frac{1}{0.29 \text{ cm}} + \frac{1}{14.5 \text{ cm}} = 3.52;$$

so $f_o = 0.28$ cm.
(c) We just calculated $d_o = 0.29$ cm, which is very close to f_o.

*34–10 Aberrations of Lenses and Mirrors

Earlier in this chapter, we developed a theory of image formation by a thin lens. We found, for example, that all rays from each point on an object are brought to a single point as the image point. This, and other results, were based on approximations such as that all rays make small angles with one another and we can use $\sin\theta \approx \theta$. Because of these approximations, we expect deviations from the simple theory and these are referred to as **lens aberrations**. There are several types of aberration; we will briefly discuss each of them separately but all may be present at one time.

Consider an object at any point (even at infinity) on the axis of a lens. Rays from this point that pass through the outer regions of the lens are brought to a focus at a different point from those that pass through the center of the lens. *Spherical aberration* This is called **spherical aberration**, and is shown exaggerated in Fig. 34–36. Consequently, the image seen on a piece of film (for example) will not be a point but a tiny circular patch of light. If the film is placed at the point C, as indicated, the circle will have its smallest diameter, which is referred to as the **circle of least confusion**. Spherical aberration is present whenever spherical surfaces are used. It can be corrected by using nonspherical lens surfaces, but to grind such lenses is difficult and expensive. It can be minimized with spherical surfaces by choosing the curvatures so that equal amounts of bending occur at each lens surface; a lens can be designed like this for only one particular object distance. Spherical aberration is usually corrected (by which we mean reduced greatly) by the use of several lenses in combination, and by using only the central part of lenses.

FIGURE 34–36 Spherical aberration (exaggerated). Circle of least confusion is at C.

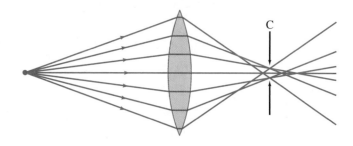

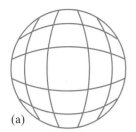 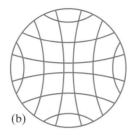

(a)   (b)

FIGURE 34–37 Distortion. Lenses may image a square grid of perpendicular lines to produce (a) barrel distortion or (b) pincushion distortion. These distortions can be seen in the photographs of Fig. 34–1d.

For object points off the lens axis, additional aberrations occur. Rays passing through the different parts of the lens cause spreading of the image that is noncircular. We won't go into the details but merely point out that there are two effects: **coma** (because the image of a point is comet-shaped rather than a circle) and **off-axis astigmatism.**[†] Furthermore, the image points for objects off the axis but at the same distance from the lens do not fall on a flat plane but on a curved surface—that is, the focal plane is not flat. (We expect this because the points on a flat plane, such as the film in a camera, are not equidistant from the lens.) This aberration is known as **curvature of field** and is obviously a problem in cameras and other devices where the film is placed in a flat plane. In the eye, however, the retina is curved, which compensates for this effect. Another aberration, known as **distortion**, is a result of variation of magnification at different distances from the lens axis. Thus a straight line object some distance from the axis may form a curved image. A square grid of lines may be distorted to produce "barrel distortion," or "pincushion distortion," Fig. 34–37. The latter is common in extreme wide-angle lenses.

Off-axis aberrations

All the above aberrations occur for monochromatic light and hence are referred to as *monochromatic aberrations*. Normal light is not monochromatic, and there will also be **chromatic aberration**. This aberration arises because of dispersion—the variation of index of refraction of transparent materials with wavelength (Section 33–6). For example, blue light is bent more than red light by glass. So if white light is incident on a lens, the different colors are focused at different points, Fig. 34–38, and there will be colored fringes in the image. Chromatic aberration can be eliminated for any two colors (and reduced greatly for all others) by the use of two lenses made of different materials with different indices of refraction and dispersion. Normally one lens is converging and the other diverging, and they are often cemented together (Fig. 34–39). Such a lens combination is called an **achromatic doublet** (or "color-corrected" lens).

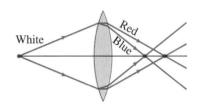

FIGURE 34–38 Chromatic aberration. Different colors are focused at different points.

It is not possible to fully correct all aberrations. Combining two or more lenses together can reduce them. High-quality lenses used in cameras, microscopes, and other devices are **compound lenses** consisting of many simple lenses (referred to as **elements**). A typical high-quality camera lens may contain six to eight (or more) elements.

FIGURE 34–39 Achromatic doublet.

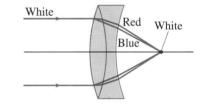

For simplicity we will normally indicate lenses in diagrams as if they were simple lenses. But it must be remembered that good-quality lenses are compound.

The human eye is also subject to aberrations, but they are minimal. Spherical aberration, for example, is minimized because (1) the cornea is less curved at the edges than at the center, and (2) the lens is less dense at the edges than at the center. Both effects cause rays at the outer edges to be bent less strongly, and thus help to reduce spherical aberration. Chromatic aberration is partially compensated for because the lens absorbs the shorter wavelengths appreciably and the retina is less sensitive to the blue and violet wavelengths. This is just the region of the spectrum where dispersion—and thus chromatic aberration—is greatest (Fig. 33–26).

Spherical mirrors (Section 33–4) also suffer aberrations including spherical aberration (see Fig. 33–13). Mirrors can be ground in a parabolic shape to correct for spherical aberration, but they are much harder to make and therefore very expensive. Spherical mirrors do not, however, exhibit chromatic aberration because the light does not pass through them.

[†] Although the effect is the same as for astigmatism in the eye (Section 34–6), the cause is different. Off-axis astigmatism is no problem in the eye because objects are clearly seen only at the fovea, on the lens axis.

Summary

A lens uses refraction to produce a real or virtual image. Parallel rays of light are focused to a point, called the **focal point**, by a converging lens. The distance of the focal point from the lens is called the **focal length** f of the lens. After parallel rays pass through a diverging lens, they appear to diverge from a point, its focal point; and the corresponding focal length is considered negative.

The **power** P of a lens, which is $P = 1/f$, is given in diopters, which are units of inverse meters (m^{-1}).

For a given object, the position and size of the image formed by a lens can be found by ray tracing. Algebraically, the relation between image and object distances, d_i and d_o, and the focal length f, is given by the **lens equation**:

$$\frac{1}{d_o} + \frac{1}{d_i} = \frac{1}{f}.$$

The ratio of image height to object height, which equals the magnification m, is

$$m = \frac{h_i}{h_o} = -\frac{d_i}{d_o}.$$

When using the various equations of geometric optics, it is important to remember the **sign conventions** for all quantities involved: carefully review them when doing problems.

A **camera** lens forms an image on film by allowing light in through a shutter. The lens is focused by moving it relative to the film, and its f-stop (or lens opening) must be adjusted for the brightness of the scene and the chosen shutter speed. The f-stop is defined as the ratio of the focal length to the diameter of the lens opening.

The human **eye** also adjusts for the available light—by opening and closing the iris. It focuses not by moving the lens, but by adjusting the shape of the lens to vary its focal length. The image is formed on the retina, which contains an array of receptors known as rods and cones.

Diverging eyeglass or contact lenses are used to correct the defect of a nearsighted eye, which cannot focus well on distant objects. Converging lenses are used to correct for defects in which the eye cannot focus on close objects.

A **simple magnifier** is a converging lens that forms a virtual image of an object placed at (or within) the focal point. The **angular magnification**, when viewed by a relaxed normal eye, is

$$M = \frac{N}{f},$$

where f is the focal length of the lens and N is the near point of the eye (25 cm for a "normal" eye).

An **astronomical telescope** consists of an **objective** lens or mirror and an **eyepiece** that magnifies the real image formed by the objective. The **magnification** is equal to the ratio of the objective and eyepiece focal lengths, and the image is inverted:

$$M = -\frac{f_o}{f_e}.$$

A compound **microscope** also uses objective and eyepiece lenses, and the final image is inverted. The total magnification is the product of the magnifications of the two lenses and is approximately

$$M \approx \left(\frac{N}{f_e}\right)\left(\frac{l}{f_o}\right),$$

where l is the distance between the lenses, N is the near point of the eye, and f_o and f_e are the focal lengths of objective and eyepiece, respectively.

Microscopes, telescopes, and other optical instruments are limited in the formation of sharp images by **lens aberrations**. These include **spherical aberration**, in which rays passing through the edge of a lens are not focused at the same point as those that pass near the center; and **chromatic aberration**, in which different colors are focused at different points. Compound lenses, consisting of several elements, can largely correct for aberrations.

Questions

1. Where must the film be placed if a camera lens is to make a sharp image of an object far away?

2. A photographer moves closer to his subject and then refocuses. Does the camera lens move farther away from or nearer to the film?

3. Can a diverging lens form a real image under any circumstances? Explain.

4. Show that a real image formed by a thin lens is always inverted, whereas a virtual image is always upright if the object is real.

5. Light rays are said to be "reversible." Is this consistent with the lens equation?

6. Can real images be projected on a screen? Can virtual images? Can either be photographed? Discuss carefully.

7. A thin converging lens is moved closer to a nearby object. Does the real image formed change (a) in position, (b) in size? If yes, describe how.

8. Compare the mirror equation with the lens equation. Discuss similarities and differences, especially the sign conventions for the quantities involved.

9. A lens is made of a material with an index of refraction $n = 1.30$. In air, it is a converging lens. Will it still be a converging lens if placed in water? Explain, using a ray diagram.

10. A dog stands facing a converging lens with its tail in the air. If the nose and the tail are each focused on a screen in turn, which will have the greater magnification?

11. A cat stands facing a converging lens with its tail in the air. Under what circumstances (if any) would the image of the nose be virtual and the image of the tail be real? Where would the image of the rest of the cat be?

12. Why, in Example 34–5, must the converging lens have a shorter focal length than the diverging lens if the latter's focal length is to be determined by combining them?

13. An unsymmetrical lens (say, planoconvex) forms an image of a nearby object. Does the image point change if the lens is turned around?

14. The thicker a double convex lens is in the center as compared to its edges, the shorter its focal length for a given lens diameter. Explain.

15. Does the focal length of a lens depend on the fluid in which it is immersed? What about the focal length of a spherical mirror? Explain.

16. An underwater lens consists of a carefully shaped thin-walled plastic container filled with air. What shape should it have in order to be (a) converging (b) diverging? Use ray diagrams to support your answer.

* 17. Why is the depth of field greater, and the image sharper, when a camera lens is "stopped down" to a larger f-number? Ignore diffraction.

* 18. Explain why swimmers with good eyes see distant objects as blurry when they are underwater. Use a diagram and also show why goggles correct this problem.

* 19. Will a nearsighted person who wears corrective lenses be able to see clearly underwater when wearing glasses? Use a diagram to show why or why not.

* 20. You can tell whether a person is nearsighted or farsighted by looking at the width of the face through their glasses. If the person's face appears narrower through the glasses, is the person farsighted or nearsighted?

* 21. The human eye is much like a camera—yet, when a camera shutter is left open and the camera moved, the image will be blurred; but when you move your head with your eyes open, you still see clearly. Explain.

* 22. In attempting to discern distant details, people will sometimes squint. Why does this help?

* 23. Is the image formed on the retina of the human eye upright or inverted? Discuss the implications of this for our perception of objects.

* 24. Reading glasses use converging lenses. A simple magnifier is also a converging lens. Are reading glasses therefore magnifiers? Discuss the similarities and differences between converging lenses as used for these two different purposes.

* 25. Spherical aberration in a thin lens is minimized if rays are bent equally by the two surfaces. If a planoconvex lens is used to form a real image of an object at infinity, which surface should face the object? Use ray diagrams to show why.

Problems

Sections 34–1 and 34–2

1. (I) A sharp image is located 88.0 mm behind a 65.0-mm-focal-length converging lens. Find the object distance (a) using a ray diagram, (b) by calculation.

2. (I) A certain lens focuses an object 2.85 m away as an image 48.3 cm on the other side of the lens. What type of lens is it and what is its focal length? Is the image real or virtual?

3. (I) (a) What is the power of a 27.5-cm-focal-length lens? (b) What is the focal length of a −6.25-diopter lens? Are these lenses converging or diverging?

4. (II) A stamp collector uses a converging lens with focal length 24 cm to view a stamp 18 cm in front of the lens. (a) Where is the image located? (b) What is the magnification?

5. (II) How large is the image of the Sun on the film used in a camera with (a) a 28-mm-focal-length lens, (b) a 50-mm-focal-length lens, and (c) a 200-mm-focal-length lens? The Sun's diameter is 1.4×10^6 km and it is 1.5×10^8 km away.

6. (II) A 35-mm slide (picture size is actually 24 by 36 mm) is to be projected on a screen 1.80 by 2.70 m placed 8.00 m from the projector. What focal-length lens should be used if the image is to cover the screen?

7. (II) An 80-mm-focal-length lens is used to focus an image on the film of a camera. The maximum distance allowed between the lens and the film plane is 120 mm. (a) How far ahead of the film should the lens be if the object to be photographed is 10.0 m away? (b) 3.0 m away? (c) 1.0 m away? (d) What is the closest object this lens could photograph sharply?

8. (II) A −6.0-diopter lens is held 12.5 cm from an ant 1.0 mm high. What is the position, type, and height of the image?

9. (II) It is desired to magnify reading material by a factor of 2.5× when a book is placed 8.0 cm behind a lens. (a) Draw a ray diagram and describe the type of image this would be. (b) What type of lens is needed for this? (c) What is the power of the lens in diopters?

10. (II) (a) How far from a 50.0-mm-focal-length lens must an object be placed if its image is to be magnified 2.00× and be real? (b) What if the image is to be virtual and magnified 2.00×?

11. (II) (a) An object 37.5 cm in front of a certain lens is imaged 8.20 cm in front of that lens (on the same side as the object). What type of lens is this and what is its focal length? Is the image real or virtual? (b) If the image were located, instead, 46.0 cm in front of the lens, what type of lens would it be and what focal length would it have?

12. (II) (a) A 2.20-cm-high insect is 1.10 m from a 135-mm-focal-length lens. Where is the image, how high is it, and what type is it? (b) What if $f = -135$ mm?

13. (II) A bright object and a viewing screen are separated by a distance of 76.0 cm. At what location(s) between the object and the screen should a lens of focal length 16.0 cm be placed in order to produce a crisp image on the screen? [Hint: First draw a diagram.]

14. (II) How far apart are an object and an image formed by a 75-cm-focal-length converging lens if the image is 2.75× larger than the object and is real?

15. (II) Show analytically that the image formed by a converging lens is real and inverted if the object is beyond the focal point $(d_o > f)$, and is virtual and upright if the object is within the focal point $(d_o < f)$. Describe the image if the object is itself an image formed by another lens, so its position is beyond the lens (on the side of the lens opposite from the incoming light), for which $-d_o > f$, and for which $0 < -d_o < f$.

16. (III) (a) Show that the lens equation can be written in the *Newtonian form*:

$$xx' = f^2,$$

where x is the distance of the object from the focal point on the front side of the lens, and x' is the distance of the image to the focal point on the other side of the lens. Calculate the location of an image if the object is placed 45.0 cm in front of a convex lens with a focal length of 32.0 cm using (b) the standard form of the thin lens formula, and (c) the Newtonian form, derived above.

Section 34–3

17. (II) Two 27.0-cm-focal-length converging lenses are placed 16.5 cm apart. An object is placed 35.0 cm in front of one. Where will the final image formed by the second lens be located? What is the total magnification?

18. (II) A diverging lens with $f = -31.5$ cm is placed 14.0 cm behind a converging lens with $f = 20.0$ cm. Where will an object at infinity be focused?

19. (II) The two converging lenses of Example 34–4 are placed so they are now only 20.0 cm apart. The object is still 60.0 cm in front of the first lens as in Fig. 34–13. In this case, determine (a) the position of the final image, and (b) the overall magnification. (c) Sketch the ray diagram for this system.

20. (II) A 31.0-cm-focal-length converging lens is 24.0 cm behind a diverging lens. Parallel light strikes the diverging lens. After passing through the converging lens, the light is again parallel. What is the focal length of the diverging lens? [Hint: First draw a ray diagram.]

21. (II) A diverging lens is placed next to a converging lens of focal length f_C, as in Fig. 34–14. If f_T represents the focal length of the combination, show that the focal length of the diverging lens, f_D, is given by

$$\frac{1}{f_D} = \frac{1}{f_T} - \frac{1}{f_C}.$$

Section 34–4

22. (I) A double concave lens has surface radii of 33.4 cm and 23.8 cm. What is the focal length if $n = 1.58$?

23. (I) Both surfaces of a double convex lens have radii of 31.0 cm. If the focal length is 28.9 cm, what is the index of refraction of the lens material?

24. (I) Show that if the lens of Example 34–7 is reversed so the light enters the curved face, the focal length is unchanged.

25. (I) A planoconvex lens (Fig. 34–1a) is to have a focal length of 17.5 cm. If made from fused quartz, what must be the radius of curvature of the convex surface?

26. (I) A glass $(n = 1.50)$ planoconcave lens has a focal length of -21.5 cm. What is the radius of the concave surface?

27. (II) A prescription for a corrective lens calls for +2.50 diopters. The lensmaker grinds the lens from a "blank" with $n = 1.56$ and a preformed convex front surface of radius of curvature of 20.0 cm. What should be the radius of curvature of the other surface?

28. (II) An object is placed 90.0 cm from a glass lens $(n = 1.56)$ with one concave surface of radius 22.0 cm and one convex surface of radius 18.5 cm. Where is the final image? What is the magnification?

29. (III) A glass lens $(n = 1.50)$ in air has a power of +4.5 diopters. What would its power be if it were submerged in water?

* Section 34–5

* 30. (I) A 45-mm-focal-length lens has f-stops ranging from $f/1.4$ to $f/22$. What is the corresponding range of lens diaphragm diameters?

* 31. (I) A television camera lens has a 14-cm focal length and a lens diameter of 6.0 cm. What is its f-number?

* 32. (I) A properly exposed photograph is taken at $f/16$ and $\frac{1}{60}$ s. What lens opening would be required if the shutter speed were $\frac{1}{1000}$ s?

* 33. (II) A "pinhole" camera uses a tiny pinhole instead of a lens. Show, using ray diagrams, how reasonably sharp images can be formed using such a pinhole camera. In particular, consider two point objects 2.0 cm apart that are 1.0 m from a 1.0-mm-diameter pinhole. Show that on a piece of film 7.0 cm behind the pinhole the two objects produce two separate circles that do not overlap.

* 34. (II) Suppose that a correct exposure is $\frac{1}{250}$ s at $f/11$. Under the same conditions, what exposure time would be needed for a pinhole camera if the pinhole diameter is 1.0 mm and the film is 7.0 cm from the hole?

* 35. (II) A nature photographer wishes to photograph a 32-m tall tree from a distance of 55 m. What focal-length lens should be used if the image is to fill the 24-mm height of the film?

Section 34–6

36. (I) A human eyeball is about 2.0 cm long and the pupil has a maximum diameter of about 5.0 mm. What is the "speed" of this lens?

37. (II) A person struggles to read by holding a book at arm's length, a distance of 50 cm away. What power of reading glasses should be prescribed for him, assuming they will be placed 2.0 cm from the eye and he wants to read at the "normal" near point of 25 cm?

38. (II) Reading glasses of what power are needed for a person whose near point is 120 cm, so that he can read a computer screen at 50 cm? Assume a lens–eye distance of 1.8 cm.

39. (II) Show that if the nearsighted person in Example 34–10 wore contact lenses corrected for the far point ($= \infty$), that the near point would be 41 cm. (Would glasses be better in this case?)

40. (II) A person's left eye is corrected by a −4.0-diopter lens, 2.0 cm from the eye. (a) Is this person near- or farsighted? (b) What is this person's far point without glasses?

41. (II) A person's right eye can see objects clearly only if they are between 25 cm and 75 cm away. (a) What power of contact lens is required so that objects far away are sharp? (b) What will be the near point with the lens in place?

42. (II) About how much longer is the nearsighted eye in Example 34–10 than the 2.0 cm of a normal eye?

43. (II) One lens of a nearsighted person's eyeglasses has a focal length of −25.0 cm and the lens is 1.8 cm from the eye. If the person switches to contact lenses that are placed directly on the eye, what should be the focal length of the corresponding contact lens?

44. (II) What is the focal length of the eye lens system when viewing an object (a) at infinity, and (b) 30 cm from the eye? Assume that the lens–retina distance is 2.0 cm.

45. (II) A nearsighted person has near and far points of 10.0 and 20.0 cm respectively. If she puts on a pair of glasses with power $P = -4.0$ D, what are her new near and far points? Neglect the distance between her eye and the lens.

Section 34–7

46. (I) What is the magnification of a lens used with a relaxed eye if its focal length is 11 cm?

47. (I) What is the focal length of a magnifying glass of 3.0× magnification for a relaxed normal eye?

48. (I) A magnifier is rated at 2.5× for a normal eye focusing on an image at the near point. (a) What is its focal length? (b) What is its focal length if the 2.5× refers to a relaxed eye?

49. (II) Sherlock Holmes is using an 8.50-cm-focal-length lens as his magnifying glass. To obtain maximum magnification, where must the object be placed (assume a normal eye), and what will be the magnification?

50. (II) A 3.70-mm-wide bolt is viewed with a 9.00-cm-focal-length lens. A normal eye views the image at its near point. Calculate (a) the angular magnification, (b) the width of the image, and (c) the object distance from the lens.

51. (II) A small insect is placed 5.35 cm from a +6.00-cm-focal-length lens. Calculate (a) the position of the image, and (b) the angular magnification.

52. (II) A magnifying glass with a focal length of 9.5 cm is used to read print placed at a distance of 8.5 cm. Calculate: (a) the position of the image; (b) the linear magnification; and (c) the angular magnification.

53. (II) A magnifying glass is rated at 3.0× for a normal eye that is relaxed. What would be the magnification for a relaxed eye whose near point is (a) 60 cm, and (b) 18 cm? Explain the differences.

* Section 34–8

* 54. (I) What is the magnification of an astronomical telescope whose objective lens has a focal length of 72 cm, and whose eyepiece has a focal length of 2.8 cm? What is the overall length of the telescope when adjusted for a relaxed eye?

* 55. (I) The overall magnification of an astronomical telescope is desired to be 25×. If an objective of 80 cm focal length is used, what must be the focal length of the eyepiece? What is the overall length of the telescope when adjusted for use by the relaxed eye?

* 56. (I) An 8.0× binocular has 3.0-cm-focal-length eyepieces. What is the focal length of the objective lenses?

* 57. (II) An astronomical telescope has an objective with focal length 95 cm and a +35 D eyepiece. What is the total magnification?

* 58. (II) An astronomical telescope has its two lenses spaced 76.0 cm apart. If the objective lens has a focal length of 74.5 cm, what is the magnification of this telescope? Assume a relaxed eye.

* 59. (II) A Galilean telescope adjusted for a relaxed eye is 33.0 cm long. If the objective lens has a focal length of 36.0 cm, what is the magnification?

* 60. (II) What is the magnifying power of an astronomical telescope using a reflecting mirror whose radius of curvature is 6.2 m and an eyepiece whose focal length is 2.8 cm?

* 61. (II) The Moon's image appears to be magnified 120× by a reflecting astronomical telescope with an eyepiece having a focal length of 3.3 cm. What are the focal length and radius of curvature of the main mirror?

* 62. (II) A 130× astronomical telescope is adjusted for a relaxed eye when the two lenses are 1.25 m apart. What is the focal length of each lens?

* 63. (III) A 7.0× pair of binoculars has an objective focal length of 26 cm. If the binoculars are focused on an object 4.0 m away (from the objective), what is the magnification? (The 7.0× refers to objects at infinity; Eq. 34–7 holds only for objects at infinity and not for nearby ones.)

* Section 34–9

* 64. (I) A microscope uses an eyepiece with a focal length of 1.6 cm. Using a normal eye with a final image at infinity, the tube length is 17.5 cm and the focal length of the objective lens is 0.65 cm. What is the magnification of the microscope?

* 65. (I) A 650× microscope uses a 0.40-cm-focal-length objective lens. If the tube length is 17.5 cm, what is the focal length of the eyepiece? Assume a normal eye and that the final image is at infinity.

*66. (II) A microscope has a 12.0× eyepiece and a 56.0× objective 20.0 cm apart. Calculate (a) the total magnification, (b) the focal length of each lens, and (c) where the object must be for a normal relaxed eye to see it in focus.

*67. (II) A microscope has a 1.8-cm-focal-length eyepiece and 0.80-cm objective. Assuming a relaxed normal eye calculate (a) the position of the object if the distance between the lenses is 16.0 cm, and (b) the total magnification.

*68. (II) Repeat Problem 69 assuming that the final image is located 25 cm from the eyepiece (near point of a normal eye).

*69. (II) The eyepiece of a compound microscope has a focal length of 2.70 cm and the objective has $f = 0.740$ cm. If an object is placed 0.790 cm from the objective lens, calculate (a) the distance between the lenses when the microscope is adjusted for a relaxed eye, and (b) the total magnification.

* Section 34–10

*70. (II) An achromatic lens is made of two very thin lenses placed in contact that have focal lengths of $f_1 = -28$ cm and $f_2 = +23$ cm. (a) Is the combination converging or diverging? (b) What is the net focal length?

*71. (III) Let's examine spherical aberration in a particular situation. A planoconvex lens of index of refraction 1.50 and a radius of curvature $R = 12.0$ cm is shown in Fig. 34–40. Consider an incoming ray parallel to the principle axis and a height h above it as shown. Determine the distance d, from the flat face of the lens, to where this ray crosses the principle axis if (a) $h = 1.0$ cm, and (b) $h = 6.0$ cm. (c) How far apart are these "focal points"? (d) How large is the "circle of confusion" produced by the $h = 6.0$ cm ray at the "focal point" for $h = 1.0$ cm?

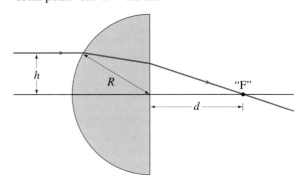

FIGURE 34–40 Problem 71.

General Problems

72. If a 135-mm telephoto lens is designed to cover object distances from 1.40 m to ∞, over what distance must the lens move relative to the plane of the film?

73. A 200-mm-focal-length lens can be adjusted so that it is 200.0 mm to 206.0 mm from the film. For what range of object distances can it be adjusted?

74. Show that for objects very far away (assume infinity), the magnification of a camera lens is proportional to its focal length.

75. For a camera equiped with a 50-mm-focal-length lens, what is the object distance if the image height equals the object height? How far is the object from the image on the film?

76. What is the magnifying power of a +8.0-D lens used as a magnifier? Assume a relaxed normal eye.

77. A lucite planoconvex lens (Fig. 34–1a) has one flat surface and one with $R = 18.4$ cm. It is used to view an object, located 66.0 cm away from the lens, which is a mixture of red and yellow. The index of refraction of the lucite is 1.5106 for red light and 1.5226 for yellow light. What are the locations of the red and yellow images formed by the lens?

78. A movie star catches a reporter shooting pictures of her at home. She claims the reporter was trespassing and to prove her point, she gives as evidence the film she seized. Her 1.75-m height is 8.25 mm high on the film, and the focal length of the camera lens was 200 mm. How far away from the subject was the reporter standing?

79. In a slide or movie projector, the film acts as the object whose image is projected on a screen (Fig. 34–41). If a 100-mm-focal-length lens is to project an image on a screen 7.50 m away, how far from the lens should the slide be? If the slide is 36 mm wide, how wide will the picture be on the screen?

Slide   Lens   Screen

FIGURE 34–41 Slide projector, Problem 79.

80. A converging lens with focal length of 12.0 cm is placed in contact with a diverging lens with a focal length of 20.0 cm. What is the focal length of the combination, and is the combination converging or diverging?

81. Show analytically that a diverging lens can never form a real image of a real object. Can you describe a situation in which a diverging lens can form a real image?

82. A lighted candle is placed 30 cm in front of a converging lens of focal length $f_1 = 15$ cm, which in turn is 50 cm in front of another converging lens of focal length $f_2 = 10$ cm (see Fig. 34–42). (a) Draw a ray diagram and estimate the location and the size of the final image. (b) Calculate the position and size of the final image.

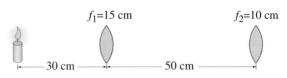

$f_1 = 15$ cm $f_2 = 10$ cm

— 30 cm —   — 50 cm —

FIGURE 34–42 Problem 82.

83. A bright object is placed on one side of a converging lens of focal length f, and a white screen for viewing the image is on the opposite side. The distance $d_T = d_i + d_o$ between the object and the screen is kept fixed, but the lens can be moved. Show that (a) if $d_T > 4f$, there will be two positions where the lens can be placed and a sharp image will be produced on the screen, and (b) if $d_T < 4f$, there will be no lens position where a sharp image is formed. (c) Determine the distance between the two lens positions in part (a), and the ratio of the image sizes.

84. (a) Show that if two thin lenses of focal lengths f_1 and f_2 are placed in contact with each other, the focal length of the combination is given by $f_T = f_1 f_2/(f_1 + f_2)$. ($b$) Show that the power P of the combination of two lenses is the sum of their separate powers, $P = P_1 + P_2$.

85. A lens whose index of refraction is n is submerged in a material whose index of refraction is n' ($n' \neq 1$). Derive the equivalent of Eqs. 34–2, 34–3 and 34–4 for this lens.

86. Sam purchases +2.5-diopter eyeglasses which correct his faulty vision to put his near point at 25 cm. (Assume he wears the lenses 2.0 cm from his eyes.) (a) Is Sam nearsighted or farsighted? (b) Calculate the focal length of Sam's glasses. (c) Calculate Sam's near point without glasses. (d) Pam, who has normal eyes with near point at 25 cm, puts on Sam's glasses. Calculate Pam's near point with Sam's glasses on.

87. A 50-year-old man uses +2.5-diopter lenses to be able to read a newspaper 25 cm away. Ten years later, he finds that he must hold the paper 35 cm away to see clearly with the same lenses. What power lenses does he need now? (Distances are measured from the lens.)

88. A woman can see clearly with her right eye only when objects are between 40 cm and 150 cm away. Prescription bifocals should have what powers so that she can see distant objects clearly (upper part) and be able to read a book 25 cm away (lower part)? Assume that the glasses will be 2.0 cm from the eye.

89. A child has a near point of 15 cm. What is the maximum magnification the child can obtain using an 8.0-cm-focal-length magnifier? Compare to that for a normal eye.

* 90. As early morning passed toward midday, and the sunlight got more intense, a photographer who was taking repeated shots of the same subject noted that, if she kept her shutter speed constant, she had to change the f-number from $f/5.6$ to $f/22$. By how much had the sunlight intensity increased during that time?

* 91. A person tries to make a telescope using the lenses from reading glasses. They have powers of +2.0 D and +5.0 D, respectively. (a) What maximum magnification telescope is possible? (b) Which lens should be used as the eyepiece?

* 92. Exposure times must be increased for pictures taken at very short distances, because of the increased distance of the lens from the film for a focused image. (a) Show that when the object is so close to the camera that the image height equals the object height, the exposure time must be four times longer than when the object is a long distance away (say, ∞), given the same illumination and f-stop. (b) Show that if d_o is at least four or five times the focal length f of the lens, the exposure time is increased negligibly relative to the same object being a great distance away.

* 93. The objective lens and the eyepiece of a telescope are spaced 85 cm apart. If the eyepiece is +25 diopters, what is the total magnification of the telescope?

The wave nature of light nicely explains how light reflected from the front and back surfaces of this very thin film of soapy water interferes constructively to produce the bright colors. Which color we see depends on the thickness of the film at that point—the film is very thin at the top and gets thicker going down toward the bottom (because of gravity). Can you use the clues (colors of light) to determine the thickness of the film at any point? You should be able to after reading this chapter.

CHAPTER 35

The Wave Nature of Light; Interference

That light carries energy is obvious to anyone who has focused the Sun's rays with a magnifying glass on a piece of paper and burned a hole in it. But how does light travel, and in what form is this energy carried? In our discussion of waves in Chapter 15, we noted that energy can be carried from place to place in basically two ways: by particles or by waves. In the first case, material bodies or particles can carry energy, such as an avalanche or rushing water. In the second case, water waves and sound waves, for example, can carry energy over long distances even though mass itself does not travel these distances. Then, what can we say about the nature of light: does light travel as a stream of particles away from its source, or does it travel in the form of waves that spread outward from the source?

Historically, this question has turned out to be a difficult one. For one thing, light does not reveal itself in any obvious way as being made up of tiny particles nor do we see tiny light waves passing by as we do water waves. The evidence seemed to favor first one side and then the other until about 1830, when most physicists had accepted the wave theory. By the end of the nineteenth century, light was considered to be an *electromagnetic wave* (Chapter 32). In the early twentieth century, light was shown to have a particle nature as well, as we shall discuss in Chapter 38. Nonetheless, the wave theory of light remains valid and has proved very successful. We now investigate the evidence for the wave theory and how it has explained a wide range of phenomena.

35-1 | Huygens' Principle and Diffraction

The Dutch scientist Christiaan Huygens (1629–1695), a contemporary of Newton, proposed a wave theory of light that had much merit. Still useful today is a technique he developed for predicting the future position of a wave front when an earlier position is known. This is known as **Huygens' principle** and can be stated as follows: *Every point on a wave front can be considered as a source of tiny wavelets that spread out in the forward direction at the speed of the wave itself. The new wave front is the envelope of all the wavelets—that is, the tangent to all of them.*

As a simple example of the use of Huygens' principle, consider the wave front AB in Fig. 35–1, which is traveling away from a source S. We assume the medium is *isotropic*—that is, the speed v of the waves is the same in all directions. To find the wave front a short time t after it is at AB, tiny circles are drawn with radius $r = vt$. The centers of these tiny circles are on the original wave front AB and the circles represent Huygens' (imaginary) wavelets. The tangent to all these wavelets, the line CD, is the new position of the wave front.

Huygens' principle is particularly useful when waves impinge on an obstacle and the wave fronts are partially interrupted. Huygens' principle predicts that waves bend in behind an obstacle, as shown in Fig. 35–2. This is just what water waves do, as we saw in Chapter 15 (Figs. 15–32 and 15–33). The bending of waves behind obstacles into the "shadow region" is known as **diffraction**. Since diffraction occurs for waves, but not for particles, it can serve as one means for distinguishing the nature of light.

Does light actually exhibit diffraction? In the mid-seventeenth century, a Jesuit priest, Francesco Grimaldi (1618–1663), had observed that when sunlight entered a darkened room through a tiny hole in a screen, the spot on the opposite wall was larger than would be expected from geometric rays. He also observed that the border of the image was not clear but was surrounded by colored fringes. Grimaldi attributed this to the diffraction of light.

Note that the ray model (Chapter 33) cannot account for diffraction, and it is important to be aware of the limitations of the ray model. Geometric optics using rays is so successful in its limited sphere because normal openings and obstacles are much larger than the wavelength of the light, and so relatively little diffraction or bending occurs.

Huygens' principle

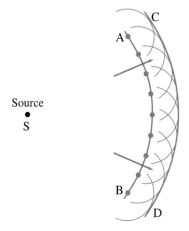

FIGURE 35–1 Huygens' principle used to determine the wave front CD when the wave front AB is given.

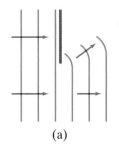 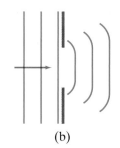 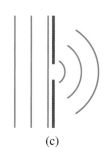

(a) (b) (c)

FIGURE 35–2 Huygens' principle is consistent with diffraction (a) around the edge of an obstacle, (b) through a large hole, (c) through a small hole whose size is on the order of the wavelength of the wave.

35-2 | Huygens' Principle and the Law of Refraction

The laws of reflection and refraction were well known in Newton's time. The law of reflection could not distinguish between the two theories: waves versus particles, that we discussed on the previous page. For when waves reflect from an obstacle, the angle of incidence equals the angle of reflection (Fig. 15–22). The same is true of particles—think of a tennis ball without spin striking a flat surface.

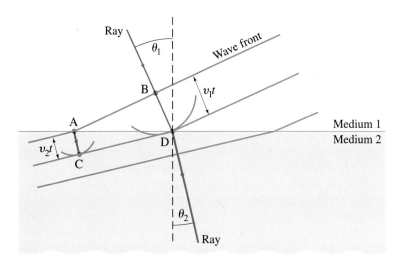

FIGURE 35–3 Refraction explained, using Huygens' principle.

The law of refraction is another matter. Consider light entering a medium where it is bent toward the normal, as when it travels from air into water. As shown in Fig. 35–3, this effect can be constructed using Huygens' principle if we assume the speed of light is less in the second medium $(v_2 < v_1)$. In time t, the point B on wave front AB goes a distance $v_1 t$ to reach point D. Point A, on the other hand, travels a distance $v_2 t$ to reach point C, and $v_2 t < v_1 t$. Huygens' principle is applied to points A and B to obtain the curved wavelets shown at C and D. The wave front is tangent to these two wavelets, so the new wave front is the line CD. Hence the rays, which are perpendicular to the wave fronts, bend toward the normal if $v_2 < v_1$, as drawn.[†] Newton favored a particle theory of light which predicted the opposite result, that the speed of light would be greater in the second medium $(v_2 > v_1)$. Thus the wave theory predicts that the speed of light in water, for example, is less than in air; and Newton's particle theory predicts the reverse. An experiment to actually measure the speed of light in water was performed in 1850 by the French physicist Jean Foucault, and it confirmed the wave-theory prediction. By then, however, the wave theory was already fully accepted, as we shall see in the next Section.

It is easy to show that Snell's law of refraction follows directly from Huygens' principle, given that the speed of light v in any medium is related to the speed in a vacuum, c, and the index of refraction, n, by Eq. 33–1, $v = c/n$. From the Huygens' construction of Fig. 35–3, angle ADC is equal to θ_2 and angle BAD is equal to θ_1. Then for the two triangles that have the common side AD, we have

$$\sin \theta_1 = \frac{v_1 t}{\text{AD}}, \qquad \sin \theta_2 = \frac{v_2 t}{\text{AD}}.$$

We divide these two equations and obtain:

$$\frac{\sin \theta_1}{\sin \theta_2} = \frac{v_1}{v_2}.$$

Then, since $v_1 = c/n_1$ and $v_2 = c/n_2$,

$$n_1 \sin \theta_1 = n_2 \sin \theta_2,$$

which is Snell's law of refraction, Eq. 33–5. (The law of reflection can be derived from Huygens' principle in a similar way, and this is given as Problem 1 at the end of the chapter.)

[†] This is basically the same as the discussion around Fig. 15–31.

When a light wave travels from one medium to another, its frequency does not change, but its wavelength does. This can be seen from Fig. 35–3, where we assume each of the blue lines representing a wave front corresponds to a crest (peak) of the wave. Then

Wavelength depends on n

$$\frac{\lambda_2}{\lambda_1} = \frac{v_2 t}{v_1 t} = \frac{v_2}{v_1} = \frac{n_1}{n_2},$$

where, in the last step, we used Eq. 33–1, $v = c/n$. If medium 1 is a vacuum (or air), so $n_1 = 1$, $v_1 = c$, and we call λ_1 simply λ, then the wavelength in another medium of index of refraction $n (= n_2)$ will be

$$\lambda_n = \frac{\lambda}{n}.$$

(35–1)

This result is consistent with the frequency f being unchanged since $c = f\lambda$. Combining this with $v = f\lambda_n$ in a medium where $v = c/n$ gives $\lambda_n = v/f = c/nf = f\lambda/nf = \lambda/n$, which checks.

Wave fronts can be used to explain how mirages are produced by refraction of light. Let us explain why, for example, on a hot day motorists sometimes see a mirage of water on the highway ahead of them, with distant vehicles seemingly reflected in it (Fig. 35–4a). On a hot day, there can be a layer of very hot air next to the roadway (made hot by the sun beating on the road). Hot air is less dense than cooler air, so the index of refraction is slightly lower in the hot air. In Fig. 35–4b, we see a diagram of light coming from one point on a distant car (on the right) heading left toward the observer. Wave fronts and two rays are shown. Ray A heads directly at the observer and follows a straight-line path, and represents the normal view of the car. Ray B is a ray initially directed slightly downward. But it does not hit the ground. Instead, ray B is bent slightly as it moves through layers of air of different index of refraction. The wave fronts, shown in blue in Fig. 35–4b, move slightly faster in the layers of air nearer the ground (see Fig. 35–3, and also the soldier analogy in Fig. 15–31). Thus ray B is bent as shown, and seems to the observer to be coming from below (dashed line) as if reflected off the road. Hence the mirage.

➡ PHYSICS APPLIED

Highway mirages

FIGURE 35–4 (a) A highway mirage. (b) Drawing (greatly exaggerated) showing wave fronts and rays to explain highway mirages. Note how the sections of the wavefronts near the ground move faster and so are further apart.

(a)

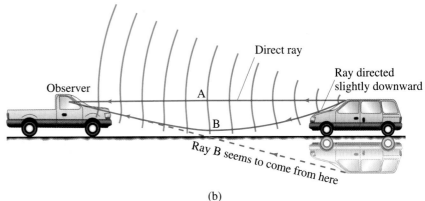

(b)

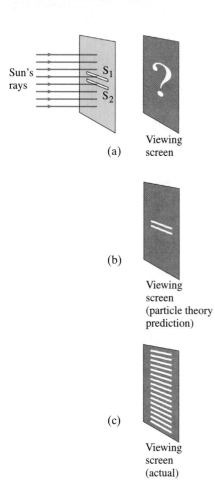

Sun's rays

S₁
S₂

?

(a)

Viewing screen

(b)

Viewing screen (particle theory prediction)

(c)

Viewing screen (actual)

FIGURE 35–5 (a) Young's double-slit experiment. (b) If light consists of particles, we would expect to see two bright lines on the screen behind the slits. (c) Young observed many lines.

35–3 Interference—Young's Double-Slit Experiment

In 1801, the Englishman Thomas Young (1773–1829) obtained convincing evidence for the wave nature of light and was even able to measure the wavelengths for visible light. Figure 35–5a shows a schematic diagram of Young's famous double-slit experiment. Light from a single source (Young used the Sun) falls on a screen containing two closely spaced slits S_1 and S_2. If light consists of tiny particles, we might expect to see two bright lines on a screen placed behind the slits as in (b). But Young observed instead a series of bright lines as in (c). Young was able to explain this result as a **wave-interference** phenomenon. To see this, imagine plane waves of light of a single wavelength—called **monochromatic**, meaning "one color"—falling on the two slits as shown in Fig. 35–6. Because of diffraction, the waves leaving the two small slits spread out as shown. This is equivalent to the interference pattern produced when two rocks are thrown into a lake (Fig. 15–24), or when sound from two loudspeakers interferes (Fig. 16–16).

FIGURE 35–6 If light is a wave, light passing through one of two slits should interfere with light passing through the other slit.

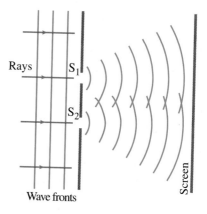

Rays

S₁

S₂

Wave fronts

Screen

To see how an interference pattern is produced on the screen, we make use of Fig. 35–7. Waves of wavelength λ are shown entering the slits S_1 and S_2, which are a distance d apart. The waves spread out in all directions after passing through the slits, but they are shown only for three different angles θ. In Fig. 35–7a, the waves reaching the center of the screen are shown ($\theta = 0$). The waves from the two slits travel the same distance, so they are in phase: a crest of one wave arrives at the same time as a crest of the other wave. Hence the amplitudes of the two waves add to

FIGURE 35–7 How the wave theory explains the pattern of lines seen in the double-slit experiment. (a) At the center of the screen the waves from each slit have traveled the same distance and are in phase. (b) At this angle θ, the lower wave travels an extra distance of one whole wavelength, and the waves are in phase; note from the shaded triangle that the extra distance equals $d \sin \theta$. (c) For this angle θ, the lower wave travels an extra distance equal to one-half wavelength, so the two waves arrive at the screen fully out of phase. (d) A more detailed diagram showing the geometry for parts (b) and (c).

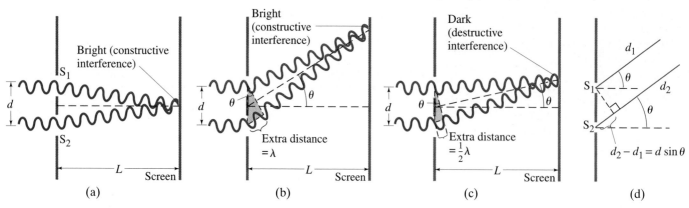

Bright (constructive interference)

S₁

d

S₂

L

Screen

(a)

Bright (constructive interference)

d

θ

θ

Extra distance = λ

L

Screen

(b)

Dark (destructive interference)

d

θ

θ

Extra distance = ½λ

L

Screen

(c)

d₁

θ

S₁

d₂

θ

S₂

d₂ − d₁ = d sin θ

(d)

form a larger amplitude as shown in Fig. 35–8a. This is **constructive interference**, and there is a bright spot at the center of the screen. Constructive interference also occurs when the paths of the two rays differ by one wavelength (or any whole number of wavelengths), as shown in Fig. 35–7b. But if one ray travels an extra distance of one-half wavelength (or $\frac{3}{2}\lambda$, $\frac{5}{2}\lambda$, and so on), the two waves are exactly out of phase when they reach the screen: the crests of one wave arrive at the same time as the troughs of the other wave, and so they add to produce zero amplitude (Fig. 35–8b). This is **destructive interference**, and the screen is dark, Fig. 35–7c. Thus, there will be a series of bright and dark lines (or **fringes**) on the viewing screen.

To determine exactly where the bright lines fall, first note that Fig. 35–7 is somewhat exaggerated; in real situations, the distance d between the slits is very small compared to the distance L to the screen. The rays from each slit for each case will therefore be essentially parallel and θ is the angle they make with the horizontal, as shown in Fig. 35–7d. From the right triangle shown in Fig. 35–7, we can see that the extra distance traveled by the lower ray is $d \sin \theta$. Constructive interference will occur, and a bright fringe will appear on the screen, when the *path difference*, $d \sin \theta$, equals a whole number of wavelengths:

$$d \sin \theta = m\lambda, \qquad m = 0, 1, 2, \cdots. \qquad \left[\begin{array}{c}\text{constructive}\\\text{interference}\end{array}\right] \quad \textbf{(35–2a)}$$

The value of m is called the **order** of the interference fringe. The first order ($m = 1$), for example, is the first fringe on each side of the central fringe (at $\theta = 0$). Destructive interference occurs when the path difference, $d \sin \theta$, is $\frac{1}{2}$, $\frac{3}{2}$, and so on, wavelengths:

$$d \sin \theta = \left(m + \tfrac{1}{2}\right)\lambda, \qquad m = 0, 1, 2, \cdots. \qquad \left[\begin{array}{c}\text{destructive}\\\text{interference}\end{array}\right] \quad \textbf{(35–2b)}$$

The intensity of the bright fringes is greatest for the central fringe ($m = 0$) and decreases for higher orders, as shown in Fig. 35–9.

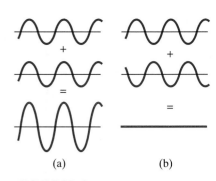

FIGURE 35–8
(a) Constructive interference.
(b) Destructive interference. (See also Section 15–8.)

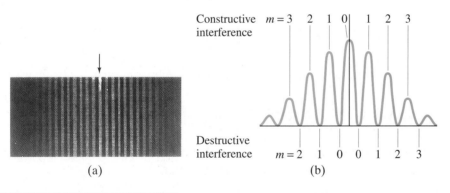

(a)

Constructive interference $\quad m = 3 \quad 2 \quad 1 \quad 0| \quad 1 \quad 2 \quad 3$

Destructive interference $\quad m = 2 \quad 1 \quad 0 \quad 0 \quad 1 \quad 2 \quad 3$

(b)

FIGURE 35–9 (a) Interference fringes produced by a double-slit experiment and detected by photographic film placed on the viewing screen. The arrow marks the central fringe. (b) Intensity of light in the interference pattern. Also shown are values of m for Eq. 35–2a (constructive interference) and Eq. 35–2b (destructive interference).

CONCEPTUAL EXAMPLE 35–1 **Interference pattern lines.** (*a*) Will there be an infinite number of points on the viewing screen where constructive and destructive interference occur, or only a finite number of points? (*b*) Are neighboring points of constructive interference uniformly spaced, or is the spacing between neighboring points of constructive interference not uniform?

RESPONSE (*a*) When you look at Eqs. 35–2a and b you might be tempted to say, given the statement $m = 0, 1, 2, \cdots$ beside the equations, that there are an infinite number of points of constructive and destructive interference. However, recall that $\sin \theta$ cannot exceed 1. Thus, there is an upper limit to the values of m that can be used in these equations. For Eq. 35–2a, the maximum value of m is the integer closest in value but smaller than d/λ. So there are a finite number of points of constructive and destructive interference spread out over an even infinitely tall screen. (*b*) The spacing between neighboring points of constructive or destructive interference is not uniform: The spacing gets larger as θ gets larger, and you can verify this statement mathematically. For small values of θ the spacing is nearly uniform as you will see in Example 35–2.

FIGURE 35–10 For small angles, the interference fringes occur at distance $x = \theta L$ above the center fringe $(m = 0)$; θ_1 and x_1 are for the first order fringe $(m = 1)$, θ_2 and x_2 are for $m = 2$.

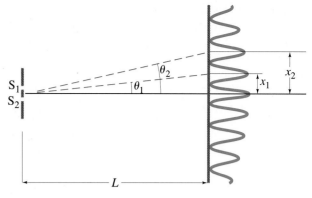

EXAMPLE 35–2 **Line spacing for double-slit interference.** A screen containing two slits 0.100 mm apart is 1.20 m from the viewing screen. Light of wavelength $\lambda = 500$ nm falls on the slits from a distant source. Approximately how far apart will the bright interference fringes be on the screen?

SOLUTION Given $d = 0.100$ mm $= 1.00 \times 10^{-4}$ m, $\lambda = 500 \times 10^{-9}$ m, and $L = 1.20$ m, the first-order fringe $(m = 1)$ occurs at an angle θ given by

$$\sin\theta_1 = \frac{m\lambda}{d} = \frac{(1)(500 \times 10^{-9}\,\text{m})}{1.00 \times 10^{-4}\,\text{m}} = 5.00 \times 10^{-3}.$$

This is a very small angle, so we can take $\sin\theta = \theta$, with θ in radians. The first-order fringe will occur a distance x_1 above the center of the screen (see Fig. 35–10) given by $x_1/L = \tan\theta_1 = \theta_1$, so

$$x_1 = L\theta_1 = (1.20\,\text{m})(5.00 \times 10^{-3}) = 6.00\,\text{mm}.$$

The second-order fringe $(m = 2)$ will occur at

$$x_2 = L\theta_2 = L\frac{2\lambda}{d} = 12.0\,\text{mm}$$

above the center, and so on. Thus the lower order fringes are 6.00 mm apart.

CONCEPTUAL EXAMPLE 35–3 **Changing the wavelength.** (*a*) What happens to the interference pattern shown in Fig. 35–10, Example 35–2, if the incident light (500 nm) is replaced by light of wavelength 700 nm? (*b*) What happens instead if the slits are moved farther apart?

RESPONSE (*a*) When λ increases in Eq. 35–2a but d stays the same, then the angle θ for maxima increases and the interference pattern spreads out. (*b*) Increasing the slit spacing d reduces θ for each order, so the lines are closer together.

From Eqs. 35–2 we can see that, except for the zeroth-order fringe at the center, the position of the fringes depends on wavelength. Consequently, when white light falls on the two slits, as Young found in his experiments, the central fringe is white, but the first- (and higher-) order fringes contain a spectrum of colors like a rainbow; θ was found to be smallest for violet light and largest for red. By measuring the position of these fringes, Young was the first to determine the wavelengths of visible light (using Eqs. 35–2). In doing so, he showed that what distinguishes different colors physically is their wavelength, an idea put forward earlier by Grimaldi in 1665.

Wavelength (or frequency) determines color

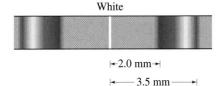

EXAMPLE 35-4 **Wavelengths from double-slit interference.** White light passes through two slits 0.50 mm apart and an interference pattern is observed on a screen 2.5 m away. The first-order fringe resembles a rainbow with violet and red light at either end. The violet light falls about 2.0 mm and the red 3.5 mm from the center of the central white fringe (Fig. 35–11). Estimate the wavelengths of the violet light and the red light.

SOLUTION We use Eq. 35–2a with $m = 1$ and $\sin\theta = \theta$. Then for violet light, $x = 2.0$ mm, so (see also Fig. 35–10)

$$\lambda = \frac{d\theta}{m} = \frac{d}{m}\frac{x}{L} = \left(\frac{5.0 \times 10^{-4}\,\text{m}}{1}\right)\left(\frac{2.0 \times 10^{-3}\,\text{m}}{2.5\,\text{m}}\right) = 4.0 \times 10^{-7}\,\text{m},$$

or 400 nm. For red light, $x = 3.5$ mm, so

$$\lambda = \frac{d}{m}\frac{x}{L} = \left(\frac{5.0 \times 10^{-4}\,\text{m}}{1}\right)\left(\frac{3.5 \times 10^{-3}\,\text{m}}{2.5\,\text{m}}\right) = 7.0 \times 10^{-7}\,\text{m} = 700\,\text{nm}.$$

FIGURE 35–11 Example 35–4.

35-4 Coherence

The two slits in Fig. 35–7 act as if they were two sources of radiation. They are called **coherent sources** because the waves leaving them bear the same phase relationship to each other at all times. This happens because the waves come from a single source to the left of the two slits in Fig. 35–7. An interference pattern is observed only when the sources are coherent. If two tiny lightbulbs replaced the two slits, an interference pattern would not be seen. The light emitted by one lightbulb would have a random phase with respect to the second bulb, and the screen would be more or less uniformly illuminated. Two such sources, whose output waves have phases that bear no fixed relationship to each other over time, are called **incoherent sources**.

Coherent and incoherent sources

The subject of coherence is rather complicated, and we discuss it only briefly. Two light beams do not have to be in phase to be coherent and produce an interference pattern. For example, suppose a piece of glass were placed in front of the lower slit in Fig. 35–7 and suppose the glass is just thick enough to slow down the light so that it enters the lower slit a half wavelength behind the light entering the upper slit. The two beams would be a constant 180° out of phase, but there would still be an interference pattern on the screen. (Can you guess what it would look like? Hint: The central point would be dark instead of bright.) Two beams can be coherent whether they are in phase or out of phase; the important thing is that they have a *constant* phase relation to each other over time.

Coherent sources of water or sound waves are easier to obtain than are coherent sources of light—two loudspeakers receiving the same pure frequency signal from an amplifier will be coherent sources. And two antennas connected to the same LC oscillator can be coherent sources of low-frequency electromagnetic waves. But LC oscillators at the high frequencies of visible light (10^{15} Hz) don't exist since L and C can't be made small enough. For sources of visible light we have to rely on the oscillations (or acceleration) of electric charge within atoms. In an incandescent light bulb, for example, the atoms in the filament are excited by heating, and give off "wave trains" of light, each of which lasts only about 10^{-8} s. The light we see is the sum of a great many such wave trains that bear a random phase relation to each other. Two light bulbs are thus not coherent, and an interference pattern would not be seen. It wasn't until the 1950s that a really coherent source of light was developed, the *laser*. Because of its coherence, laser light on a double slit produces a very "clean" interference pattern.

Coherence is a relative concept. Perfect coherence of a beam would correspond to light that is perfectly sinusoidal (of one frequency) for all times, whereas complete incoherence occurs when there are waves whose phase relations are completely random over time. A measure of the "relative coherence" can be defined in terms of the sharpness of a two-slit interference pattern.

35–5 Intensity in the Double-Slit Interference Pattern

We saw in Section 35–3 that the interference pattern produced by the coherent light from two slits, S_1 and S_2 (Figs. 35–7 and 35–9), produces a series of bright and dark fringes. If the two monochromatic waves of wavelength λ are in phase at the slits, the maxima (brightest points) occur at angles θ given by

$$d \sin \theta = m\lambda,$$

and the minima (darkest points) when

$$d \sin \theta = \left(m + \tfrac{1}{2}\right)\lambda,$$

where m is an integer ($m = 0, 1, 2, \ldots$).

We now determine the intensity of the light at all points in the pattern (that is, for all θ). For simplicity, let us assume that if either slit were covered, the light passing through the other would diffract sufficiently to illuminate a large portion of the screen uniformly. The intensity I of the light at any point is proportional to the square of its wave amplitude (Section 15–3). Treating light as an electromagnetic wave, I is proportional to the square of the electric field E or magnetic field B (Section 32–7). Since E and B are proportional to each other, it doesn't matter which we use, but it is conventional to use E and write the intensity as $I \propto E^2$. The electric field $\mathbf{E}$ at any point P (see Fig. 35–12) will be the sum of the electric field vectors of the waves coming from each of the two slits, $\mathbf{E}_1$ and $\mathbf{E}_2$. Since $\mathbf{E}_1$ and $\mathbf{E}_2$ are essentially parallel (on a screen far away compared to the slit separation), the magnitude of the electric field at angle θ (that is, at point P) will be

$$E_\theta = E_1 + E_2.$$

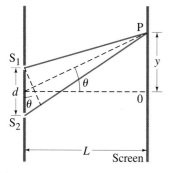

FIGURE 35–12 Determining the intensity in a double-slit interference pattern. Not to scale: in fact $L \gg d$, and the two rays become essentially parallel.

Both E_1 and E_2 vary sinusoidally with frequency $f = c/\lambda$, but they differ in phase, depending on their different travel distances from the slits. The electric field at P can then be written for the light from each of the two slits, using $\omega = 2\pi f$, as

$$E_1 = E_{10} \sin \omega t$$
$$E_2 = E_{20} \sin(\omega t + \delta) \tag{35–3}$$

where E_{10} and E_{20} are their respective amplitudes and δ is the phase difference. The value of δ depends on the angle θ, so let us now determine δ as a function of θ.

At the center of the screen (point 0), $\delta = 0$. If the difference in path length from P to S_1 and S_2 is $d \sin \theta = \lambda/2$, the two waves are exactly out of phase so $\delta = \pi$ (or 180°). If $d \sin \theta = \lambda$, the two waves differ in phase by $\delta = 2\pi$. In general, then, δ is related to θ by

$$\frac{\delta}{2\pi} = \frac{d \sin \theta}{\lambda}$$

or

$$\delta = \frac{2\pi}{\lambda} d \sin \theta. \tag{35–4}$$

To determine $E_\theta = E_1 + E_2$, we add the two scalars E_1 and E_2 which are sine functions differing by the phase δ. One way to determine the sum of E_1 and E_2 is to use a **phasor diagram**. (We used this technique before, in Chapter 31.) As shown in

Fig. 35–13, we draw an arrow of length E_{10} to represent the amplitude of E_1 (Eq. 35–3); and the arrow of length E_{20}, which we draw to make a fixed angle δ with E_{10}, represents the amplitude of E_2. When the diagram rotates at angular frequency ω about the origin, the projections of E_{10} and E_{20} on the vertical axis represent E_1 and E_2 as a function of time (see Eq. 35–3). We let $E_{\theta 0}$ be the "vector" sum[†] of E_{10} and E_{20}; it is the amplitude of the sum $E_\theta = E_1 + E_2$, and the projection of $E_{\theta 0}$ on the vertical axis is just E_θ. If the two slits provide equal illumination, so that $E_{10} = E_{20} = E_0$, then from Fig. 35–13 the angle $\phi = \delta/2$, and we can write

$$E_\theta = E_{\theta 0} \sin\left(\omega t + \frac{\delta}{2}\right). \qquad (35\text{–}5a)$$

From Fig. 35–13 we can also see that

$$E_{\theta 0} = 2E_0 \cos\phi = 2E_0 \cos\frac{\delta}{2}. \qquad (35\text{–}5b)$$

Combining Eqs. 35–5a and b, we obtain

$$E_\theta = 2E_0 \cos\frac{\delta}{2} \sin\left(\omega t + \frac{\delta}{2}\right), \qquad (35\text{–}5c)$$

where δ is given by Eq. 35–4. We can also derive Eq. 35–5c, when $E_{10} = E_{20}$, using the trigonometric identity

$$\sin A + \sin B = 2 \sin\left(\frac{A+B}{2}\right) \cos\left(\frac{A-B}{2}\right).$$

That is, we add the two Eqs. 35–3 and obtain Eq. 35–5c:

$$E_\theta = E_1 + E_2 = 2E_0 \sin\left(\omega t + \frac{\delta}{2}\right) \cos\frac{\delta}{2}.$$

We are not really interested in E_θ as a function of time, since for visible light the frequency $(10^{14}$ to $10^{15}\,\text{Hz})$ is much too high to be noticeable. We are interested in the average intensity, which is proportional to the amplitude squared, $E_{\theta 0}^2$. We now drop the word "average," and we let $I_\theta (I_\theta \propto E_\theta^2)$ be the intensity at any point P at an angle θ to the horizontal. We let I_0 be the intensity at point 0, the center of the screen, where $\theta = \delta = 0$, so $I_0 \propto (E_{10} + E_{20})^2 = (2E_0)^2$. Then the ratio I_θ/I_0 is equal to the ratio of the squares of the electric-field amplitudes at these two points, so

$$\frac{I_\theta}{I_0} = \frac{E_{\theta 0}^2}{(2E_0)^2} = \cos^2\frac{\delta}{2}$$

where we used Eq. 35–5b. Thus the intensity I_θ at any point is related to the maximum intensity at the center of the screen by

$$\begin{aligned} I_\theta &= I_0 \cos^2\frac{\delta}{2} \\ &= I_0 \cos^2\left(\frac{\pi d \sin\theta}{\lambda}\right) \end{aligned} \qquad (35\text{–}6)$$

where δ was given by Eq. 35–4. This is the relation we sought. From Eq. 35–6 we see that maxima occur where $\cos\delta/2 = \pm 1$, which corresponds to $\delta = 0, 2\pi, 4\pi, \cdots$; from Eq. 35–4, δ has these values when

$$d \sin\theta = m\lambda, \qquad m = 0, 1, 2, \cdots.$$

Minima occur where $\delta = \pi, 3\pi, 5\pi, \ldots$, which corresponds to

$$d \sin\theta = (m + \tfrac{1}{2})\lambda, \qquad m = 0, 1, 2, \cdots.$$

These are the same results we obtained in Section 35–3. But now we know not only the position of maxima and minima, but from Eq. 35–6 we can determine the intensity at all points.

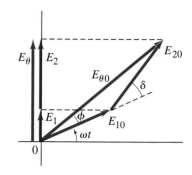

FIGURE 35–13 Phasor diagram for double-slit interference pattern.

[†] We are not really adding the actual electric field vectors; instead we are adding the "phasors" in order to get the amplitude, taking into account the phase difference of the two waves.

In the usual situation where the distance L to the screen from the slits is large compared to the slit separation d ($L \gg d$), if we consider only points P whose distance y from the center (point 0) is small compared to L ($y \ll L$)—see Fig. 35–12—then

$$\sin \theta = \frac{y}{L}.$$

From this it follows (see Eq. 35–4) that

$$\delta = \frac{2\pi}{\lambda} \frac{d}{L} y.$$

Equation 35–6 then becomes

$$I_\theta = I_0 \left[\cos \left(\frac{\pi d}{\lambda L} y \right) \right]^2. \qquad\qquad [y \ll L, d \ll L] \quad \textbf{(35–7)}$$

The intensity I_θ as a function of the phase difference δ is plotted in Fig. 35–14. In the approximation of Eq. 35–7, the horizontal axis could as well be y, the position on the screen.

The intensity pattern expressed in Eqs. 35–6 and 35–7, and plotted in Fig. 35–14, shows a series of maxima of equal height, and is based on the assumption that each slit (alone) would illuminate the screen uniformly. This is never quite true, as we shall see when we discuss diffraction in the next chapter. We will see that the center maximum is strongest and each succeeding maximum to each side is less strong.

FIGURE 35–14 Intensity I as a function of phase difference δ and position on screen y (assuming $y \ll L$).

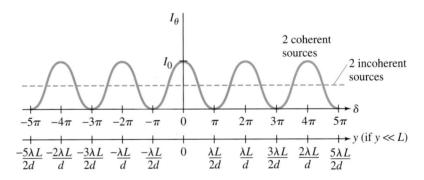

FIGURE 35–15 Example 35–5. The two dots represent the antennas.

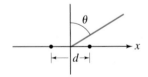

EXAMPLE 35–5 **Radar antenna intensity.** Two radio antennas are located close to each other as shown in Fig. 35–15, separated by a distance d. The antennas radiate in phase with each other, emitting waves of intensity I_0 at wavelength λ. (a) Calculate the net intensity as a function of θ for points very far from the antennas. (b) For $d = \lambda$, determine I and find in which directions I is a maximum and a minimum. (c) Repeat part b when $d = \lambda/2$.

SOLUTION (a) This setup is similar to Young's double-slit experiment. Points of constructive and destructive interference are still given by Eqs. 35–2a and b, and the net intensity as a function of position is given by Eq. 35–6 or 35–7. (b) We let $d = \lambda$ in Eq. 35–6, and find for the intensity,

$$I = I_0 \cos^2(\pi \sin \theta).$$

I is a maximum, equal to I_0, when $\sin \theta = 0$, 1 or -1, meaning $\theta = 0$, 90°, 180° and 270°. I is zero when $\sin \theta = \frac{1}{2}$ and $-\frac{1}{2}$, for which $\theta = 30°$, 150°, 210° and 330°. (c) For $d = \lambda/2$ I is maximized for $\theta = 0$ and 180° and minimized for 90° and 270°.

Antenna arrays are important when a directional signal needs to be trasmitted such as in radar. We saw in Example 35–5 b and c that the intensity is maximized along particular directions and falls off slowly to zero at other angles. A narrow beam would be useful to concentrate radar energy on a target, and can be achieved using an array of more than two antennas. If the phases of the antennas are varied relative to each other, then the radiation pattern will shift direction, without having to physically rearrange the antennas. Such an antenna array is called an *electronically steered phase array*.

* Intensity for Incoherent Sources

Let us consider, briefly, the intensity pattern if the two slits of Fig. 35–12 were replaced by two *incoherent* sources of equal strength $\left(E_{10} = E_{20} = E_0\right)$. There would be no interference pattern, but rather a uniform illumination of the screen. The phase difference δ between the two waves varies randomly at any point P. Thus we must use the time average over $\cos^2 \delta/2$ in Eq. 35–6, which is $\frac{1}{2}$. The intensity, I_{inc}, of the two incoherent sources, will be

$$I_{\text{inc}} = \tfrac{1}{2} I_{\text{coh}}$$

where I_{coh} is the intensity due to two coherent sources at a maximum (I_{coh} is the I_0 of Eq. 35–6). This result, which is shown as the dashed line in Fig. 35–14, is confirmed by experiment and can be obtained in another way: when the sources are coherent, we add their wave amplitudes and then square to get the intensity

$$I_{\text{coh}} \propto \left(E_{10} + E_{20}\right)^2.$$

But if the sources are incoherent, each wave produces an intensity unrelated to the other, and the two intensities add up:

$$I_{\text{inc}} \propto E_{10}^2 + E_{20}^2.$$

That is, we square the amplitudes and *then* add them. If $E_{10} = E_{20} = E_0$, then the two relations above become

$$I_{\text{coh}} \propto \left(2E_0\right)^2 = 4E_0^2$$

$$I_{\text{inc}} \propto E_0^2 + E_0^2 = 2E_0^2$$

so

$$I_{\text{inc}} = \tfrac{1}{2} I_{\text{coh}}$$

which is our result above.

35–6 | Interference in Thin Films

Interference of light gives rise to many everyday phenomena such as the bright colors reflected from soap bubbles and from thin oil or gasoline films on water, Fig. 35–16. In these and other cases, the colors are a result of constructive interference between light reflected from the two surfaces of the thin film.

FIGURE 35–16 Thin film interference patterns seen in (a) soap bubbles, (b) thin films of soapy water, and (c) a thin layer of gasoline on water.

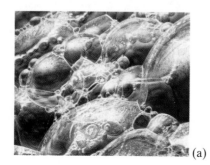
(a)

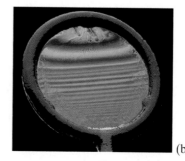
(b)

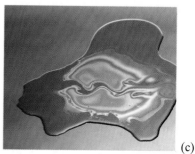
(c)

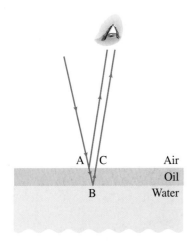

FIGURE 35–17 Light reflected from the upper and lower surfaces of a thin film of oil lying on water. Analysis assumes the light strikes the surface perpendicularly, but is shown here at a slight angle for clarity.

Newton's rings

To see how this happens, consider a smooth surface of water on top of which is a thin uniform layer of another substance, say an oil whose index of refraction is less than that of water (we'll see why we assume this in a moment); see Fig. 35–17. Assume for the moment that the incident light is of a single wavelength. Part of the incident light is reflected at A on the top surface, and part of the light transmitted is reflected at B on the lower surface. The part reflected at the lower surface must travel the extra distance ABC. If this *path difference* ABC is equal to one or a whole number of wavelengths in the film (λ_n), the two waves will reach the eye in phase and interfere constructively. Hence the region AC on the surface film will appear bright. But if ABC equals $\frac{1}{2}\lambda_n$, $\frac{3}{2}\lambda_n$, and so on, the two waves will be exactly out of phase, destructive interference occurs, and the area AC on the film will be dark. The wavelength λ_n is *the wavelength in the film:* $\lambda_n = \lambda/n$ where n is the index of refraction in the film and λ is the wavelength in vacuum.

When white light falls on such a film, the path difference ABC will equal λ_n (or $m\lambda_n$, with m = an integer) for only one wavelength at a given viewing angle. The color corresponding to λ (λ in air) will be seen as very bright. For light viewed at a slightly different angle, the path difference ABC will be longer or shorter and a different color will undergo constructive interference. Thus, for an extended (non-point) source emitting white light, a series of bright colors will be seen next to one another. Variations in thickness of the film will also alter the path difference ABC and therefore affect the color of light that is most strongly reflected.

When a curved glass surface is placed in contact with a flat glass surface, Fig. 35–18, a series of concentric rings is seen when illuminated from above by monochromatic light. These are called **Newton's rings**[†] and they are due to interference between rays reflected by the top and bottom surfaces of the very thin *air gap* between the two pieces of glass. Because this gap (which is equivalent to a thin film) increases in width from the central contact point out to the edges, the extra path length for the lower ray (equal to BCD) varies; where it equals $0, \frac{1}{2}\lambda, \lambda, \frac{3}{2}\lambda, 2\lambda$, and so on, it corresponds to constructive and destructive interference; and this gives rise to the series of bright and dark lines seen in Fig. 35–18b.

The point of contact of the two glass surfaces (A in Fig. 35–18a) is dark in Fig. 35–18b. Since the path difference is zero here, we expect the rays reflected from each surface to be in phase and so this central point ought to be bright. But it is dark, which tells us the two rays must be completely out of phase. This can happen only because one of the waves undergoes a change in phase of 180° upon reflection, corresponding to $\frac{1}{2}$ cycle. Indeed, this and other experiments reveal that *a beam of light reflected by a material whose index of refraction is greater than that of the material in which it is traveling, changes phase by 180° or $\frac{1}{2}$ cycle.*

[†]Although Newton gave an elaborate description of them, they had been first observed and described by his contemporary, Robert Hooke.

FIGURE 35–18 Newton's rings.

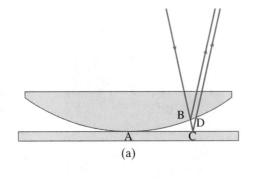

(a)

(b)

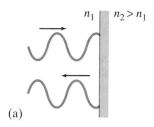
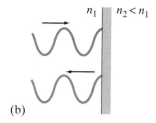

(a)   (b)

FIGURE 35–19 (a) Reflected ray changes phase by 180° or $\frac{1}{2}$ cycle if $n_2 > n_1$, but (b) does not if $n_2 < n_1$.

See Fig. 35–19. If the refractive index is less than that of the material in which the light is traveling, no phase change occurs.[†] [This can be derived from Maxwell's equations. It corresponds to the reflection of a wave traveling along a rope when it reaches the end; as we saw in Fig. 15–19, if the end is tied down, the wave changes phase and the pulse flips over, but if the end is free, no phase change occurs.] Thus the ray reflected by the curved surface above the air gap in Fig. 35–18a undergoes no change in phase. The ray reflected at the lower surface, where the beam in air strikes the glass, undergoes a 180° ($\frac{1}{2}\lambda$) phase change. Thus the two rays reflected at the point of contact A of the two glass surfaces (where the air gap approaches zero thickness) will be 180° ($\frac{1}{2}\lambda$) out of phase, and a dark spot occurs. Other dark bands will occur when the path difference BCD in Fig. 35–18a is equal to an integral number of wavelengths. Bright bands will occur when the path difference is $\frac{1}{2}\lambda, \frac{3}{2}\lambda$, and so on, because the phase change at one surface effectively adds another $\frac{1}{2}\lambda$.

FIGURE 35–20 (a) Light rays reflected from the upper and lower surfaces of a thin wedge of air interfere to produce bright and dark bands. (b) Pattern observed when glass plates are optically flat; (c) pattern when plates are not so flat. See Example 35–6.

EXAMPLE 35–6 Thin film of air, wedge-shaped. A very fine wire 7.35×10^{-3} mm in diameter is placed between two flat glass plates as in Fig. 35–20a. Light whose wavelength in air is 600 nm falls (and is viewed) perpendicular to the plates, and a series of bright and dark bands is seen, Fig. 35–20b. How many light and dark bands will there be in this case? Will the area next to the wire be bright or dark?

SOLUTION The thin film is the wedge of air between the two glass plates. Because of the phase change at the lower surface, there will be a dark band when the path difference is 0, λ, 2λ, 3λ, and so on. Since the light rays are perpendicular to the plates, the extra path length equals $2t$, where t is the thickness of the air gap at any point; so dark bands occur where

$$2t = m\lambda, \qquad m = 0, 1, 2, \cdots.$$

Bright bands occur when $2t = \left(m + \frac{1}{2}\right)\lambda$, where m is an integer. At the position of the wire, $t = 7.35 \times 10^{-6}$ m. At this point there will be $2t/\lambda = (2)(7.35 \times 10^{-6}\text{ m})/(6.00 \times 10^{-7}\text{ m}) = 24.5$ wavelengths. This is a "half integer," so the area next to the wire will be bright. There will be a total of 25 dark lines along the plates, corresponding to path lengths of $0\lambda, 1\lambda, 2\lambda, 3\lambda, \cdots, 24\lambda$, including the one at the point of contact A ($m = 0$). Between them, there will be 24 bright lines plus the one at the end, or 25. The bright and dark bands will be straight only if the glass plates are extremely flat. If they are not, the pattern is uneven, as in Fig. 35–20c. Thus we see a very precise way of testing a glass surface for flatness. Curved lens surfaces can be tested for precision similarly by placing the lens on a flat glass surface and observing Newton's rings (Fig. 35–18b) for perfect circularity.

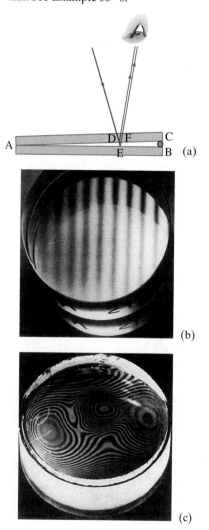

(a)

(b)

(c)

[†]Note that in Fig. 35–17, the light reflecting at both interfaces, air–oil and oil–water, underwent a phase change of $\frac{1}{2}\lambda$, since we assumed $n_{\text{water}} > n_{\text{oil}} > n_{\text{air}}$. Since the phase changes were equal, they didn't affect our analysis.

If the wedge between the two glass plates of Example 35–6 is filled with some transparent substance other than air—say, water—the pattern shifts because the wavelength of the light changes. In a material where the index of refraction is n, the wavelength is $\lambda_n = \lambda/n$ where λ is the wavelength in vacuum (see Eq. 35–1). For instance, if the thin wedge of Example 35–6 were filled with water, $\lambda_n = 600 \, \text{nm}/1.33 = 450 \, \text{nm}$; instead of 25 dark lines, there would be 33.

When white light (rather than monochromatic light) is incident on the thin wedge of Figs. 35–18a or 35–20a, a colorful series of fringes is seen. This is because constructive interference occurs in the reflected light at different locations along the wedge for different wavelengths. Such a difference in thickness is part of the reason bright colors appear when light is reflected from a soap bubble or a thin layer of oil or gasoline on a puddle or lake (Fig. 35–16). Which wavelengths appear brightest also depends on the viewing angle, as we saw earlier.

EXAMPLE 35–7 **Thickness of soap bubble skin.** A soap bubble appears green ($\lambda = 540 \, \text{nm}$) at the point on its front surface nearest the viewer. What is its minimum thickness? Assume $n = 1.35$.

SOLUTION The light is reflected perpendicularly from the point on a spherical surface nearest the viewer, Fig. 35–21. Therefore the path difference is $2t$, where t is the thickness of the soap film. Light reflected from the first (outer) surface undergoes a $\frac{1}{2}\lambda$ phase change (index of refraction of soap is greater than that of air), whereas that at the second (inner) surface does not. Therefore, green light is bright when the minimum path difference equals $\frac{1}{2}\lambda_n$. Thus, $2t = \lambda/2n$, so

$$t = \frac{\lambda}{4n} = \frac{(540 \, \text{nm})}{(4)(1.35)} = 100 \, \text{nm}.$$

This is the minimum thickness. The front surface would also appear green if $2t = 3\lambda/2n$, and, in general, if $2t = (2m + 1)\lambda/2n$ where m is an integer. Note that green is seen in air, so $\lambda = 540 \, \text{nm}$ (not λ/n).

➡ **PROBLEM SOLVING**

A formula is not enough: you must also check for phase changes at surfaces

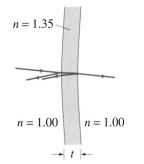

$n = 1.35$

$n = 1.00$ $n = 1.00$

$\rightarrow|\, t \,|\leftarrow$

FIGURE 35–21 Example 35–7.

➡ **PHYSICS APPLIED**

Lens coatings

An important application of thin-film interference is in the coating of glass to make it "nonreflecting," particularly for lenses. A glass surface reflects about 4 percent of the light incident upon it. Good-quality cameras, microscopes, and other optical devices may contain six to ten thin lenses. Reflection from all these surfaces can reduce the light level considerably, and multiple reflections produce a background haze that reduces the quality of the image. By reducing reflection, transmission is increased. A very thin coating on the lens surfaces can reduce reflections considerably: the thickness of the film is chosen so that light (at least for one wavelength) reflecting from the front and rear surfaces of the film destructively interferes. The amount of reflection at a boundary depends on the difference in index of refraction between the two materials. Ideally, the coating material should have an index of refraction which is the geometric mean of those for air and glass, so that the amount of reflection at each surface is about equal. Then destructive interference can occur nearly completely for one particular wavelength depending on the thickness of the coating. Nearby wavelengths will at least partially destructively interfere, but it is clear that a single coating cannot eliminate reflections for all wavelengths. Nonetheless, a single coating can reduce total reflection from 4 percent to 1 percent of the incident light. Often the coating is designed to eliminate the center of the reflected spectrum (around 550 nm). The extremes of the spectrum—red and violet—will not be reduced as much. Since a mixture of red and violet produces purple, the light seen reflected from such coated lenses is purple (Fig. 35–22). Lenses containing two or three separate coatings can more effectively reduce a wider range of reflecting wavelengths.

FIGURE 35–22 A coated lens. Note color of light reflected from the front lens surface.

EXAMPLE 35–8 **Nonreflective coating.** What is the thickness of an optical coating of MgF$_2$, whose index of refraction is $n = 1.38$, which is designed to eliminate reflected light at wavelengths centered at 550 nm when incident normally on glass for which $n = 1.50$?

SOLUTION Figure 35–23 shows an incoming ray and two rays reflected from the front and rear surfaces of the coating on the lens. The rays are drawn not quite perpendicular to the lens so we can see each of them. To eliminate reflection, we want the reflected rays 1 and 2 to be $\frac{1}{2}$ wavelength out of phase with each other so they destructively interfere. Rays 1 and 2 *both* undergo a change of phase by $\frac{1}{2}\lambda$ when they reflect, respectively, from the front and rear surfaces of the coating. Therefore, we want the extra distance traveled by ray 2 ($= 2t$) to be a half integral number of wavelengths. That is, $2t = (m + \frac{1}{2})\lambda_n$, where m is an integer and λ_n is the wavelength inside the MgF$_2$ coating. The minimum thickness ($m = 0$) is usually chosen because destructive interference will then occur over the widest angle. Then

$$t = \frac{\lambda_n}{4} = \frac{\lambda}{4n} = \frac{(550 \text{ nm})}{(4)(1.38)} = 99.6 \text{ nm}.$$

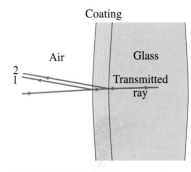

FIGURE 35–23 Incident ray of light is partially reflected at the front surface of a lens coating (ray 1) and again partially reflected at the rear surface of the coating (ray 2), with most of the energy passing as the transmitted ray into the glass.

---

* 35–7 Michelson Interferometer

A useful instrument involving wave interference is the **Michelson interferometer** (Fig. 35–24),[†] invented by the American Albert A. Michelson (Section 33–2). Monochromatic light from a single point on an extended source is shown striking a half-silvered mirror M$_S$. This **beam splitter** mirror M$_S$ has a thin layer of silver that reflects only half the light that hits it, so that half of the beam passes through to a fixed mirror M$_2$, where it is reflected back. The other half is reflected by M$_S$ up to a mirror M$_1$ that is movable (by a fine-thread screw), where it is also reflected back. Upon its return, part of beam 1 passes through M$_S$ and reaches the eye; and part of beam 2, on its return, is reflected by M$_S$ into the eye. If the two path lengths are identical, the two coherent beams entering the eye constructively interfere and brightness will be seen. If the movable mirror is moved a distance $\lambda/4$, one beam will travel an extra distance equal to $\lambda/2$ (because it travels back and forth over the distance $\lambda/4$). In this case, the two beams will destructively interfere and darkness will be seen. As M$_1$ is moved farther, brightness will recur (when the path difference is λ), then darkness, and so on.

Very precise length measurements can be made with an interferometer. The motion of mirror M$_1$ by only $\frac{1}{4}\lambda$ produces a clear difference between brightness and darkness. For $\lambda = 400$ nm, this means a precision of 100 nm or 10^{-4} mm! If mirror M$_1$ is tilted very slightly, the bright or dark spots are seen instead as a series of bright and dark lines or "fringes." By counting the number of fringes, or fractions thereof, extremely precise length measurements can be made.

Michelson saw that the interferometer could be used to determine the length of the standard meter in terms of the wavelength of a particular light. In 1960, that standard was chosen to be a particular orange line in the spectrum of krypton-86 (krypton atoms with atomic mass 86). Careful repeated measurements of the old standard meter (the distance between two marks on a platinum–iridium bar kept in Paris) were made to establish 1 meter as being 1,650,763.73 wavelengths of this light, which was *defined* to be the meter. In 1983, the meter was redefined in terms of the speed of light (Section 1–4).

FIGURE 35–24 Michelson interferometer.

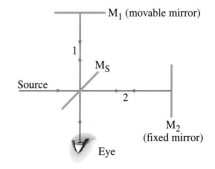

[†]There are other types of interferometer, but Michelson's is the best known.

Although the *intensity* of light, as for any electromagnetic wave, is measured by the Poynting vector in W/m^2, and the total power output of a source can be measured in watts (the *radiant flux*), these are not adequate for measuring the visual sensation we call brightness. The reason is that we are really interested here only in the visible spectrum, whereas the two quantities just mentioned would take into account all wavelengths present. It is also important to take into account the eye's sensitivity to different wavelengths—the eye is most sensitive in the central, 550-nm (yellow), portion of the spectrum; so a yellow source would appear brighter than a red or blue source of the same power output.

These factors are taken into account in the quantity **luminous flux**, F_l, whose unit is the **lumen** (lm). One lumen is defined experimentally as the brightness of $\frac{1}{60}$ cm^2 of platinum surface at its melting temperature (1770°C). It is equivalent to $\frac{1}{683}$ watts of 555-nm light.

Since the luminous flux from a source may not be uniform over all directions, we define the **luminous intensity** I_l as the luminous flux per unit solid angle (steradian). Its unit is the **candela** (cd) where $1 \text{ cd} = 1 \text{ lm/sr}$.

Candela (unit)

The **illuminance**, E_l, is the luminous flux incident on a surface per unit area of the surface: $E_l = F_l/A$. Its unit is the lumen per square meter (lm/m^2) and is a measure of the illumination falling on a surface.[†]

We will not go into this subject in any more detail. We have introduced it here for completeness since the *luminous intensity*, as measured by the candela in SI units, is one of the seven basic quantities in the SI. (See Section 1–4 and Table 1–5.) The other six (and their units) we have already met: length (m), time (s), mass (kg), electric current (A), temperature (K), and amount of substance (mol).

EXAMPLE 35–9 **Light bulb illuminance.** The brightness of a particular type of 100-W light bulb is rated at 1700 lm. Determine (*a*) the luminous intensity and (*b*) the illuminance at a distance of 2.0 m. Assume the light output is uniform in all directions.

SOLUTION (*a*) A full sphere corresponds to 4π sr. Hence, $I_l = 1700 \text{ lm}/4\pi \text{ sr} = 135 \text{ cd}$. It does not depend on distance. (*b*) At $d = 2.0$ m from the source, the luminous flux per unit area is

$$E_l = \frac{F_l}{4\pi d^2} = \frac{1700 \text{ lm}}{(4\pi)(2.0 \text{ m})^2} = 34 \text{ lm/m}^2.$$

The illuminance decreases as the square of the distance.

PROBLEM	Interference
SOLVING	

1. Interference effects depend on the simultaneous arrival of two or more waves at the same point in space.

2. Constructive interference occurs when waves arrive in phase with each other: a crest of one wave arrives at the same time as a crest of the other wave. The amplitudes of the waves then add to form a larger amplitude. Constructive interference also occurs when the phase difference is exactly one full wavelength or any integer multiple of a full wavelength: $1\lambda, 2\lambda, 3\lambda, \cdots$.

3. Destructive interference occurs when a crest of one wave arrives at the same time as a trough of the other

wave. The amplitudes add, but they are of opposite sign, so the total amplitude is reduced to zero if the two amplitudes are equal. Destructive interference occurs whenever the phase difference is a half-integral number of wavelengths. Thus, the total amplitude will be zero if two identical waves arrive one-half wavelength out of phase, or $\left(m + \frac{1}{2}\right)\lambda$ out of phase where m is an integer.

4. For thin-film interference, don't forget an extra one-half wavelength phase shift that occurs when light reflects from an optically more dense medium (going from a medium of lesser toward greater index of refraction).

[†]The British unit is the foot-candle, or lumen per square foot.

Summary

The wave theory of light is strongly supported by the observations that light exhibits **interference** and **diffraction**. Wave theory also explains the refraction of light and the fact that light travels more slowly in transparent solids and liquids than it does in air. The wavelength of light in a medium with index of refraction n is

$$\lambda_n = \frac{\lambda}{n},$$

where λ is the wavelength in vacuum. The frequency is not changed.

Young's double-slit experiment clearly demonstrated the interference of light. The observed bright spots of the interference pattern were explained as constructive interference between the beams coming through the two slits, where the beams differ in path length by an integral number of wavelengths. The dark areas in between are due to destructive interference when the path lengths differ by $\frac{1}{2}\lambda, \frac{3}{2}\lambda$, and so on. The angles θ at which **constructive interference** occurs are given by

$$\sin\theta = m\frac{\lambda}{d},$$

where λ is the wavelength of the light, d the separation of the slits, and m an integer $(0, 1, 2, \cdots)$. **Destructive interference** occurs at angles θ given by

$$\sin\theta = \left(m + \tfrac{1}{2}\right)\frac{\lambda}{d}$$

where m is an integer $(0, 1, 2, \cdots)$.

The light intensity I_θ at any point in a double-slit interference pattern can be calculated using a phasor diagram, which predicts that

$$I_\theta = I_0 \cos^2\frac{\delta}{2}$$

where I_0 is the intensity at $\theta = 0$ and the phase angle δ is

$$\delta = \frac{2\pi d}{\lambda} \sin\theta.$$

Two sources of light are perfectly **coherent** if the waves leaving them are sinusoidal, of the same single frequency, and maintain the same phase relationship at all times. If the light waves from the two sources have a random phase with respect to each other over time (as for two incandescent light bulbs) the two sources are **incoherent**.

Light reflected from the front and rear surfaces of a thin film of transparent material can interfere constructively or destructively, depending on the optical path difference. A phase change of 180° or $\frac{1}{2}\lambda$ occurs when the light reflects at a surface where the index of refraction increases. Such **thin-film interference** has many practical applications, such as lens coatings and using Newton's rings to check the uniformity of glass surfaces.

Questions

1. Does Huygens' principle apply to sound waves? To water waves?

2. What is the evidence that light carries energy?

3. Why is light sometimes described as rays and sometimes as waves?

4. We can hear sounds around corners but we cannot see around corners, yet both sound and light are waves. Explain the difference.

5. Can the wavelength of light be determined from reflection or refraction measurements?

6. Monochromatic red light is incident on a double slit and the interference pattern is viewed on a screen some distance away. Explain how the fringe pattern would change if the red light source is replaced by a blue light source.

7. Suppose white light falls on the two slits of Fig. 35–7, but one slit is covered by a red filter (700 nm) and the other by a blue (450 nm) filter. Describe the pattern on the screen.

8. Compare a double-slit experiment for sound waves to that for light waves. Discuss the similarities and differences.

9. Two rays of light from the same source destructively interfere if their path lengths differ by how much?

10. If Young's double-slit experiment were submerged in water, how would the fringe pattern be changed?

11. Why doesn't the light from the two headlights of a distant car produce an interference pattern?

12. Why are interference fringes noticeable only for a *thin* film like a soap bubble and not for a thick piece of glass, say?

13. Why are Newton's rings (Fig. 35–18) closer together farther from the center?

14. Some coated lenses appear greenish yellow when seen by reflected light. What wavelengths do you suppose the coating is designed to eliminate completely?

15. A drop of oil on a pond appears bright at its edges where its thickness is much less than the wavelengths of visible light. What can you say about the index of refraction of the oil?

* 16. Describe how a Michelson interferometer could be used to measure the index of refraction of air.

Problems

1. (II) Derive the law of reflection—namely, that the angle of incidence equals the angle of reflection from a flat surface—using Huygens' principle for waves.

Section 35–3

2. (I) Monochromatic light falling on two slits 0.016 mm apart produces the fifth-order fringe at a 9.8° angle. What is the wavelength of the light used?

3. (I) The third-order fringe of 610 nm light is observed at an angle of 18° when the light falls on two narrow slits. How far apart are the slits?

4. (II) Monochromatic light falls on two very narrow slits 0.048 mm apart. Successive fringes on a screen 5.00 m away are 6.5 cm apart near the center of the pattern. What is the wavelength and frequency of the light?

5. (II) A parallel beam of light from a He-Ne laser, with a wavelength 656 nm, falls on two very narrow slits 0.060 mm apart. How far apart are the fringes in the center of the pattern if the screen is 3.6 m away?

6. (II) Light of wavelength 680 nm falls on two slits and produces an interference pattern in which the fourth-order fringe is 38 mm from the central fringe on a screen 2.0 m away. What is the separation of the two slits?

7. (II) If 720-nm and 660-nm light passes through two slits 0.58 mm apart, how far apart are the second-order fringes for these two wavelengths on a screen 1.0 m away?

8. (II) Suppose a thin piece of glass were placed in front of the lower slit in Fig. 35–7 so that the two waves enter the slits 180° out of phase (Fig. 35–25). Describe in detail the interference pattern on the screen.

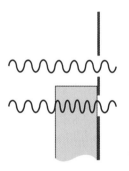

FIGURE 35–25
Problem 8.

9. (II) In a double-slit experiment it is found that blue light of wavelength 460 nm gives a second-order maximum at a certain location on the screen. What wavelength of visible light would have a minimum at the same location?

10. (II) Light of wavelength 480 nm in air falls on two slits 6.00×10^{-2} mm apart. The slits are immersed in water, as is a viewing screen 40.0 cm away. How far apart are the fringes on the screen?

11. (III) A very thin sheet of plastic ($n = 1.60$) covers one slit of a double-slit apparatus illuminated by 640-nm light. The center point on the screen, instead of being a maximum, is dark. What is the (minimum) thickness of the plastic?

Section 35–5

12. (I) If one slit in Fig. 35–12 is covered, by what factor does the intensity at the center of the screen change?

13. (II) Show that the angular full width at half maximum of the central peak in a double-slit interference pattern is given by $\Delta\theta = \lambda/2d$ if $\lambda \ll d$.

14. (II) Suppose that one slit of a double-slit apparatus is wider than the other so that the intensity of light passing through it is twice as great. Determine the intensity I as a function of position (θ) on the screen for coherent light.

15. (III) (a) Consider three equally spaced and equal-intensity coherent sources of light (such as adding a third slit to the two slits of Fig. 35–12). Use the phasor method to obtain the intensity as a function of the phase difference δ (Eq. 35–4). (b) Determine the positions of maxima and minima.

16. (III) Apply the phasor method to four parallel slits separated by equal distances d. Assume coherent light and determine the intensity as a function of position on the screen and find the positions of maxima and minima.

Section 35–6

17. (I) If a soap bubble is 120 nm thick, what color will appear at the center when illuminated normally by white light? Assume that $n = 1.34$.

18. (I) How far apart are the dark fringes in Example 35–6 if the glass plates are each 26.5 cm long?

19. (II) What is the minimum thickness (>0) of a soap film ($n = 1.34$) that would appear black if illuminated with 480-nm light? Assume there is air on both sides of the soap film.

20. (II) A lens appears greenish yellow ($\lambda = 570$ nm is strongest) when white light reflects from it. What minimum thickness of coating ($n = 1.28$) do you think is used on such a (glass) lens, and why?

21. (II) A total of 28 bright and 28 dark Newton's rings (not counting the dark spot at the center) are observed when 650-nm light falls normally on a planoconvex lens resting on a flat glass surface (Fig. 35–18). How much thicker is the center than the edges?

22. (II) A fine metal foil separates one end of two pieces of optically flat glass, as in Fig. 35–20. When light of wavelength 670 nm is incident normally, 25 dark lines are observed (with one at each end). How thick is the foil?

23. (II) How thick (minimum) should the air layer be between two flat glass surfaces if the glass is to appear bright when 480-nm light is incident normally? What if the glass is to appear dark?

24. (II) A thin film of alcohol ($n = 1.36$) lies on a flat glass plate ($n = 1.51$). When monochromatic light, whose wavelength can be changed, is incident normally, the reflected light is a minimum for $\lambda = 512$ nm and a maximum for $\lambda = 640$ nm. What is the thickness of the film?

25. (II) When a Newton's ring apparatus (Fig. 35–18) is immersed in a liquid, the diameter of the eighth dark ring decreases from 2.92 cm to 2.60 cm. What is the refractive index of the liquid?

26. (II) A planoconvex lucite lens 3.4 cm in diameter is placed on a flat piece of glass as in Fig. 35–18. When 580-nm light is incident normally, 48 bright rings are observed, the last one right at the edge. What is the radius of curvature of the lens surface, and the focal length of the lens?

27. (II) Show that the radius r of the mth dark Newton's ring, as viewed from directly above (Fig. 35–18), is given by $r = \sqrt{m\lambda R}$ where R is the radius of curvature of the curved glass surface and λ is the wavelength of light used. Assume that the thickness of the air gap is much less than R at all points and that $r \ll R$.

28. (III) Use the result of Problem 27 to show that the distance between adjacent dark Newton's rings is

$$\Delta r \approx \sqrt{\frac{\lambda R}{4\,m}}$$

for the mth ring, assuming $m \gg 1$.

29. (III) A single optical coating reduces reflection to zero for $\lambda = 550$ nm. By what factor is the intensity reduced by the coating for $\lambda = 450$ nm and $\lambda = 650$ nm as compared to no coating? Assume normal incidence.

* Section 35–7

* 30. (II) What is the wavelength of the light entering an interferometer if 344 bright fringes are counted when the movable mirror moves 0.125 mm?

* 31. (II) How far must the mirror M_1 in a Michelson interferometer be moved if 750 fringes of 589-nm light are to pass by a reference line?

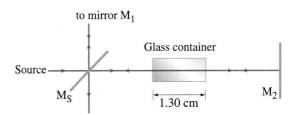

FIGURE 35–26 Problem 32.

* 32. (III) One of the beams of an interferometer (Fig. 35–26) passes through a small glass container containing a cavity 1.30 cm deep. When a gas is allowed to slowly fill the container, a total of 186 dark fringes are counted to move past a reference line. The light used has a wavelength of 610 nm. Calculate the index of refraction of the gas at its final density, assuming that the interferometer is in vacuum.

* 33. (III) The yellow sodium D lines have wavelengths of 589.0 and 589.6 nm. When they are used to illuminate a Michelson interferometer, it is noted that the interference fringes disappear and reappear periodically as the mirror M_1 is moved. Why does this happen? How far must the mirror move between one disappearance and the next?

* Section 35–8

* 34. (I) The illuminance of direct sunlight on Earth is about 10^5 lm/m². Estimate the luminous flux and luminous intensity of the Sun.

* 35. (II) The *luminous efficiency* of a light bulb is the ratio of luminous flux to electric power input. (a) What is the luminous efficiency of a 100-W, 1700-lm bulb? (b) How many 40-W, 60-lm/W fluorescent lamps would be needed to provide an illuminance of 250 lm/m² on a factory floor of area 25 m × 30 m? Assume the lights are 10 m above the floor and that half their flux reaches the floor.

General Problems

36. Light of wavelength λ strikes a screen containing two slits a distance d apart at an angle θ_i to the normal. Determine the angle θ_m at which the mth-order maximum occurs.

37. Television and radio waves can reflect from nearby mountains or from airplanes, and the reflections can interfere with the direct signal from the station. (a) Determine what kind of interference will occur when 75-MHz television signals arrive at a receiver directly from a distant station, and are reflected from an airplane 118 m directly above the receiver. (Assume $\frac{1}{2}\lambda$ change in phase of the signal upon reflection.) (b) What kind of interference will occur if the plane is 22 m closer to the receiver?

38. A radio station operating at 102.1 MHz broadcasts from two identical antennae at the same elevation but separated by a 7.0-m horizontal distance d, Fig. 35–27. A maximum signal is found along the midline, perpendicular to d at its midpoint and extending horizontally in both directions. If the midline is taken as 0°, at what other angle(s) θ is a maximum signal detected? A minimum signal? Assume all measurements are made much farther than 7 m from the antenna towers.

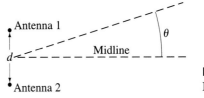

FIGURE 35–27
Problem 38.

39. Light of wavelength 690 nm passes through two narrow slits 0.60 mm apart. The screen is 1.50 m away. A second source of unknown wavelength produces its second-order fringe 1.13 mm closer to the central maximum than the 690-nm light. What is the wavelength of the unknown light?

40. What is the index of refraction of a clear material if a minimum of 150 nm thickness of it, when laid on glass, is needed to reduce reflection to nearly zero when light of 600 nm is incident normally upon it? Do you have a choice for an answer?

41. Monochromatic light of variable wavelength is incident normally on a thin sheet of plastic film in air. The reflected light is a minimum only for $\lambda = 510$ nm and $\lambda = 680$ nm in the visible spectrum. What is the thickness of the film ($n = 1.58$)?

42. In a water tank experiment, water waves are generated with their crests 3.5 cm apart and parallel. They pass through two openings 6.0 cm apart in a long wooden board. If the end of the tank is 2.0 m beyond the boards, where would you stand, relative to the "straight-through" direction, so that you received little or no wave action?

43. Compare the minimum thickness needed for an antireflective coating ($n = 1.38$) applied to a glass lens in order to eliminate (a) blue (450 nm), or (b) red (700 nm) reflections for light at normal incidence.

44. Suppose you viewed the light *transmitted* through a thin film on a flat piece of glass. Draw a diagram, similar to Fig. 35–17 or 35–23, and describe the conditions required for maxima and minima. Consider all possible values of index of refraction. Discuss the relative size of the minima compared to the maxima and to zero.

45. Stealth aircraft are designed to not reflect radar, whose wavelength is typically 2 cm, by using an antireflecting coating. Ignoring any change in wavelength in the coating, estimate its thickness.

46. Very highly reflective mirrors for a particular wavelength can be made by alternating many layers of *transparent* materials of indices of refraction n_1 and $n_2 (1 < n_1 < n_2)$. What should the minimum thicknesses d_1 and d_2 of Fig. 35–28 be, in terms of the incident wavelength λ, to maximize reflection?

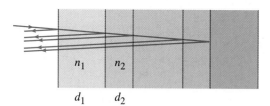

FIGURE 35–28   Problem 46.

47. When a charged particle, such as a proton, travels through a transparent medium at a speed v_p faster than the speed of light in that medium ($v = c/n$), it emits electromagnetic radiation (light) called **Čerenkov radiation**. It is the electromagnetic equivalent of a shock wave (see Section 16–8), and is confined to a particular angle that depends on v_p and v. Determine this angle for a proton traveling 2.21×10^8 m/s in a plastic whose index of refraction is 1.52.

48. What is the minimum (non-zero) thickness for the air layer between two flat glass surfaces if the glass is to appear dark when 640-nm light is incident normally? What if the glass is to appear bright?

49. *Lloyd's mirror* provides one way of obtaining a double-slit interference pattern from a single source so the light is coherent; as shown in Fig. 35–29, the light that reflects from the plane mirror appears to come from the virtual image of the slit. Describe in detail the interference pattern on the screen.

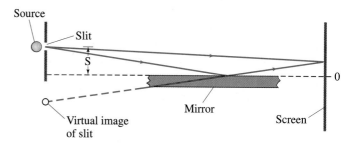

FIGURE 35–29   Problem 49.

50. Consider the antenna array of Example 35–5, Fig. 35–15. Let $d = \lambda/2$, and suppose that the two antennas are now 180° out of phase with each other. Find the directions for constructive and destructive interference, and compare with the case when the sources are in phase. (These results illustrate the basis for directional antennas.)

*51. Suppose the mirrors in a Michelson interferometer are perfectly aligned and the path lengths to mirrors M_1 and M_2 are identical. With these initial conditions, an observer sees a bright maximum at the center of the viewing area. Now one of the mirrors is moved a distance x. Determine a formula for the intensity at the center of the viewing area as a function of x, the distance the movable mirror is moved from the initial conditions.

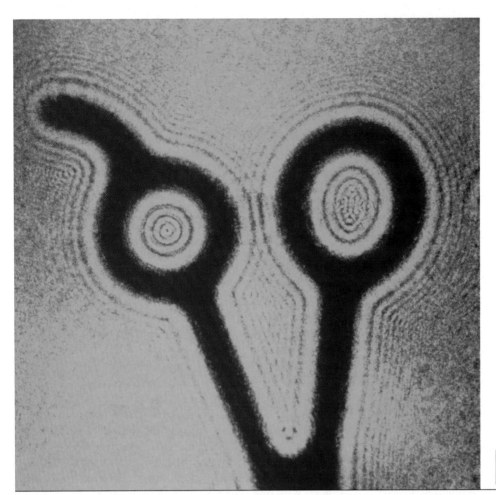

Parallel coherent light from a laser, which acts as nearly a point source, illuminates these shears. Instead of a clean shadow, there is a dramatic diffraction pattern, which is a strong confirmation of the wave theory of light. Diffraction patterns are washed out when typical extended sources of light are used, and hence are not seen, although a careful examination of shadows will reveal fuzziness. We will examine diffraction by a single slit, and how it affects the double-split pattern, as well as diffraction gratings and x-rays. We will see how diffraction affects the resolution of optical instruments, and that the ultimate resolution can never be greater than the wavelength of the radiation used. Finally we study the polarization of light.

CHAPTER 36

Diffraction and Polarization

Young's double-slit experiment put the wave theory of light on a firm footing. But full acceptance came only with studies on diffraction more than a decade later.

We have already discussed diffraction briefly with regard to water waves (Section 15–11) as well as for light (Section 35–1) and we have seen that it refers to the spreading or bending of waves around edges. Now we look at diffraction more closely, including its important practical effects of limiting the amount of detail, or *resolution*, that can be obtained with any optical instrument including telescopes, cameras, and the eye.

A part of the history of the wave theory of light belongs to Augustin Fresnel (1788–1827) who in 1819 presented to the French Academy a wave theory of light that predicted and explained interference and diffraction effects. Almost immediately Siméon Poisson (1781–1840) pointed out a counter-intuitive inference: according to Fresnel's wave theory, if light from a point source were to fall on a solid disk, part of the incident light would be diffracted around the edges and would constructively interfere at the center of the shadow (Fig. 36–1). That prediction seemed very unlikely. But when the experiment was actually carried out by François Arago, the bright spot was seen at the very center of the shadow (Fig. 36–2a). This was strong evidence for the wave theory.

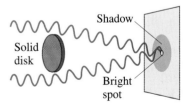

FIGURE 36–1 If light is a wave, a bright spot will appear at the center of the shadow of a solid disk illuminated by a point source of monochromatic light.

Figure 36–2a is a photograph of the shadow cast by a coin using a (nearly) point source of light, a laser in this case. The bright spot is clearly present at the center. Note that there also are bright and dark fringes beyond the shadow. These resemble the interference fringes of a double slit. Indeed, they are due to interference of waves diffracted around the disk, and the whole is referred to as a **diffraction pattern**. A diffraction pattern exists around any sharp-edged object illuminated by a point source, as shown in Fig. 36–2b and c. We are not always aware of diffraction because most sources of light in everyday life are not points, so light from different parts of the source washes out the pattern.

36–1 | Diffraction by a Single Slit

To see how a diffraction pattern arises, we will analyze the important case of monochromatic light passing through a narrow slit. We will assume that parallel rays (plane waves) of light fall on the slit of width a, and pass through to a viewing screen very far away. If the viewing screen is not far away, lenses can be used to make the rays parallel.[†] As we know from studying water waves and from Huygens' principle, the waves passing through the slit spread out in all directions. We will now examine how the waves passing through different parts of the slit interfere with each other.

Parallel rays of monochromatic light pass through the narrow slit as shown in Fig. 36–3a. The light falls on a screen which is assumed to be very far away, so the rays heading for any point are essentially parallel. First we consider rays that pass straight through as in Fig. 36–3a. They are all in phase, so there will be a central bright spot on the screen. In Fig. 36–3b, we consider rays moving at an angle θ such that the ray from the top of the slit travels exactly one wavelength farther than the ray from the bottom edge. The ray passing through the very center of the slit will travel one-half wavelength farther than the ray at the bottom of the slit. These two rays will be exactly out of phase with one another and so will destructively interfere. Similarly, a ray slightly above the bottom one will cancel a ray that is the same distance above the central one. Indeed, each ray passing through the lower half of the slit will cancel with a corresponding ray passing through the upper half. Thus, all the

[†] Such a diffraction pattern, involving parallel rays, is called *Fraunhofer diffraction*. If the screen is close and no lenses are used, it is called *Fresnel diffraction*. The analysis in the latter case is rather involved, so we consider only the limiting case of Fraunhofer diffraction.

FIGURE 36–2 Diffraction pattern of (a) a circular disk (a coin), (b) a razor blade, (c) a single slit, each illuminated by a (nearly) point source of monochromatic light.

(a)   (b)   (c)

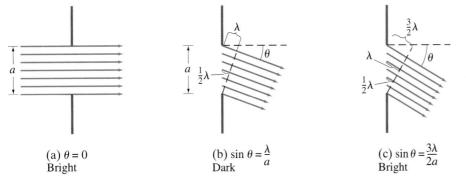

(a) $\theta = 0$
Bright

(b) $\sin\theta = \dfrac{\lambda}{a}$
Dark

(c) $\sin\theta = \dfrac{3\lambda}{2a}$
Bright

(d) $\sin\theta = \dfrac{2\lambda}{a}$
Dark

FIGURE 36–3 Analysis of diffraction pattern formed by light passing through a narrow slit.

rays destructively interfere in pairs, and so no light will reach the viewing screen at this angle. The angle θ at which this takes place can be seen from the diagram to occur when $\lambda = a\sin\theta$, so

$$\sin\theta = \frac{\lambda}{a}.$$ [first minimum]   **(36–1)**

Diffraction equation (angular width of central spot)

The light intensity is a maximum at $\theta = 0°$ and decreases to a minimum (intensity = zero) at the angle θ given by Eq. 36–1.

Now consider a larger angle θ such that the top ray travels $\frac{3}{2}\lambda$ farther than the bottom ray, as in Fig. 36–3c. In this case, the rays from the bottom third of the slit will cancel in pairs with those in the middle third because they will be $\lambda/2$ out of phase. However, light from the top third of the slit will still reach the screen, so there will be a bright spot centered near $\sin\theta \approx 3\lambda/2a$, but it will not be nearly as bright as the central spot at $\theta = 0°$. For an even larger angle θ such that the top ray travels 2λ farther than the bottom ray, Fig. 36–3d, rays from the bottom quarter of the slit will cancel with those in the quarter just above it because the path lengths differ by $\lambda/2$. And the rays through the quarter of the slit just above center will cancel with those through the top quarter. At this angle there will again be a minimum of zero intensity in the diffraction pattern. A plot of the intensity as a function of angle is shown in Fig. 36–4. This corresponds well with the photo of Fig. 36–2c. Notice that minima (zero intensity) occur at

$$a\sin\theta = m\lambda, \qquad m = 1, 2, 3, \cdots,$$ [minima]   **(36–2)**

but *not* at $m = 0$ where there is the strongest maximum. Between the minima, smaller intensity maxima occur at approximately (but not exactly) $m \approx \frac{3}{2}\lambda, \frac{5}{2}\lambda, \cdots$. [Note that the *minima* for a diffraction pattern, Eq. 36–2, satisfy a criterion very similar to that for the *maxima* (bright spots) for double-slit interference, Eq. 35–2a.]

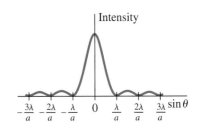

FIGURE 36–4 Intensity in the diffraction pattern of a single slit as a function of $\sin\theta$. Note that the central maximum is not only much higher than the maxima to each side, but it is also twice as wide ($2\lambda/a$ wide) as any of the others (only λ/a wide each).

Single slit diffraction minima

FIGURE 36–5 Example 36–1.

| **EXAMPLE 36–1** | **Single-slit diffraction maximum.** Light of wavelength |

EXAMPLE 36–1 **Single-slit diffraction maximum.** Light of wavelength 750 nm passes through a slit 1.0×10^{-3} mm wide. How wide is the central maximum (*a*) in degrees, and (*b*) in centimeters, on a screen 20 cm away?

SOLUTION   (*a*) The first minimum occurs at

$$\sin\theta = \frac{\lambda}{a} = \frac{7.5 \times 10^{-7}\,\text{m}}{1.0 \times 10^{-6}\,\text{m}} = 0.75.$$

So $\theta = 49°$. This is the angle between the center and the first minimum, Fig. 36–5. The angle subtended by the whole central maximum, between the minima above and below the center, is twice this, or 98°.

(*b*) The width of the central maximum is $2x$, where $\tan\theta = x/20$ cm. So $2x = 2(20\,\text{cm})(\tan 49°) = 46$ cm. A large width of the screen will be illuminated, but it will not normally be very bright since the amount of light that passes through such a small slit will be small and it is spread over a large area.

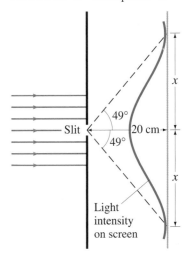

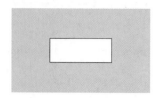

FIGURE 36–6 Example 36–2.

CONCEPTUAL EXAMPLE 36–2 **Diffraction spreads.** Light shines through a rectangular hole that is narrower in the vertical direction than the horizontal, Fig. 36–6. (a) Would you expect the diffraction pattern to be more spread out in the vertical direction or in the horizontal direction? (b) Should a rectangular loudspeaker horn at a stadium be high and narrow, or wide and flat?

RESPONSE (a) From Eq. 36–2 we can see that if we make the slit (width a) narrower, the pattern spreads out more. This is consistent with our earlier study of waves in Chapter 15. Hence the diffraction through the rectangular hole will be wider vertically, since the aperture in that direction is smaller.
(b) For the loudspeaker, the sound pattern desired is one spread out horizontally, so the horn should be tall and narrow (rotate Fig. 36–6 by 90°).

36–2 Intensity in Single-Slit Diffraction Pattern

We have determined the positions of the minima in the diffraction pattern produced by light passing through a single slit, Eq. 36–2. We now discuss a method for predicting the amplitude and intensity at any point in the pattern using the phasor technique already discussed in Section 35–5.

Let us consider the slit divided into N very thin strips of width Δy as indicated in Fig. 36–7. Each strip sends light in all directions toward a screen on the right. Again we take the rays heading for any particular point on the distant screen to be parallel, all making an angle θ with the horizontal as shown. We choose the strip width Δy to be much smaller than the wavelength λ of the monochromatic light falling on the slit, so all the light from a given strip is in phase. The strips are of equal size, and if the whole slit is uniformly illuminated, we can take the electric field wave amplitudes ΔE_0 from each thin strip to be equal as long as θ is not too large. However, the separate amplitudes from the different strips will differ in phase. The phase difference in the light coming from adjacent strips will be (see Section 35–5, Eq. 35–4)

$$\Delta \beta = \frac{2\pi}{\lambda} \Delta y \sin \theta \qquad (36\text{–}3)$$

since the difference in path length is $\Delta y \sin \theta$.

The total amplitude on the screen at any angle θ will be the sum of the separate wave amplitudes due to each strip. These wavelets have the same amplitude ΔE_0 but differ in phase. To obtain the total amplitude, we can use a phasor diagram as we did in Section 35–5 (Fig. 35–13). The phasor diagrams for four different angles θ are shown in Fig. 36–8. At the center of the screen, $\theta = 0$, the waves from each strip are all in phase ($\Delta \beta = 0$, Eq. 36–3), so the arrows representing each ΔE_0 line up as shown in Fig. 36–8a. The total amplitude of the light arriving at the center of the screen is then $E_0 = N \Delta E_0$. At a small angle θ, for a point on the distant screen not far from the center, Fig. 36–8b shows how the wavelets of amplitude ΔE_0 add up to give E_θ, the total amplitude on the screen at this angle θ. Note that each wavelet differs in phase from the adjacent one by $\Delta \beta$. The phase difference between the wavelets from the top and bottom edges of the slit is

$$\beta = N \Delta \beta = \frac{2\pi}{\lambda} N \Delta y \sin \theta = \frac{2\pi}{\lambda} a \sin \theta \qquad (36\text{–}4)$$

where $a = N \Delta y$ is the total width of the slit. Although the "arc" in Fig. 36–8b has length $N \Delta E_0$, and so equals E_0 (total amplitude at $\theta = 0$), the amplitude of the total wave E_θ at angle θ is the *vector* sum of each wavelet amplitude and so is equal to the length of the chord as shown. The chord is shorter than the arc, so $E_\theta < E_0$.

FIGURE 36–7 Slit of width a divided into N strips of width Δy.

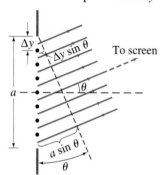

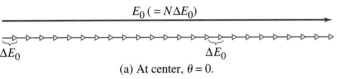

$E_0 \, (= N\Delta E_0)$

ΔE_0 ΔE_0

(a) At center, $\theta = 0$.

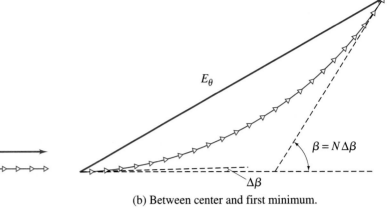

E_θ

$\beta = N\Delta\beta$

$\Delta\beta$

(b) Between center and first minimum.

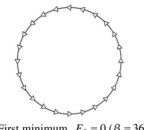

(c) First minimum, $E_\theta = 0$ ($\beta = 360°$).

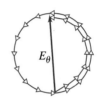

E_θ

(d) Near secondary maximum.

FIGURE 36–8 Phasor diagram for single-slit diffraction, giving the total amplitude E_θ at four different angles θ.

For greater θ, we eventually come to the case, illustrated in Fig. 36–8c, where the chain of arrows closes on itself. In this case the vector sum is zero, so $E_\theta = 0$ for this angle θ. This corresponds to the first minimum. Since $\beta = N\,\Delta\beta$ is 360° or 2π in this case, we have from Eq. 36–3,

$$2\pi = N\,\Delta\beta = N\left(\frac{2\pi}{\lambda}\,\Delta y\sin\theta\right)$$

or, since the slit width $a = N\,\Delta y$,

$$\sin\theta = \frac{\lambda}{a}.$$

Thus the first minimum $(E_\theta = 0)$ occurs where $\sin\theta = \lambda/a$, which is the same result we obtained in the previous Section, Eq. 36–1.

For even greater values of θ, the chain of arrows spirals beyond 360°. Figure 36–8d shows the case near the secondary maximum next to the first minimum. Here $\beta = N\,\Delta\beta \approx 360° + 180° = 540°$ or 3π. (Note that although β may be 540°, θ can still be a very small angle, depending on the values of a and λ.) When $\beta = 4\pi$, we have a double circle and again a minimum, where $\sin\theta = 2\lambda/a$, corresponding to $m = 2$ in Eq. 36–2. When greater angles θ are considered, new maxima and minima occur. But since the total length of the coil remains constant, equal to $N\,\Delta E_0 (= E_0)$, each succeeding maximum is smaller and smaller as the coil winds in on itself.

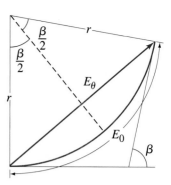

FIGURE 36–9 Determining amplitude E_θ as a function of θ for single-slit diffraction.

To obtain a quantitative expression for the amplitude (and intensity) for any point on the screen (that is, for any angle θ), we now consider the limit $N \to \infty$ so Δy become the infinitesimal width dy. In this case, the diagrams of Fig. 36–8 become smooth curves, one of which is shown in Fig. 36–9. For any angle θ, the wave amplitude on the screen is E_θ, equal to the chord in Fig. 36–9. The length of the arc is E_0, as before. If r is the radius of curvature of the arc, then

$$\frac{E_\theta}{2} = r \sin \frac{\beta}{2}.$$

Using radian measure for $\beta/2$, we also have

$$\frac{E_0}{2} = r \frac{\beta}{2}.$$

We combine these to obtain

$$E_\theta = E_0 \frac{\sin \beta/2}{\beta/2}. \tag{36–5}$$

The angle β is the phase difference between the waves from the top and bottom edges of the slit. The path difference for these two rays is $a \sin \theta$ (see Fig. 36–7 as well as Eq. 36–4), so

$$\beta = \frac{2\pi}{\lambda} a \sin \theta. \tag{36–6}$$

Intensity is proportional to the square of the wave amplitude, so the intensity I_θ at any angle θ is, from Eq. 36–5,

$$I_\theta = I_0 \left(\frac{\sin \beta/2}{\beta/2} \right)^2 \tag{36–7}$$

where $I_0 (\propto E_0^2)$ is the intensity at $\theta = 0$ (the central maximum). We can combine Eqs. 36–7 and 36–6 (although it is often simpler to leave them as separate equations) to obtain

Intensity in single-slit diffraction pattern

$$I_\theta = I_0 \left[\frac{\sin \left(\dfrac{\pi a \sin \theta}{\lambda} \right)}{\left(\dfrac{\pi a \sin \theta}{\lambda} \right)} \right]^2. \tag{36–8}$$

According to Eq. 36–8, minima $(I_\theta = 0)$ occur where $\sin(\pi a \sin \theta/\lambda) = 0$, which means $\pi a \sin\theta/\lambda$ must be $\pi, 2\pi, 3\pi$, and so on, or

$$a \sin\theta = m\lambda, \qquad m = 1, 2, 3, \cdots \qquad \text{[minima]}$$

which is what we have obtained previously, Eq. 36–2. Notice that m cannot be zero: when $\beta/2 = \pi a \sin\theta/\lambda = 0$, the denominator as well as the numerator in Eqs. 36–7 or 36–8 vanishes. We can evaluate the intensity in this case by taking the limit as $\theta \to 0$ (or $\beta \to 0$); for very small angles, $\sin\beta/2 \approx \beta/2$, so $(\sin\beta/2)/(\beta/2) \to 1$ and $I_\theta = I_0$, the maximum at the center of the pattern.

The intensity I_θ as a function of θ, as given by Eq. 36–8, corresponds to the diagram of Fig. 36–4.

EXAMPLE 36–3 **ESTIMATE** **Intensity at secondary maxima.** Estimate the intensities of the first two secondary maxima to either side of the central maximum.

SOLUTION The secondary maxima occur close to halfway between the minima, at about

$$\frac{\beta}{2} = \frac{\pi a \sin\theta}{\lambda} \approx \left(m + \tfrac{1}{2}\right)\pi. \qquad m = 1, 2, 3, \cdots.$$

The actual maxima are not quite at these points—their positions can be determined by differentiating Eq. 36–7 (see Problem 10)—but we are only seeking an estimate. Using these values for β in Eq. 36–7 or 36–8, with $\sin\left(m + \tfrac{1}{2}\right)\pi = 1$, gives

$$I_\theta = \frac{I_0}{\left(m + \tfrac{1}{2}\right)^2 \pi^2}. \qquad m = 1, 2, 3, \cdots.$$

For $m = 1$ and 2, we get

$$I_\theta = \frac{I_0}{22.2} = 0.045\, I_0 \qquad\qquad\qquad [m = 1]$$

$$I_\theta = \frac{I_0}{61.7} = 0.016\, I_0. \qquad\qquad\qquad [m = 2]$$

The first maximum to the side of the central peak has only 1/22, or 4.5 percent, the intensity of the central intensity, and succeeding ones are smaller still, just as we can see in Fig. 36–4 and the photo of Fig. 36–2c.

Diffraction by a circular opening produces a similar pattern (though circular rather than rectangular) and is of great practical importance, since lenses are essentially circular apertures through which light passes. We will discuss this in Section 36–4 and see how diffraction limits the resolution (or sharpness) of images.

---

* 36–3 | Diffraction in the Double-Slit Experiment

When we analyzed Young's double-slit experiment in Section 35–5, we assumed that the central portion of the screen was uniformly illuminated. This is equivalent to assuming the slits are infinitesimally narrow, so that the central diffraction peak is spread out over the whole screen. This can never be the case for real slits; diffraction reduces the intensity of the bright interference fringes to the side of center so they are not all of the same height as they were shown in Fig. 35–14. (They were shown more correctly in Fig. 35–9b.)

To calculate the intensity in a double-slit interference pattern, including diffraction, let us assume the slits have equal widths a and their centers are separated by a distance d. Since the distance to the screen is large compared to the slit separation d, the wave amplitude due to each slit is essentially the same at each point on the screen. Then the total wave amplitude at any angle θ will no longer be

$$E_{\theta 0} = 2E_0 \cos \frac{\delta}{2},$$

as was given by Eq. 35–5b. Rather, it must be modified, because of diffraction, by Eq. 36–5, so that

$$E_{\theta 0} = 2E_0 \left(\frac{\sin \beta/2}{\beta/2} \right) \cos \frac{\delta}{2}.$$

(Note that up to now in this chapter we have dealt only with the amplitude of E, and so we merely wrote E_θ rather than $E_{\theta 0}$ as was done in Chapter 35.) Thus the intensity will be given by

$$I_\theta = I_0 \left(\frac{\sin \beta/2}{\beta/2} \right)^2 \left(\cos \frac{\delta}{2} \right)^2 \qquad \textbf{(36–9)}$$

where, from Eqs. 36–6 and 35–4,

$$\frac{\beta}{2} = \frac{\pi}{\lambda} a \sin \theta \qquad \text{and} \qquad \frac{\delta}{2} = \frac{\pi}{\lambda} d \sin \theta.$$

The first term in parentheses in Eq. 36–9 is sometimes called the "diffraction factor" and the second one the "interference factor." These two factors are plotted in Fig. 36–10a and b for the case when $d = 6a$, and $a = 10\lambda$. (Figure 36–10b is essentially the same as Fig. 35–14.) Figure 36–10c shows the product of these two curves, times I_0, which is the actual intensity as a function of θ (or as a function of position on the screen for θ not too large) as given by Eq. 36–9. As indicated by the dashed lines in Fig. 36–10c, the diffraction factor acts as a sort of envelope that limits the interference peaks.

EXAMPLE 36–4 **Diffraction plus interference.** Show why the central diffraction peak in Fig. 36–10c contains 11 interference fringes.

SOLUTION The first minimum in the diffraction pattern occurs where

$$\sin \theta = \frac{\lambda}{a}.$$

Since $d = 6a$,

$$d \sin \theta = 6a \left(\frac{\lambda}{a} \right) = 6\lambda.$$

From Eq. 35–2a, interference peaks (maxima) occur for $d \sin \theta = m\lambda$ where m can be $0, 1, \cdots$ or any integer. Thus the diffraction minimum ($d \sin \theta = 6\lambda$) coincides with $m = 6$ in the interference pattern, so the $m = 6$ peak won't appear. Hence the central diffraction peak encloses the central interference peak ($m = 0$) and five peaks ($m = 1$ to 5) on each side for a total of 11. Since the sixth order doesn't appear, it is said to be a "missing order."

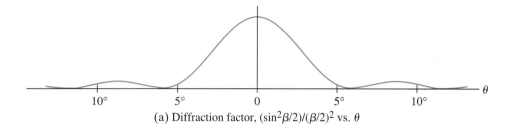

(a) Diffraction factor, $(\sin^2\beta/2)/(\beta/2)^2$ vs. θ

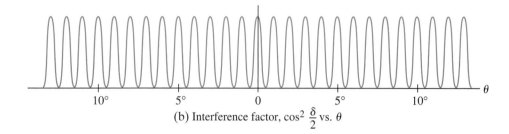

(b) Interference factor, $\cos^2\dfrac{\delta}{2}$ vs. θ

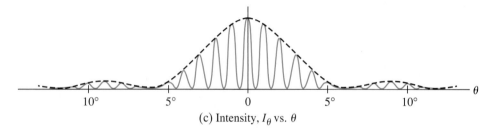

(c) Intensity, I_θ vs. θ

FIGURE 36–10 (a) Diffraction factor, (b) interference factor, and (c) the resultant intensity I_θ, plotted as a function of θ for $d = 6a = 60\lambda$.

Notice from this last Example that the number of interference fringes in the central diffraction peak depends only on the ratio d/a. It does not depend on wavelength λ. The actual spacing (in angle, or in position on the screen) does depend on λ. For the case illustrated, $a = 10\lambda$, and so the first diffraction minimum occurs at $\sin\theta = \lambda/a = 0.10$ or about 6°.

The decrease in intensity of the interference fringes away from the center, as graphed in Fig. 36–10, is shown in Fig. 36–11.

The patterns due to interference and diffraction arise from the same phenomenon—the superposition of coherent waves of different phase. The distinction between them is thus not so much physical as for convenience of description, as in this Section where we analyzed the two-slit pattern in terms of interference and diffraction separately. In general, we use the word "diffraction" when referring to an analysis by superposition of many infinitesimal and usually contiguous sources, such as when we subdivide a source into infinitesimal parts. We use the term "interference" when we superpose the wave from a finite (and usually small) number of coherent sources.

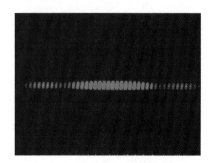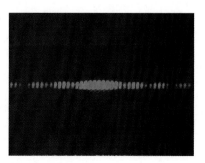

FIGURE 36–11 Photographs of a double-slit interference pattern showing effects of diffraction. In both cases $d = 0.50$ mm, whereas $a = 0.040$ mm in (a) and 0.080 mm in (b).

(a)

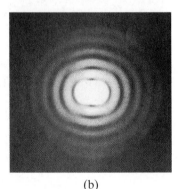

(b)

FIGURE 36–12 Photographs of images (greatly magnified) formed by a lens, showing the diffraction pattern of an image for: (a) a single point object; (b) two point objects whose images are barely resolved.

The ability of a lens to produce distinct images of two point objects very close together is called the **resolution** of the lens. The closer the two images can be and still be seen as distinct (rather than overlapping blobs), the higher the resolution. The resolution of a camera lens, for example, is often specified as so many lines per millimeter,[†] and can be determined by photographing a standard set of parallel lines on fine-grain film. The minimum spacing of lines distinguishable on film using the lens gives the resolution.

Two principal factors limit the resolution of a lens. The first is lens aberrations. As we saw, because of spherical and other aberrations, a point object is not a point on the image but a tiny blob. Careful design of compound lenses can reduce aberrations significantly, but they cannot be eliminated entirely. The second factor that limits resolution is *diffraction*, which cannot be corrected for because it is a natural result of the wave nature of light. We discuss it now.

In Section 36–1, we saw that because light travels as a wave, light from a point source passing through a slit is spread out into a diffraction pattern (Figs. 36–2 and 36–4). A lens, because it has edges, acts like a slit. When a lens forms the image of a point object, the image is actually a tiny diffraction pattern. Thus *an image would be blurred even if aberrations were absent.*

In the analysis that follows, we assume that the lens is free of aberrations, so that we can focus our attention on diffraction effects and how much they limit the resolution of a lens. In Fig. 36–4 we saw that the diffraction pattern produced by light passing through a rectangular slit has a central maximum in which most of the light falls. This central peak falls to a minimum on either side of its center at an angle $\theta \approx \sin\theta = \lambda/a$ (this is Eq. 36–1) where a is the width of the slit, λ is the wavelength of light used, and we assume θ is small. There are also low-intensity fringes beyond. For a lens, or any circular hole, the image of a point object will consist of a *circular* central peak (called the *diffraction spot* or *Airy disk*) surrounded by faint circular fringes, as shown in Fig. 36–12a. The central maximum has an angular half width given by

$$\theta = \frac{1.22\lambda}{D}$$

where D is the diameter of the circular opening.

This formula differs from that for a slit (Eq. 36–1) by the factor 1.22. This factor appears because the width of a circular hole is not uniform (like a rectangular slit) but varies from its diameter D to zero. A careful analysis shows that the "average" width is $D/1.22$. Hence we get the equation above rather than Eq. 36–1.

[†]This may be specified at the center of the field of view as well as at the edges, where it is usually less because of off-axis aberrations.

FIGURE 36–13 Intensity of light across the diffraction pattern of a circular hole.

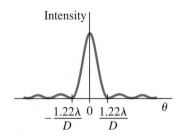

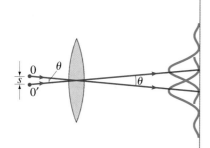

FIGURE 36–14 The *Rayleigh criterion.* Two images are just resolvable when the center of the diffraction peak of one is directly over the first minimum in the diffraction pattern of the other. The two point objects 0 and 0′ subtend an angle θ at the lens; only one ray is drawn for each object, to indicate the center of the diffraction pattern of its image.

The intensity of light in the diffraction pattern of light from a point source passing through a circular opening is shown in Fig. 36–13. The image for a non-point source is a superposition of such patterns. For most purposes we need consider only the central spot, since the concentric rings are so much dimmer.

If two point objects are very close, the diffraction patterns of their images will overlap as shown in Fig. 36–12b. As the objects are moved closer, a separation is reached where you can't tell if there are two overlapping images or a single image. The separation at which this happens may be judged differently by different observers. However, a generally accepted criterion is one proposed by Lord Rayleigh (1842–1919). This **Rayleigh criterion** states that *two images are just resolvable when the center of the diffraction disk of one image is directly over the first minimum in the diffraction pattern of the other.* This is shown in Fig. 36–14. Since the first minimum is at an angle $\theta = 1.22\lambda/D$ from the central maximum, Fig. 36–14 shows that two objects can be considered just resolvable if they are separated by this angle θ:

$$\theta = \frac{1.22\lambda}{D}. \tag{36–10}$$

Rayleigh criterion (resolution limit)

This is the limit on resolution set by the wave nature of light due to diffraction.

EXAMPLE 36–5 **Hubble space telescope.** The Hubble Space Telescope (HST) is a reflecting telescope that was placed in orbit above the Earth's atmosphere, so its resolution would not be limited by turbulence in the atmosphere (Fig. 36–15). Its objective diameter is 2.4 m. For visible light, say $\lambda = 550$ nm, estimate the improvement in resolution the Hubble offers over Earth-bound telescopes, which are limited in resolution by movement of the Earth's atmosphere to about half an arc second (each degree is divided into 60 minutes each containing 60 seconds, so $1° = 3600$ arc seconds).

SOLUTION Earth-bound telescopes are limited to an angular resolution of

$$\theta = \tfrac{1}{2}\left(\frac{1}{3600}\right)°\left(\frac{2\pi \text{ rad}}{360°}\right) = 2.4 \times 10^{-6} \text{ rad.}$$

The Hubble, on the other hand, is limited by diffraction (Eq. 36–10) which for $\lambda = 550$ nm is

$$\theta = \frac{1.22(550 \times 10^{-9} \text{ m})}{2.4 \text{ m}} = 2.8 \times 10^{-7} \text{ rad,}$$

which is almost ten times better resolution.

FIGURE 36–15 Hubble Space Telescope, with Earth in the background. The flat orange panels are solar cells that collect energy from the Sun.

36–5 Resolution of Telescopes and Microscopes; the λ Limit

You might think that a microscope or telescope could be designed to produce any desired magnification, depending on the choice of focal lengths and quality of the lenses. But this is not possible, because of diffraction. An increase in magnification above a certain point merely results in magnification of the diffraction patterns. This could be highly misleading since we might think we are seeing details of an object when we are really seeing details of the diffraction pattern. To examine this problem, we apply the Rayleigh criterion: two objects (or two nearby points on one object) are just resolvable if they are separated by an angle θ (Fig. 36–14) given by Eq. 36–10:

Resolution

$$\theta = \frac{1.22\lambda}{D}.$$

This is valid for either a microscope or a telescope, where D is the diameter of the objective lens. For a telescope, the resolution is specified by stating θ as given by this equation.[†]

For a microscope, it is more convenient to specify the actual distance, s, between two points that are just barely resolvable, Fig. 36–14. Since objects are normally placed near the focal point of the microscope objective, $\theta = s/f$, or $s = f\theta$. If we combine this with Eq. 36–10, we obtain for the **resolving power (RP)**:

Resolving power (microscope)

$$\text{RP} = s = f\theta = \frac{1.22\lambda f}{D}. \tag{36-11}$$

This distance s is called the resolving power of the lens because it is the minimum separation of two object points that can just be resolved, assuming the highest quality lens since this limit is imposed by the wave nature of light.

FIGURE 36–16 The 300-meter radiotelescope in Arecibo, Puerto Rico, uses radio waves (Fig. 32–12) instead of visible light.

EXAMPLE 36–6 **Telescope resolution (radio wave vs. visible light).** What is the theoretical minimum angular separation of two stars that can just be resolved by: (*a*) the 200-inch telescope on Palomar Mountain (Fig. 34–33c); and (*b*) the Arecibo radiotelescope (Fig. 36–16), whose diameter is 300 m and whose radius of curvature is also 300 m. Assume $\lambda = 550$ nm for the visible-light telescope in part (*a*), and $\lambda = 4$ cm (the shortest wavelength at which the radiotelescope has been operated) in part (*b*).

SOLUTION (*a*) Since $D = 200$ inch $= 5.1$ m, we have from Eq. 36–10 that

$$\theta = \frac{1.22\lambda}{D} = \frac{(1.22)(5.50 \times 10^{-7}\,\text{m})}{(5.1\,\text{m})} = 1.3 \times 10^{-7}\,\text{rad},$$

or 0.75×10^{-5} deg. (Note that this is equivalent to resolving two points less than 1 cm apart from a distance of 100 km!) This is the limit set by diffraction. The resolution is not this good because of aberrations and, more importantly, turbulence in the atmosphere. In fact, large-diameter objectives are not justified by increased resolution, but by their greater light-gathering ability—they allow more light in, so fainter objects can be seen.

(*b*) Radiotelescopes are not hindered by atmospheric turbulence, and for radio waves with $\lambda = 0.04$ m the resolution is

$$\theta = \frac{(1.22)(0.04\,\text{m})}{(300\,\text{m})} = 1.6 \times 10^{-4}\,\text{rad}.$$

[†]Telescopes with large-diameter objectives are usually limited not by diffraction but by other effects such as turbulence in the atmosphere. The resolution of a high-quality microscope, on the other hand, normally *is* limited by diffraction because microscope objectives are complex compound lenses containing many elements of small diameter (since f is small).

Diffraction sets an ultimate limit on the detail that can be seen on any object. In Eq. 36–11 we note that the focal length of a lens cannot be made less than (approximately) the radius of the lens, and even that is very difficult—see the lens-maker's equation (Eq. 34–4). In this best case, Eq. 36–11 gives, with $f \approx D/2$,

$$RP \approx \frac{\lambda}{2}. \qquad\qquad (36\text{–}12)$$

Thus we can say, to within a factor of 2 or so, that

it is not possible to resolve detail of objects smaller than the wavelength of the radiation being used.

Resolution limited to λ

This is an important and useful rule of thumb.

Compound lenses are now designed so well that the actual limit on resolution is often set by diffraction—that is, by the wavelength of the light used. To obtain greater detail, one must use radiation of shorter wavelength. The use of UV radiation can increase the resolution by a factor of perhaps 2. Far more important, however, was the discovery in the early twentieth century that electrons have wave properties (Chapter 38) and that their wavelengths can be very small. The wave nature of electrons is utilized in the electron microscope (Section 38–7), which can magnify 100 to 1000 times more than a visible-light microscope because of the much shorter wavelengths. X-rays, too, have very short wavelengths and are often used to study objects in great detail (Section 36–10).

* 36–6 Resolution of the Human Eye and Useful Magnification

The resolution of the human eye is limited by several factors, all of roughly the same order of magnitude. The resolution is best at the fovea, where the cone spacing is smallest, about $3 \, \mu m$ (= 3000 nm). The diameter of the pupil varies from about 0.1 cm to about 0.8 cm. So for $\lambda = 550$ nm (where the eye's sensitivity is greatest), the diffraction limit is about $\theta \approx 1.22\lambda/D \approx 8 \times 10^{-5}$ rad to 6×10^{-4} rad. Since the eye is about 2 cm long, this corresponds to a resolving power of $s \approx (8 \times 10^{-5} \, \text{rad})(2 \times 10^{-2} \, \text{m}) \approx 2 \, \mu m$ at best, to about $15 \, \mu m$ at worst (pupil small). Spherical and chromatic aberration also limit the resolution to about $10 \, \mu m$. The net result is that the eye can resolve objects whose angular separation is about 5×10^{-4} rad at best. This corresponds to objects separated by 1 cm at a distance of about 20 m.

The typical near point of a human eye is about 25 cm. At this distance, the eye can just resolve objects that are $(25 \, \text{cm})(5 \times 10^{-4} \, \text{rad}) \approx 10^{-4} \, \text{m} = \frac{1}{10} \, \text{mm}$ apart. Since the best light microscopes can resolve objects no smaller than about 200 nm at best (Eq. 36–12 for violet light, $\lambda = 400$ nm), the useful magnification $\left[= (\text{resolution by naked eye})/(\text{resolution by microscope})\right]$ is limited to about

$$\frac{10^{-4} \, \text{m}}{200 \times 10^{-9} \, \text{m}} = 500\times.$$

In practice, magnifications of about $1000\times$ are often used to minimize eye-strain. Any greater magnification would simply make visible the diffraction pattern produced by the microscope objective.

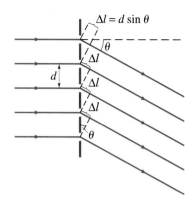

FIGURE 36–17 Diffraction grating.

A large number of equally spaced parallel slits is called a **diffraction grating**, although the term "interference grating" might be as appropriate. Gratings can be made by precision machining of very fine parallel lines on a glass plate. The untouched spaces between the lines serve as the slits. Photographic transparencies of an original grating serve as inexpensive gratings. Gratings containing 10,000 lines per centimeter are common today, and are very useful for precise measurements of wavelengths. A diffraction grating containing slits is called a **transmission grating. Reflection gratings** are also used, which are made by ruling fine lines on a metallic or glass surface from which light is reflected and analyzed. The analysis is basically the same as for a transmission grating, which we now discuss.

The analysis of a diffraction grating is much like that of Young's double-slit experiment. We assume parallel rays of light are incident on the grating as shown in Fig. 36–17. We also assume that the slits are narrow enough so that diffraction by each of them spreads light over a very wide angle on a distant screen behind the grating, and interference can occur with light from all the other slits. Light rays that pass through each slit without deviation ($\theta = 0°$) interfere constructively to produce a bright line at the center of the screen. Constructive interference also occurs at an angle θ such that rays from adjacent slits travel an extra distance of $\Delta l = m\lambda$, where m is an integer. Thus, if d is the distance between slits, then we see from Fig. 36–17 that $\Delta l = d\sin\theta$, and

Diffraction grating maxima (m = order)

$$\sin\theta = \frac{m\lambda}{d}, \qquad m = 0, 1, 2, \cdots \qquad \text{[principal maxima]} \quad \textbf{(36–13)}$$

is the criterion to have a brightness maximum. This is the same equation as for the double-slit situation, and again m is called the order of the pattern.

There is an important difference between a double-slit and a multiple-slit pattern, however. The bright maxima are much *sharper* and *narrower* for a grating. Why this happens can be seen as follows. Suppose that the angle θ is increased just slightly beyond that required for a maximum. In the case of only two slits, the two waves will be only slightly out of phase, so nearly full constructive interference occurs. This means the maxima are wide (see Fig. 35–9). For a grating, the waves from two adjacent slits will also not be significantly out of phase. But waves from one slit and those from a second one a few hundred slits away may be exactly out of phase; all or nearly all the light will cancel in pairs in this way. For example, suppose the angle θ is different from its first-order maximum so that the extra path length for a pair of adjacent slits is not exactly λ but rather 1.0010λ. The wave through one slit and another one 500 slits below will be out of phase by 1.5000λ, or exactly $1\frac{1}{2}$ wavelengths, so the two will cancel. A pair of slits, one below each of these, will also cancel. That is, the light from slit 1 cancels with that from slit 501; light from slit 2 cancels with that from slit 502, and so on. Thus even for a tiny angle[†] corresponding to an extra path length of $\frac{1}{1000}\lambda$, there is much destructive interference, and so the maxima are very narrow. The more lines there are in a grating, the sharper will be the peaks (see Fig. 36–18). Because a grating produces much sharper (and brighter) lines than two slits alone, it is a far more precise device for measuring wavelengths.

Suppose the light striking a diffraction grating is not monochromatic, but rather consists of two or more distinct wavelengths. Then for all orders other than $m = 0$, each wavelength will produce a maximum at a different angle (Fig. 36–19a), just as for a double slit. If white light strikes a grating, the central ($m = 0$) maximum will be a sharp white peak. But for all other orders, there will be a distinct

Why more slits yield sharper peaks

FIGURE 36–18 Intensity as a function of viewing angle θ (or position on the screen) for (a) two slits, (b) six slits. For a diffraction grating, the number of slits is very large ($\sim 10^4$) and the peaks are narrower still.

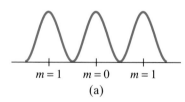

(a)

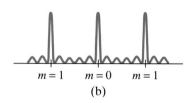

(b)

[†]Depending on the total number of slits, there may or may not be complete cancellation for such an angle, so there will be very tiny peaks between the main maxima (see Fig. 36–18b), but they are usually much too small to be seen.

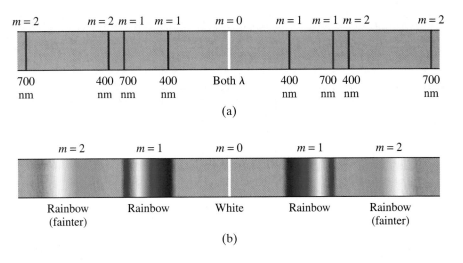

| $m = 2$ | | $m = 2$ | $m = 1$ | $m = 1$ | $m = 0$ | $m = 1$ | $m = 1$ | $m = 2$ | | $m = 2$ |

| 700 nm | | 400 nm | 700 nm | 400 nm | Both λ | 400 nm | 700 nm | 400 nm | | 700 nm |

(a)

| $m = 2$ | $m = 1$ | $m = 0$ | $m = 1$ | $m = 2$ |

| Rainbow (fainter) | Rainbow | White | Rainbow | Rainbow (fainter) |

(b)

FIGURE 36–19 Spectrum produced by a grating: (a) two wavelengths, 400 nm and 700 nm; (b) white light. The second order will normally be dimmer than the first order. (Higher orders are not shown.) If the grating spacing is small enough, the second and higher orders will be missing.

spectrum of colors spread out over a certain angular width, Fig. 36–19b. Because a diffraction grating spreads out light into its component wavelengths, the resulting pattern is called a **spectrum**.

EXAMPLE 36–7 **Diffraction grating: lines.** Calculate the first- and second-order angles for light of wavelength 400 nm and 700 nm if the grating contains 10,000 lines/cm.

SOLUTION The grating contains 10^4 lines/cm $= 10^6$ lines/m, which means the separation between slits is $d = (1/10^6)\,\text{m} = 1.0 \times 10^{-6}\,\text{m}$. In first order ($m = 1$), the angles are

$$\sin\theta_{400} = \frac{m\lambda}{d} = \frac{(1)(4.0 \times 10^{-7}\,\text{m})}{1.0 \times 10^{-6}\,\text{m}} = 0.400$$

$$\sin\theta_{700} = 0.700$$

so $\theta_{400} = 23.6°$ but $\theta_{700} = 44.4°$. In second order,

$$\sin\theta_{400} = \frac{(2)(4.0 \times 10^{-7}\,\text{m})}{1.0 \times 10^{-6}\,\text{m}} = 0.800$$

$$\sin\theta_{700} = 1.40$$

so $\theta_{400} = 53.1°$, but the second order does not exist for $\lambda = 700\,\text{nm}$ because $\sin\theta$ cannot exceed 1. No higher orders will appear.

The diffraction grating is the essential component of a spectroscope, a device for precise measurement of wavelengths, and we discuss it next.

*36–8 The Spectrometer and Spectroscopy

A **spectrometer** or **spectroscope**, Fig. 36–20, is a device[†] to measure wavelengths accurately using a diffraction grating, or a prism, to separate different wavelengths of light. Light from a source passes through a narrow slit S in the collimator. The slit is at the focal point of the lens L, so parallel light falls on the grating. The movable telescope can bring the rays to a focus. Nothing will be seen in the viewing telescope unless it is positioned at an angle θ that corresponds to a diffraction peak (first order is usually used) of a wavelength emitted by the source. The angle θ can be measured

FIGURE 36–20 Spectrometer or spectroscope.

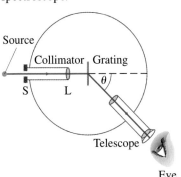

[†] If the spectrum of a source is recorded (say, on film) rather than viewed by the eye, the device is called a **spectrometer** or spectrograph, as compared to a **spectroscope**, which is for viewing only; but these terms are often used interchangeably. Devices that can also measure the intensity of light of a given wavelength are called **spectrophotometers**.

Atomic hydrogen

Mercury

Sodium

Solar absorption spectrum

FIGURE 36–21 Line spectra for the gases indicated, and spectrum from the Sun showing absorption lines.

to very high accuracy, so the wavelength of a line can be determined to high accuracy using Eq. 36–13:

$$\lambda = \frac{d}{m} \sin \theta,$$

where m is an integer representing the order, and d is the distance between grating lines. The line you see in a spectrometer corresponding to each wavelength is actually an image of the slit S. The narrower the slit, the narrower—but dimmer—the line is, and the more precisely we can measure its angular position. If the light contains a continuous range of wavelengths, then a continuous spectrum is seen in the spectroscope.

Line spectra An important use of a spectrometer is for the identification of atoms or molecules. When a gas is heated or a large electric current is passed through it, the gas emits a characteristic **line spectrum**. That is, only certain discrete wavelengths of light are emitted, and these are different for different elements and compounds. Figure 36–21 shows the line spectra for a number of elements in the gas state. Line spectra occur only for gases at high temperatures and low pressure. The light from heated solids, such as a lightbulb filament, and even from a dense gaseous object such as the Sun, produces a **continuous spectrum** including a wide range of wavelengths.

Figure 36–21 also shows the Sun's "continuous spectrum," which contains a number of *dark* lines (only the most prominent are shown), called **absorption lines**. Atoms and molecules can absorb light at the same wavelengths at which they emit light. The Sun's absorption lines are due to absorption by atoms and molecules in the cooler outer atmosphere of the Sun, as well as by atoms and molecules in the Earth's atmosphere. A careful analysis of all these thousands of lines reveals that at least two-thirds of all elements are present in the Sun's atmosphere. The presence of elements in the atmosphere of other planets, in interstellar space, and in stars is also determined by spectroscopy.

*36-9 Peak Widths and Resolving Power for a Diffraction Grating

We now look at the pattern of maxima produced by a multiple-slit grating using phasor diagrams. We can determine a formula for the width of each peak, and we will see why there are tiny maxima between the principal maxima, as indicated in Fig. 36–18b. First of all, it should be noted that the two-slit and six-slit patterns shown in Fig. 36–18 were drawn assuming very narrow slits so that diffraction does not limit the height of the peaks. For real diffraction gratings, this is not normally the case: the slit width a is often not much smaller than the slit separation d, and diffraction thus limits the intensity of the peaks so the central peak ($m = 0$) is brighter than the side peaks. We won't worry about this effect on intensity except to note that if a diffraction minimum coincides with a particular order of the interference pattern, that order will not appear. (For example, if $d = 2a$, all the even orders, $m = 2, 4, \cdots$, will be missing. Can you see why? Hint: See Example 36–4.)

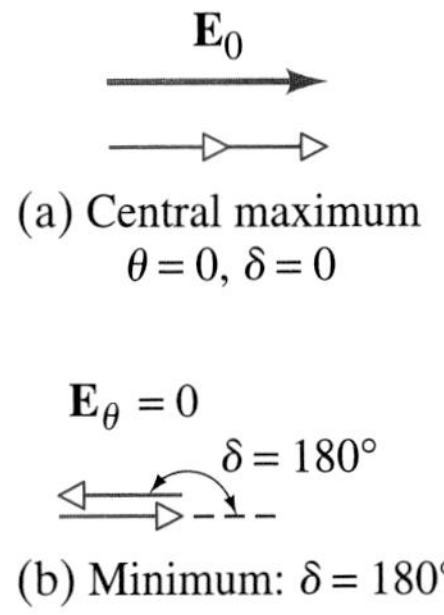

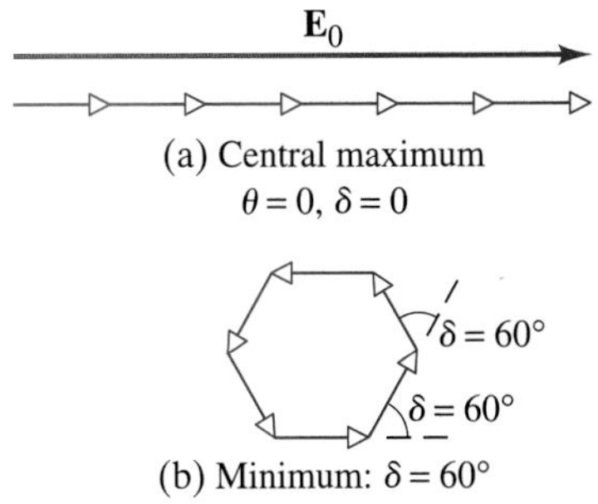

Figures 36–22 and 36–23 show phasor diagrams for a two-slit and a six-slit grating, respectively. Each short arrow represents the amplitude of a wave from a single slit, and their vector sum (as phasors) represents the total amplitude for a given viewing angle θ. Part (a) of each Figure shows the phasor diagram at $\theta = 0°$, at the center of the pattern, which is the central maximum ($m = 0$). Part (b) of each Figure shows the condition for the adjacent minimum: where the arrows first close on themselves (add to zero) so the amplitude E_θ is zero. For two slits, this occurs when the two separate amplitudes are 180° out of phase. For six slits, it occurs when each amplitude makes a 60° angle with its neighbor. For two slits, the minimum occurs when the phase between slits is $2\pi/2$ (in radians); for six slits it occurs when the phase δ is $2\pi/6$; and in the general case of N slits, the minimum occurs for a phase difference between adjacent slits of

$$\delta = \frac{2\pi}{N}. \tag{36-14}$$

What does this correspond to in θ? First note that δ is related to θ by

$$\frac{\delta}{2\pi} = \frac{d \sin \theta}{\lambda} \quad \text{or} \quad \delta = \frac{2\pi}{\lambda} d \sin \theta \tag{36-15}$$

just as in Eq. 35–4. Let us call $\Delta\theta_0$ the angular position of the minimum next to the peak at $\theta = 0$. This corresponds to an extra path length between adjacent slits (see Fig. 36–17) of $\Delta l = d \sin \Delta\theta_0$, so

$$\frac{\delta}{2\pi} = \frac{\Delta l}{\lambda} = \frac{d \sin \Delta\theta_0}{\lambda}.$$

We insert Eq. 36–14 for δ and find

$$\sin \Delta\theta_0 = \frac{\lambda}{Nd}. \tag{36-16a}$$

Since $\Delta\theta_0$ is usually small (N is usually very large for a grating), $\sin \Delta\theta_0 \approx \Delta\theta_0$, so in the small angle limit we can write

$$\Delta\theta_0 = \frac{\lambda}{Nd}. \tag{36-16b}$$

Peak half width ($m = 0$)

It is clear from either of the last two relations that the larger N is, the narrower will be the central peak. (For $N = 2$, $\sin \Delta\theta_0 = \lambda/2d$, which is what we obtained earlier for the double slit, Eq. 35–2b with $m = 0$.)

FIGURE 36–22 Phasor diagram for two slits (a) at the central maximum, (b) at the nearest minimum.

FIGURE 36–23 Phasor diagram for six slits (a) at the central maximum, (b) at the nearest minimum.

FIGURE 36–24 Phasor diagram for the secondary peak.

Either of Eqs. 36–16 shows why the peaks become narrower for larger N. The origin of the small secondary maxima between the principal peaks (see Fig. 36–18b) can be deduced from the diagram of Fig. 36–24. This is just a continuation of Fig. 36–23b (where $\delta = 60°$); but now the phase δ has been increased to almost $90°$, where E_θ is a relative maximum. Note that E_θ is much less than E_0 (Fig. 36–23a), so the intensity in this secondary maximum is much smaller than in a principal peak. As δ (and θ) is increased further, E_θ again decreases to zero (a "double circle"), then reaches another tiny maximum, and so on. Eventually the diagram unfolds again and when $\delta = 360°$, all the amplitudes again lie in a straight line (as in Fig. 35–23a) corresponding to the next principal maximum ($m = 1$ in Eq. 36–13).

Equation 36–16b gives the half width of the central ($m = 0$) peak. To determine the half width of higher order peaks, $\Delta\theta_m$ for order m, we differentiate Eq. 36–15 so as to relate the change $\Delta\delta$ in δ, to the change $\Delta\theta$ in the angle θ:

$$\Delta\delta \approx \frac{d\delta}{d\theta}\Delta\theta = \frac{2\pi d}{\lambda}\cos\theta\,\Delta\theta.$$

If $\Delta\theta_m$ represents the half width of a peak of order $m\,(m = 1, 2, \cdots)$—that is, the angle between the peak maximum and the minimum to either side—then $\Delta\delta = 2\pi/N$ as given by Eq. 36–14. We insert this into the above relation and find

Peak half width (order m)

$$\Delta\theta_m = \frac{\lambda}{Nd\cos\theta_m}, \tag{36–17}$$

where θ_m is the angular position of the mth peak as given by Eq. 36–13. This derivation is valid, of course, only for small $\Delta\delta\,(= 2\pi/N)$ which is indeed the case for real gratings since N is on the order of 10^4 or more.

An important property of any diffraction grating used in a spectroscope is its ability to resolve two very closely spaced wavelengths. The **resolving power** R of a grating is defined as

Resolving power (grating)

$$R = \frac{\lambda}{\Delta\lambda}. \tag{36–18}$$

With a little work, using Eq. 36–17, we can show that $\Delta\lambda = \lambda N/m$ where N is the total number of grating lines and m is the order. Then we have

$$R = Nm. \tag{36–19}$$

The larger the value of R, the closer two wavelengths can be and still be resolvable. If R is given, the minimum separation $\Delta\lambda$ between two wavelengths near λ, is (Eq. 36–18)

$$\Delta\lambda = \frac{\lambda}{R}.$$

EXAMPLE 36–8 **Resolving two close lines.** Yellow sodium light, which consists of two wavelengths, $\lambda_1 = 589.00\,\text{nm}$ and $\lambda_2 = 589.59\,\text{nm}$, falls on a 7500-line/cm diffraction grating. Determine (a) the maximum order m that will be present for sodium light, (b) the width of grating necessary to resolve the two sodium lines, (c) the angular width of each sodium line.

SOLUTION (a) Since $d = 1\,\text{cm}/7500 = 1.33 \times 10^{-6}\,\text{m}$, then the maximum value of m for $\lambda = 589\,\text{nm}$ can be found from Eq. 35–13 with $\sin\theta \leq 1$:

$$m = \frac{d}{\lambda}\sin\theta \leq \frac{1.33 \times 10^{-6}\,\text{m}}{5.89 \times 10^{-7}\,\text{m}} = 2.25,$$

so $m = 2$ is the maximum order present.

(b) The resolving power needed is

$$R = \frac{\lambda}{\Delta\lambda} = \frac{589 \text{ nm}}{0.59 \text{ nm}} = 1000.$$

From Eq. 36–19, the total number N of lines needed is $N = R/m = 1000/2 = 500$, so the grating need only be $500/7500 \text{ cm}^{-1} = 0.0667 \text{ cm}$ wide. A typical grating is a few centimeters wide, and so will easily resolve the two lines.

(c) We need to use Eq. 36–17 for $m = 2$, but what is $\cos\theta_2$? From Eq. 36–13,

$$\sin\theta_2 = m\frac{\lambda}{d} = \frac{2(589 \times 10^{-9}\text{ m})}{1.33 \times 10^{-6}\text{ m}} = 0.886.$$

Then $\cos\theta_2 = (1 - \sin^2\theta_2)^{\frac{1}{2}} = 0.464$. For a grating with the minimum 500 lines, the angular width (Eq. 36–17) for $m = 2$ is

$$\Delta\theta_2 = \frac{\lambda}{Nd\cos\theta_2} = \frac{589 \times 10^{-9}\text{ m}}{(500)(1.33 \times 10^{-6}\text{ m})(0.464)} = 0.0019 \text{ rad},$$

or $0.11°$.

*36–10 X-Rays and X-Ray Diffraction

In 1895, W. C. Roentgen (1845–1923) discovered that when electrons were accelerated by a high voltage in a vacuum tube and allowed to strike a glass (or metal) surface inside the tube, fluorescent minerals some distance away would glow, and photographic film would become exposed. Roentgen attributed these effects to a new type of radiation (different from cathode rays). They were given the name **X-rays** after the algebraic symbol x, meaning an unknown quantity. He soon found that X-rays penetrated through some materials better than through others, and within a few weeks he presented the first X-ray photograph (of his wife's hand). The production of X-rays today is usually done in a tube (Fig. 36–25) similar to Roentgen's, using voltages of typically 30 kV to 150 kV.

Investigations into the nature of X-rays indicated they were not charged particles (such as electrons) since they could not be deflected by electric or magnetic fields. It was suggested that they might be a form of invisible light. However, they showed no diffraction or interference effects using ordinary gratings. Of course, if their wavelengths were much smaller than the typical grating spacing of $10^{-6}\text{ m} (= 10^3 \text{ nm})$, no effects would be expected. Around 1912, it was suggested by Max von Laue (1879–1960) that if the atoms in a crystal were arranged in a regular array (see Fig. 17–2a), such a crystal might serve as a diffraction grating for very short wavelengths on the order of the spacing between atoms, estimated to be about $10^{-10}\text{ m} (= 10^{-1} \text{ nm})$. Experiments soon showed that X-rays scattered from a crystal did indeed show the peaks and valleys of a diffraction pattern (Fig. 36–26). Thus it was shown, in a single blow, that X-rays have a wave nature and that atoms are arranged in a regular way in crystals. Today, X-rays are recognized as electromagnetic radiation with wavelengths in the range of about 10^{-2} nm to 10 nm, the range readily produced in an X-ray tube.

X-rays

FIGURE 36–25 X-ray tube. Electrons emitted by a heated filament in a vacuum tube are accelerated by a high voltage. When they strike the surface of the anode, the "target," X-rays are emitted.

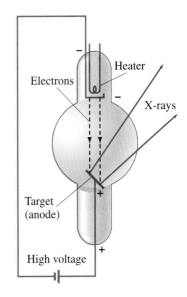

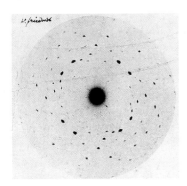

FIGURE 36–26 This X-ray diffraction pattern is one of the first observed by Max von Laue in 1912 when he aimed a beam of X-rays at a zinc sulfide crystal. The diffraction pattern was detected directly on a photographic plate.

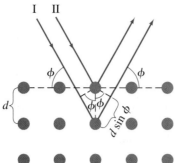

FIGURE 36–27 X-ray diffraction by a crystal.

We saw in Section 36–5 that light of shorter wavelength provides greater resolution when we are examining an object microscopically. Since X-rays have much shorter wavelengths than visible light, in principle they should offer much greater resolution. However, there seems to be no effective material to use as lenses for the very short wavelengths of X-rays. Instead, the clever but complicated technique of **X-ray diffraction** (or **crystallography**) has proved very effective for examining the microscopic world of atoms and molecules. In a simple crystal such as NaCl, the atoms are arranged in an orderly cubical fashion, Fig. 36–27, with atoms spaced a distance d apart. Suppose that a beam of X-rays is incident on the crystal at an angle ϕ to the surface, and that the two rays shown are reflected from two subsequent planes of atoms as shown. The two rays will constructively interfere if the extra distance ray I travels is a whole number of wavelengths farther than what ray II travels. This extra distance is $2d \sin \phi$. Therefore, constructive interference will occur when

$$m\lambda = 2d \sin \phi, \qquad m = 1, 2, 3, \cdots, \tag{36–20}$$

where m can be any integer. (Notice that ϕ is *not* the angle with respect to the normal to the surface.) This is called the **Bragg equation** after W. L. Bragg (1890–1971), who derived it and who, together with his father W. H. Bragg (1862–1942), developed the theory and technique of X-ray diffraction by crystals in 1912–1913. Thus, if the X-ray wavelength is known and the angle ϕ at which constructive interference occurs is measured, d can be obtained. This is the basis for X-ray crystallography.

Actual X-ray diffraction patterns are quite complicated. First of all, a crystal is a three-dimensional object, and X-rays can be diffracted from different planes at different angles within the crystal, as shown in Fig. 36–28. Although the analysis is complex, a great deal can be learned about any substance that can be put in crystalline form. If the substance is not a single crystal but a mixture of many tiny crystals—as in a metal or a powder—then instead of a series of spots, as in Fig. 36–26, a series of circles is obtained, Fig. 36–29, each corresponding to diffraction of a certain order m from a particular set of parallel planes.

X-ray diffraction has been very useful in determining the structure of biologically important molecules. For example, it was with the help of X-ray diffraction that, in 1953, J. D. Watson and F. H. C. Crick worked out the double-helix structure of DNA.

Bragg equation

FIGURE 36–28 There are many possible planes existing within a crystal from which X-rays can be diffracted.

FIGURE 36–29 (a) Diffraction of X-rays from a polycrystalline substance produces a set of circular rings as in (b), which is for polycrystalline sodium acetoacetate.

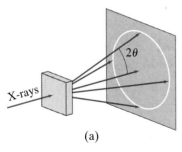

(a)   (b)

36–11 Polarization

An important and useful property of light is that it can be *polarized*. To see what this means, let us examine waves traveling on a rope. A rope can be set into oscillation in a vertical plane as in Fig. 36–30a, or in a horizontal plane as in Fig. 36–30b. In either case, the wave is said to be **linearly polarized** or **plane-polarized**—that is, the oscillations are in a plane.

If we now place an obstacle containing a vertical slit in the path of the wave, Fig. 36–31, a vertically polarized wave passes through, but a horizontally polarized wave will not. If a horizontal slit were used, the vertically polarized wave would be stopped. If both types of slit were used, both types of wave would be stopped. Note that polarization can exist *only* for *transverse waves*, and not for longitudinal waves such as sound. The latter oscillate only along the direction of motion, and neither orientation of slit would stop them.

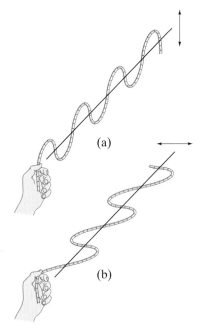

FIGURE 36–30 (above) Transverse waves on a rope polarized (a) in a vertical plane and (b) in a horizontal plane.

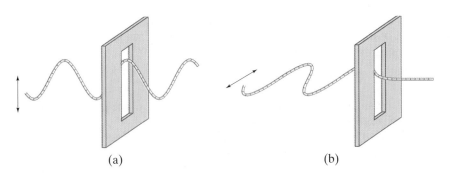

(a)   (b)

FIGURE 36–31 Vertically polarized wave passes through a vertical slit, but a horizontally polarized wave will not. [Note: This diagram would apply to the magnetic field **B** in an EM wave—see footnote below.]

Maxwell's theory of light as electromagnetic (EM) waves predicted that light can be polarized since an EM wave is a transverse wave. The direction of polarization in a plane-polarized EM wave is taken as the direction of the electric field vector **E**.

Light is not necessarily polarized. It can be **unpolarized**, which means that the source has oscillations in many planes at once, as shown in Fig. 36–32. An ordinary incandescent lightbulb emits unpolarized light, as does the Sun.

Polaroids

Plane-polarized light can be obtained from unpolarized light using certain crystals such as tourmaline. Or, more commonly today, we can use a **Polaroid sheet**. (Polaroid materials were invented in 1929 by Edwin Land.) A Polaroid sheet consists of complicated long molecules arranged parallel to one another. Such a Polaroid acts like a series of parallel slits to allow one orientation of polarization to pass through nearly undiminished (this direction is called the *axis* of the Polaroid), whereas a perpendicular polarization is absorbed almost completely.[†]

FIGURE 36–32 Oscillation of the electric field vectors in unpolarized light. The light is traveling into or out of the page.

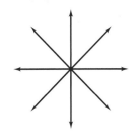

[†]How this occurs can be explained at the molecular level. An electric field **E** that oscillates parallel to the long molecules can set electrons into motion along the molecules, thus doing work on them and transferring energy. Hence, if **E** is parallel to the molecules, it gets absorbed. An electric field **E** perpendicular to the long molecules does not have this possibility of doing work and transferring its energy, and so passes through freely. When we speak of the *axis* of a Polaroid, we mean the direction for which **E** is passed, so a Polaroid axis is perpendicular to the long molecules. (If we want to think of there being slits between the parallel molecules in the sense of Fig. 36–31, then Fig. 36–31 would apply for the **B** field in the EM wave, not the **E** field.)

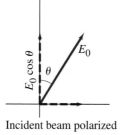

FIGURE 36–33 Vertical Polaroid transmits only the vertical component of a wave (electric field) incident upon it.

Incident beam polarized at angle θ to the vertical and has amplitude E_0.

Vertical Polaroid

Transmitted wave

If a beam of plane-polarized light strikes a Polaroid whose axis is at an angle θ to the incident polarization direction, the beam will emerge plane-polarized parallel to the Polaroid axis and its amplitude will be reduced by $\cos\theta$, Fig. 36–33. Thus, a Polaroid passes only that component of polarization (the electric field vector, **E**) that is parallel to its axis. Because the intensity of a light beam is proportional to the square of the amplitude (Sections 15–3 and 32–7), we see that the intensity of a plane-polarized beam transmitted by a polarizer is

Intensity of plane polarized wave reduced by polarizer

$$I = I_0 \cos^2 \theta, \tag{36–21}$$

where θ is the angle between the polarizer axis and the plane of polarization of the incoming wave, and I_0 is the incoming intensity.[†]

A Polaroid can be used as a **polarizer** to produce plane-polarized light from unpolarized light, since only the component of light parallel to the axis is transmitted. A Polaroid can also be used as an **analyzer** to determine (1) if light is polarized and (2) what is the plane of polarization. A Polaroid acting as an analyzer will pass the same amount of light independent of the orientation of its axis if the light is unpolarized; try rotating one lens of a pair of Polaroid sunglasses while looking through it at a lightbulb. If the light is polarized, however, when you rotate the Polaroid the transmitted light will be a maximum when the plane of polarization is parallel to the Polaroid's axis, and a minimum when perpendicular to it. If you do this while looking at the sky, preferably at right angles to the Sun's direction, you will see that skylight is polarized. (Direct sunlight is unpolarized, but don't look directly at the Sun, even through a polarizer, for damage to the eye may occur.) If the light trasmitted by an analyzer Polaroid falls to zero at one orientation, then the light is 100 percent plane-polarized. If it merely reaches a minimum, the light is *partially polarized.*

Unpolarized light consists of light with random directions of polarization. Each of these polarization directions can be resolved into components along two mutually perpendicular directions. On average, an unpolarized beam can be thought of as two plane-polarized beams of equal magnitude perpendicular to one another. When unpolarized light passes through a polarizer, one of the components is eliminated. So the intensity of the light passing through is reduced by half since half the light is eliminated, $I = \frac{1}{2} I_0$ (Fig. 36–34).

When two Polaroids are *crossed*—that is, their axes are perpendicular to one another—unpolarized light can be entirely stopped. As shown in Fig. 36–35, unpolarized light is made plane-polarized by the first Polaroid (the polarizer). The second Polaroid, the analyzer, then eliminates this component since its axis is perpendicular to the first. You can try this with Polaroid sunglasses (Fig. 36–36). Note that Polaroid sunglasses eliminate 50 percent of unpolarized light because of their polarizing property; they absorb even more because they are colored.

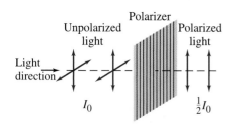

FIGURE 36–34 Unpolarized light has both vertical and horizontal components. After passing through a polarizer, one of these components is eliminated. The intensity of the light is reduced to half.

[†]Equation 36–21 is often referred to as **Malus' law**, after Etienne Malus, a contemporary of Fresnel.

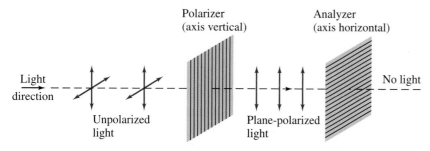

FIGURE 36–35   Crossed Polaroids completely eliminate light.

FIGURE 36–36   Crossed Polaroids. When the two polarized sunglass lenses overlap, with axes perpendicular, almost no light passes through.

EXAMPLE 36–9 **Two Polaroids at 60°.**   Unpolarized light passes through two Polaroids; the axis of one is vertical and that of the other is at 60° to the vertical. What is the orientation and intensity of the transmitted light?

SOLUTION   The first Polaroid eliminates half the light so the intensity is reduced by half: $I_1 = \frac{1}{2}I_0$. The light reaching the second polarizer is vertically polarized and so is reduced in intensity (Eq. 36–21) to

$$I_2 = I_1(\cos 60°)^2 = \tfrac{1}{4}I_1.$$

Thus, $I_2 = \frac{1}{8}I_0$. The transmitted light has an intensity one-eighth that of the original and is plane-polarized at a 60° angle to the vertical.

CONCEPTUAL EXAMPLE 36–10 | **Three Polaroids.**   We saw in Fig. 36–35 that when unpolarized light falls on two crossed Polaroids (axes at 90°), no light passes through. What happens if a third Polaroid, with axis at 45° to each of the other two, is placed between them?

RESPONSE   We start just as in Example 36–9. The first Polaroid changes the unpolarized light to plane-polarized and reduces the intensity from I_0 to $I = \frac{1}{2}I_0$. The second polarizer further reduces the intensity by $(\cos 45°)^2$, Eq. 36–21:

$$I_2 = I_1(\cos 45°)^2 = \tfrac{1}{2}I_1 = \tfrac{1}{4}I_0.$$

The light leaving the second polarizer is plane polarized at 45° (Fig. 36–37) relative to the third polarizer, so the latter reduces the intensity to

$$I_3 = I_2(\cos 45°)^2 = \tfrac{1}{2}I_2$$

or $I_3 = \frac{1}{8}I_0$. Thus $\frac{1}{8}$ of the original intensity gets transmitted. But if we don't put the 45° Polaroid in at all, zero intensity results (Fig. 36–35). The 45° Polaroid must be inserted *between* the other two if transmission is to occur. Placing it before or after the other two results in zero intensity.

FIGURE 36–37   Example 36–10.

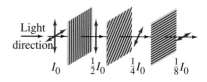

An important use of polarizers is in a **liquid crystal display** (LCD). LCDs are used in many displays, from digital watches to small TV sets. Each pixel (picture element) consists of a small liquid crystal contained between two glass plates whose outer surfaces have a thin film of polarizer. Liquid crystals are materials that, at room temperature, exist in a phase neither fully solid nor fully liquid. Useful ones are composed of rod-like molecules that interact only weakly with each other and tend to align with each other. The overall direction of alignment can be controlled using electric fields. Each pixel can be bright or dark depending on that alignment relative to the polarizers on the ends. Color displays can be obtained by using color filters.

➡ **PHYSICS APPLIED**

Liquid crystals (LCD)

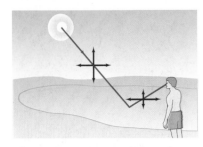

FIGURE 36–38 Light reflected from a nonmetallic surface, such as the smooth surface of water in a lake, is partially polarized parallel to the surface.

FIGURE 36–39 Photographs of a river, (a) allowing all light into the camera lens, and (b) using a polarizer which is adjusted to absorb most of the (polarized) light reflected from the water's surface, allowing the dimmer light from the bottom of the river, and any fish swimming there, to be seen more readily.

FIGURE 36–40 At θ_p the reflected light is plane-polarized parallel to the surface, and $\theta_p + \theta_r = 90°$, where θ_r is the refraction angle. (The large dots represent vibrations perpendicular to the page.)

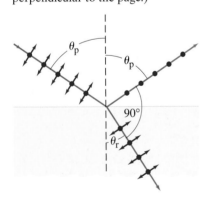

Polarization by Reflection

Another means of producing polarized light from unpolarized light is by reflection. When light strikes a nonmetallic surface at any angle other than perpendicular, the reflected beam is polarized preferentially in the plane parallel to the surface, Fig. 36–38. In other words, the component with polarization in the plane perpendicular to the surface is preferentially transmitted or absorbed. You can check this by rotating Polaroid sunglasses while looking through them at a flat surface of a lake or road. Since most outdoor surfaces are horizontal, Polaroid sunglasses are made with their axes vertical to eliminate the more strongly reflected horizontal component, and thus reduce glare. Fishermen wear Polaroids to eliminate reflected glare from the surface of a lake or stream and thus see beneath the water more clearly (Fig. 36–39).

(a)

(b)

The amount of polarization in the reflected beam depends on the angle, varying from no polarization at normal incidence to 100 percent polarization at an angle known as the **polarizing angle**, θ_p.[†] This angle is related to the index of refraction of the two materials on either side of the boundary by the equation

$$\tan \theta_p = \frac{n_2}{n_1}, \qquad (36\text{–}22a)$$

where n_1 is the index of refraction of the material in which the beam is traveling, and n_2 is that of the medium beyond the reflecting boundary. If the beam is traveling in air, $n_1 = 1$, and Eq. 36–22a becomes

$$\tan \theta_p = n. \qquad (36\text{–}22b)$$

The polarizing angle θ_p is also called **Brewster's angle**, and Eqs. 36–22 called *Brewster's law*, after the Scottish physicist David Brewster (1781–1868), who worked it out experimentally in 1812. Equations 36–22 can be derived from the electromagnetic wave theory of light. It is interesting that at Brewster's angle, the reflected and transmitted rays make a 90° angle to each other; that is, $\theta_p + \theta_r = 90°$, Fig. 36–40. This can be seen by substituting Eq. 36–22a, $n_2 = n_1 \tan \theta_p = n_1 \sin \theta_p / \cos \theta_p$, into Snell's law, $n_1 \sin \theta_p = n_2 \sin \theta_r$, and get $\cos \theta_p = \sin \theta_r$ which can only hold if $\theta_p = 90° - \theta_r$.

EXAMPLE 36–11 **Polarizing angle.** (*a*) At what incident angle is sunlight reflected from a lake plane-polarized? (*b*) What is the refraction angle?

SOLUTION (*a*) We use Eq. 36–22b with $n = 1.33$, so $\tan \theta_p = 1.33$ and $\theta_p = 53.1°$.
(*b*) $\theta_r = 90.0° - \theta_p = 36.9°$.

[†]Only a fraction of the incident light is reflected at the surface of the transparent medium. Although this reflected light is 100 percent polarized $\left(\text{if } \theta = \theta_p\right)$, the remainder of the light, which is transmitted into the new medium, is only partially polarized.

*36–12 Scattering of Light by the Atmosphere

Sunsets are red, the sky is blue, and skylight is polarized (at least partially). These phenomena can be explained on the basis of the *scattering* of light by the molecules of the atmosphere. In Fig. 36–41 we see unpolarized light from the Sun impinging on a molecule of the Earth's atmosphere. The electric field of the EM wave sets the electric charges within the molecule into motion, and the molecule absorbs some of the incident radiation. But it quickly reemits this light since the charges are oscillating. As discussed in Section 32–4, oscillating electric charges produce EM waves. The electric field of these waves is in a plane that includes the line of oscillation. The intensity is strongest along a line perpendicular to the oscillation, and drops to zero along the line of oscillation (Section 32–4). In Fig. 36–41 the motion of the charges is resolved into two components. An observer at right angles to the direction of the sunlight, as shown, will see plane-polarized light because no light is emitted along the line of the other component of the oscillation. (Another way to understand this is to note that when viewing along the line of oscillation, one doesn't see the oscillation, and hence sees no waves made by it. Furthermore, the EM wave produces no oscillation along its own direction of motion.) At other viewing angles, both components will be present; one will be stronger, however, so the light appears partially polarized. Thus, the process of scattering explains the polarization of skylight.

Scattering of light by the Earth's atmosphere depends on λ. For particles much smaller than the wavelength of light (such as molecules of air), the particles will be less of an obstruction to long wavelengths than to short ones. The scattering decreases, in fact, as $1/\lambda^4$. Blue and violet light are thus scattered much more than red and orange, which is why the sky looks blue. At sunset, the Sun's rays pass through a maximum length of atmosphere. Much of the blue has been taken out by scattering. The light that reaches the surface of the Earth, and reflects off clouds and haze, is thus lacking in blue which is why sunsets appear reddish.

The dependence of scattering on $1/\lambda^4$ is valid only if the scattering objects are much smaller than the wavelength of the light. This is valid for oxygen and nitrogen molecules whose diameters are about $0.2\,nm$. Clouds, however, contain water droplets or crystals that are much larger than λ. They scatter all frequencies of light nearly uniformly. Hence clouds appear white (or gray, if shadowed).

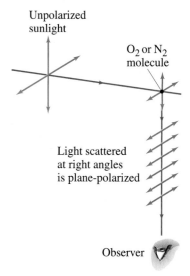

FIGURE 36–41 Unpolarized sunlight scattered by molecules of the air. An observer at right angles sees plane-polarized light, since the component of oscillation along the line of sight emits no light along that line.

➡ PHYSICS APPLIED

Why the sky is blue
Why sunsets are red

Summary

Diffraction refers to the fact that light, like other waves, bends around objects it passes, and spreads out after passing through narrow slits. This bending gives rise to a **diffraction pattern** due to interference between rays of light that travel different distances.

Light passing through a very narrow slit of width a will produce a pattern with a bright central maximum of half-width θ given by

$$\sin \theta = \frac{\lambda}{a},$$

flanked by fainter lines to either side.

The minima in the diffraction pattern occur at

$$a \sin \theta = m\lambda$$

where $m = 1, 2, 3, \cdots$, but not $m = 0$ (for which the pattern has its strongest maximum).

The **intensity** at any point in the single-slit diffraction pattern can be calculated using **phasor** diagrams. The same technique can be used to determine the intensity of the pattern produced by two slits.

The pattern for two-slit interference can be described as a series of maxima due to interference of light from the two slits, modified by an "envelope" due to diffraction at each slit.

The wave nature of light limits the sharpness or **resolution** of images. Because of diffraction, it is not possible to *discern details smaller than the wavelength* of the radiation being used. The useful magnification of a light microscope is limited by diffraction to about $1000\times$.

A **diffraction grating** consists of many parallel slits or lines, each separated from its neighbors by a distance d. The peaks of constructive interference occur at angles θ given by

$$\sin\theta = \frac{m\lambda}{d},$$

where $m = 0, 1, 2, \cdots$. The peaks are much brighter and sharper for a diffraction grating than for the simple two-slit apparatus. Peak width is inversely proportional to the total number of lines in the grating.

A diffraction grating (or a prism) is used in a **spectroscope** to separate different colors or to observe **line spectra**, since for a given order m, θ depends on λ. Precise determination of wavelength can be done with a spectroscope by careful measurement of θ.

In **unpolarized light**, the electric field vectors oscillate transversely at all angles. If the electric vector oscillates only in one plane the light is said to be **plane-polarized**. Light can also be partially polarized.

When an unpolarized light beam passes through a Polaroid sheet, the emerging beam is plane-polarized. When a light beam is polarized and passes through a Polaroid, the intensity varies as the Polaroid is rotated. Thus a Polaroid can act as a polarizer or as an analyzer.

The intensity of a plane-polarized beam incident on a Polaroid is reduced by the factor

$$\cos^2\theta$$

where θ is the angle between the axis of the Polaroid and the initial plane of polarization.

Light can also be partially or fully **polarized by reflection**. If light traveling in air is reflected from a medium of index of refraction n, the reflected beam will be *completely* plane-polarized if the incident angle θ_p is given by $\tan\theta_p = n$. The fact that light can be polarized shows that it must be a transverse wave.

☐ Questions

1. Radio waves and light are both electromagnetic waves. Why can we hear a radio behind a hill when we cannot see the transmitting antenna?

2. Hold one hand close to your eye and focus on a distant light source through a narrow slit between two fingers. (Adjust your fingers to obtain the best pattern.) Describe the pattern that you see.

3. A rectangular slit is twice as high as it is wide. Will the light spread more in the horizontal or in the vertical plane? Describe the pattern.

4. Explain why diffraction patterns are more difficult to observe with an extended light source than for a point source. Compare also a monochromatic source to white light.

5. For diffraction by a single slit, what is the effect of increasing (a) the slit width, (b) the wavelength?

6. Describe the single-slit diffraction pattern produced when white light falls on a slit having a width of (a) 50 nm, (b) 50,000 nm.

7. What happens to the diffraction pattern of a single slit if the whole apparatus is immersed in (a) water, (b) a vacuum, instead of in air.

8. Describe clearly the difference between the angles θ and β, as in Eq. 36–6.

9. In the single-slit diffraction pattern, why does the first off-center maximum not occur at exactly $\sin\theta = \frac{3}{2}\lambda/a$?

*10. Figure 36–10 shows a two-slit interference pattern for the case when d is larger than a. Can the reverse case occur, when d is less than a?

*11. When both diffraction and interference are taken into account in the double-slit experiment, discuss the effect of increasing (a) the wavelength, (b) the slit separation, (c) the slit width.

12. Discuss the similarities, and differences, of double-slit interference and single-slit diffraction.

13. Does diffraction limit the resolution of images formed by (a) spherical mirrors, (b) plane mirrors?

14. Do diffraction effects occur for virtual as well as real images?

15. What are the advantages (give at least two) for the use of large reflecting mirrors in astronomical telescopes?

16. Atoms have diameters of about 10^{-8} cm. Can visible light be used to "see" an atom? Why or why not?

17. Which color of visible light would give the best resolution in a microscope?

18. If monochromatic light were used in a microscope, would the color affect the resolution? Explain.

19. Could a diffraction grating just as well be called an interference grating? Discuss.

20. For light consisting of wavelengths between 400 nm and 700 nm, incident normally on a diffraction grating, for what orders (if any) would there be overlap in the observed spectrum? Does your answer depend on the slit width?

21. What is the difference in the interference patterns formed (i) by two slits 10^{-4} cm apart, (ii) by a diffraction grating containing 10^4 lines/cm?

22. White light strikes (a) a diffraction grating and (b) a prism. A rainbow appears on a wall just below the direction of the horizontal incident beam in each case. What is the color of the top of the rainbow in each case?

23. Explain why there are tiny peaks between the main peaks produced by a diffraction grating illuminated with monochromatic light. Why are the peaks so tiny?

24. What does polarization tell us about the nature of light?

25. How can you tell if a pair of sunglasses is polarizing or not?

*26. What would be the color of the sky if the Earth had no atmosphere?

*27. If the Earth's atmosphere were 50 times denser than it is, would sunlight still be white, or would it be some other color?

Problems

Section 36–1

1. (I) If 680-nm light falls on a slit 0.0345 mm wide, what is the angular width of the central diffraction peak?

2. (I) Monochromatic light falls on a slit that is 3.00×10^{-3} mm wide. If the angle between the first dark fringes on either side of the central maximum is 37.0° (dark fringe to dark fringe), what is the wavelength of the light used?

3. (II) Light of wavelength 550 nm falls on a slit that is 3.50×10^{-3} mm wide. Estimate how far from the central maximum the first diffraction maximum fringe is if the screen is 10.0 m away?

4. (II) Monochromatic light of wavelength 689 nm falls on a slit. If the angle between the first bright fringes on either side of the central maximum is 38°, estimate the slit width.

5. (II) If a slit diffracts 550-nm light so that the diffraction maximum is 8.0 cm wide on a screen 2.50 m away, what will be the width of the diffraction maximum for light with a wavelength of 400 nm?

6. (II) (a) For a given wavelength λ, what is the maximum slit width for which there will be no diffraction minima? (b) What is the maximum slit width so that no visible light exhibits a diffraction minimum?

7. (II) How wide is the central diffraction peak on a screen 3.50 m behind a 0.0655-mm-wide slit illuminated by 400-nm light?

8. (II) If parallel light falls on a single slit of width a at a 30° angle to the normal, describe the diffraction pattern.

Section 36–2

9. (II) If you double the width of a single slit, the intensity of the light passing through the slit is doubled. (a) Show, however, that the intensity at the center of the screen increases by a factor of 4. (b) Explain why this does not violate conservation of energy.

10. (III) (a) Explain why the secondary maxima in the single-slit diffraction pattern do not occur precisely at $\beta/2 = \left(m + \frac{1}{2}\right)\pi$ where $m = 1, 2, 3, \cdots$. (b) By differentiating Eq. 36–7, show that the secondary maxima occur when $\beta/2$ satisfies the relation $\tan(\beta/2) = \beta/2$. (c) Carefully and precisely plot the curves $y = \beta/2$ and $y = \tan\beta/2$. From their intersections, determine the values of β for the first and second secondary maxima. What is the percent difference from $\beta/2 = \left(m + \frac{1}{2}\right)\pi$?

11. (III) Determine, approximately, the angular width at half maximum $\left(\text{where } I = \frac{1}{2}I_0\right)$ of the central diffraction peak for a single slit. [*Hint*: Use graphical methods, or trial and error; the Problem cannot be solved analytically.] To be concrete, assume $\lambda = 550$ nm and $a = 2.60 \times 10^{-3}$ mm.

* Section 36–3

* 12. (II) Design a double-slit apparatus so that the central diffraction peak contains precisely fifteen fringes.

* 13. (II) If a double-slit pattern contains exactly seven fringes in the central diffraction peak, what can you say about the slit width and separation?

* 14. (II) Suppose $d = a$ in a double-slit apparatus, so that the two slits merge into one slit of width $2a$. Show that Eq. 36–9 reduces to the correct equation for single-slit diffraction.

* 15. (II) Two 0.010-mm-wide slits are 0.030 mm apart (center to center). Determine (a) the spacing between interference fringes for 550 nm light on a screen 1.0 m away and (b) the distance between the two diffraction minima on either side of the central maximum of the envelope.

* 16. (II) How many fringes are contained in the central diffraction peak for a double-slit pattern if (a) $d = 2.00a$, (b) $d = 12.0a$, (c) $d = 4.50a$, (d) $d = 7.20a$.

* 17. (III) Draw phasor diagrams (as in Fig. 36–8) for the double-slit experiment, including both interference and diffraction. Make a diagram for each of several crucial points in the pattern (see Fig. 36–10c) such as at the center, first minimum, next maximum, and at $\sin\theta = \lambda/a$.

* 18. (III) (a) Derive an expression for the intensity in the interference pattern for three equally spaced slits. Express in terms of $\delta = 2\pi d \sin\theta/\lambda$ where d is the distance between adjacent slits and assume the slit width $a \approx \lambda$. (b) Show that there is only one secondary maximum between principal peaks.

Sections 36–4 and 36–5

19. (I) What is the angular resolution limit set by diffraction for the 100-inch (mirror diameter) Mt. Wilson telescope ($\lambda = 500$ nm)?

20. (II) Two stars 10 light-years away are barely resolved by a 90 cm (mirror diameter) telescope. How far apart are the stars? Assume $\lambda = 550$ nm and that the resolution is limited by diffraction.

21. (II) The normal lens on a 35-mm camera has a focal length of 50 mm. Its aperture diameter varies from a maximum of 25 mm ($f/2$) to a minimum of 3.0 mm ($f/16$). Determine the resolution limit set by diffraction for $f/2$ and $f/16$. Specify as the number of lines per millimeter resolved on the film. Take $\lambda = 500$ nm.

22. (II) Suppose that you wish to construct a telescope that can resolve features 7.0 km across on the moon, 384,000 km away. You have a 2.0-m-focal-length objective lens whose diameter is 11.0 cm. What focal-length eyepiece is needed if your eye can resolve objects 0.10 mm apart at a distance of 25 cm? What is the resolution limit set by the size of the objective lens (that is, by diffraction)? Use $\lambda = 500$ nm.

Section 36–7

23. (I) At what angle will 440-nm light produce a third-order maximum when falling on a grating whose slits are 1.35×10^{-3} cm apart?

24. (I) How many lines per centimeter does a grating have if the third order occurs at a 13.0° angle for 650-nm light?

25. (I) A grating has 6600 lines/cm. How many spectral orders can be seen when it is illuminated by white light?

26. (I) A 3500-line/cm grating produces a third-order fringe at a 22.0° angle. What wavelength of light is being used?

27. (I) A source produces first-order lines when incident normally on a 10,000-line/cm diffraction grating at angles 29.8°, 37.7°, 39.6°, and 48.9°. What are the wavelengths?

28. (II) White light containing wavelengths from 400 nm to 750 nm falls on a grating with 7800 lines/cm. How wide is the first-order spectrum on a screen 2.80 m away?

29. (II) Show that the second- and third-order spectra of white light produced by a diffraction grating always overlap. What wavelengths overlap?

30. (II) Two first-order spectrum lines are measured by a 9550 line/cm spectroscope at angles, on each side of center, of $+26°38'$, $+41°02'$ and $-26°18'$, $-40°27'$. What are the wavelengths?

31. (II) Suppose the angles measured in Problem 30 were produced when the spectrometer (but not the source) was submerged in water. What then would be the wavelengths (in air)?

32. (II) The first-order line of 589-nm light falling on a diffraction grating is observed at a 15.5° angle. How far apart are the slits? At what angle will the third order be observed?

33. (II) Monochromatic light falls on a transmission diffraction grating at an angle ϕ to the normal. Show that Eq. 36–13 for diffraction maxima must be replaced by

$$d(\sin \phi \pm \sin \theta) = m\lambda. \qquad m = 0, 1, 2, \cdots.$$

Explain the $\pm$ sign.

* Section 36–9

* 34. (II) Missing orders occur for a diffraction grating when a diffraction minimum coincides with an interference maximum. Let a be the width of each slit and d the separation of slits and show (a) that if $d = 2a$, all even orders ($m = 2, 4, 6, \cdots$) are missing. (b) Show that there will be missing orders whenever

$$\frac{d}{a} = \frac{m_1}{m_2}$$

where m_1 and m_2 are integers. (c) Discuss the case $d = a$, the limit in which the space between slits becomes negligible.

* 35. (II) Let 580-nm light be incident normally on a diffraction grating for which $d = 3.00a = 1200$ nm. (a) How many orders (principal maxima) are present? (b) If the grating is 1.80 cm wide, what is the full angular width of each principal maximum?

* 36. (II) A 6500-line/cm diffraction grating is 3.61 cm wide. If light with wavelengths near 624 nm falls on the grating, how close can two wavelengths be if they are to be resolved in any order? What order gives the best resolution?

* 37. (II) A diffraction grating has 16,000 rulings in its 2.4 cm width. Determine (a) its resolving power in first and second orders, and (b) the minimum wavelength resolution ($\Delta\lambda$) it can yield for $\lambda \approx 410$ nm.

* 38. (II) For a fixed wavelength and diffraction angle, show that the resolving power of a diffraction grating depends on the total width, Nd, where N is the total number of slits each of width d.

* 39. (II) Determine a formula for the minimum difference in frequency, Δf, that a diffraction grating can resolve when two frequencies, $f_1 \approx f_2 = f$, are incident on it.

* Section 36–10

* 40. (II) X-rays of wavelength 0.138 nm fall on a crystal whose atoms, lying in planes, are spaced 0.265 nm apart. At what angle must the X-rays be directed if the first diffraction maximum is to be observed?

* 41. (II) First-order Bragg diffraction is observed at 26.2° from a crystal with spacing between atoms of 0.24 nm. (a) At what angle will second order be observed? (b) What is the wavelength of the X-rays?

* 42. (II) If X-ray diffraction peaks corresponding to the first three orders ($m = 1, 2$, and 3) are measured, can both the X-ray wavelength λ and lattice spacing d be determined? Prove your answer.

Section 36–11

43. (I) Two polarizers are oriented at 75° to one another. Unpolarized light falls on them. What fraction of the light intensity is transmitted?

44. (I) Two Polaroids are aligned so that the light passing through them is a maximum. At what angle should one of them be placed so the intensity is subsequently reduced by half?

45. (I) What is Brewster's angle for an air–glass ($n = 1.56$) surface?

46. (I) What is Brewster's angle for a diamond submerged in water if the light is hitting the diamond while traveling in the water?

47. (II) At what angle should the axes of two Polaroids be placed so as to reduce the intensity of the incident unpolarized light to (a) $\frac{1}{3}$, (b) $\frac{1}{10}$?

48. (II) The critical angle for total internal reflection at a boundary between two materials is 52°. What is Brewster's angle at this boundary?

49. (II) What would Brewster's angle be for reflections off the surface of water for light coming from beneath the surface? Compare to the angle for total internal reflection, and to Brewster's angle from above the surface.

50. (II) Two polarizers are oriented at 34.0° to one another. Light polarized at a 17.0° angle to each polarizer passes through both. What reduction in intensity takes place?

51. (II) Unpolarized light passes through five successive Polaroid sheets each of whose axis makes a 45° angle with the previous one. What is the intensity of the transmitted beam?

52. (II) Describe how to rotate the plane of polarization of a plane-polarized beam of light by 90° and produce only a 10 percent loss in intensity, using polarizers.

53. (III) The percent polarization P of a partially polarized beam of light is defined as

$$P = \frac{I_{max} - I_{min}}{I_{max} + I_{min}} \times 100$$

where I_{max} and I_{min} are the maximum and minimum intensities that are obtained when the light passes through a polarizer that is slowly rotated. Such light can be considered as the sum of two unequal plane-polarized beams of intensities I_{max} and I_{min} perpendicular to each other. Show that the light transmitted by a polarizer, whose axis makes an angle ϕ to the direction in which I_{max} is obtained, has intensity

$$\frac{1 + p \cos 2\phi}{1 + p} I_{max}$$

where $p = P/100$ is the "fractional polarization."

54. When violet light of wavelength 415 nm falls on a single slit, it creates a central diffraction peak that is 9.20 cm wide on a screen that is 2.55 m away. How wide is the slit?

55. A teacher stands well back from an outside doorway 0.88 m wide, and blows a whistle of frequency 750 Hz. Ignoring reflections, estimate at what angle(s) it is *not* possible to hear the whistle clearly on the playground outside the doorway.

56. The wings of a certain beetle have a series of parallel lines across them. When normally incident 460-nm light is reflected from the wing, the wing appears bright when viewed at an angle of 50°. How far apart are the lines?

57. How many lines per centimeter must a grating have if there is to be no second-order spectrum for any visible wavelength?

58. Light is incident on a diffraction grating with 7500 lines per centimeter and the pattern is viewed on a screen located 2.5 m from the grating. The incident light beam consists of two wavelengths, $\lambda_1 = 4.4 \times 10^{-7}$ m and $\lambda_2 = 6.3 \times 10^{-7}$ m. Calculate the linear distance between the first-order bright fringes of these two wavelengths on the screen.

59. If parallel light falls on a single slit of width a at a 20° angle to the normal, describe the diffraction pattern.

60. When yellow sodium light, $\lambda = 589$ nm, falls on a diffraction grating, its first-order peak on a screen 60.0 cm away falls 3.32 cm from the central peak. Another source produces a line 3.71 cm from the central peak. What is its wavelength? How many lines/cm are on the grating?

61. What is the highest spectral order that can be seen if a grating with 6000 lines per cm is illuminated with 633-nm laser light? Assume normal incidence.

62. Two and only two full spectral orders can be seen on either side of the central maximum when white light is sent through a diffraction grating. What is the maximum number of lines per cm for the grating?

63. Two of the lines of the atomic hydrogen spectrum have wavelengths of 656 nm and 410 nm. If these fall at normal incidence on a grating with 7600 lines per cm, what will be the angular separation of the two wavelengths in the first-order spectrum?

64. Light falling normally on a 9850 line/cm grating is revealed to contain three lines in the first-order spectrum at angles of 31.2°, 36.4°, and 47.5°. What wavelengths are these?

65. (a) How far away can a human eye distinguish two car headlights 2.0 m apart? Consider only diffraction effects and assume an eye diameter of 5.0 mm and a wavelength of 500 nm. (b) What is the minimum angular separation an eye could resolve when viewing two stars, considering only diffraction effects? In reality, it is about 1′ of arc. Why is it not equal to your answer in (b)?

66. At what angle above the horizon is the Sun when light reflecting off a smooth lake is polarized most strongly?

67. Unpolarized light falls on two polarizer sheets whose axes are at right angles. (a) What fraction of the incident light intensity is transmitted? (b) What fraction is transmitted if a third polarizer is placed between the first two so that its axis makes a 60° angle with the axis of the first polarizer? (c) What if the third polarizer is in front of the other two?

68. Four polarizers are placed in succession with their axes vertical, at 30° to the vertical, at 60° to the vertical, and at 90° to the vertical. (a) Calculate what fraction of the incident unpolarized light is transmitted by the four polarizers. (b) Can the transmitted light be *decreased* by removing one of the polarizers? If so, which one? (c) Can the transmitted light intensity be extinguished by removing polarizers? If so, which one(s)?

69. At what angle should the axes of two Polaroids be placed so as to reduce the intensity of the incident unpolarized light by an additional factor (after the first Polaroid cuts it in half) of (a) 25 percent, (b) 10 percent, (c) 1 percent?

70. Two polarizers are oriented at 40° to each other and plane-polarized light is incident on them. If only 15 percent of the light gets through both of them, what was the initial polarization direction of the incident light?

71. Show that if two equally intense sources of light produce light that is plane-polarized, but with their planes of polarization perpendicular to each other, then they cannot produce an interference pattern even if they are in phase at all moments.

72. Spy planes fly at extremely high altitudes (25 km) to avoid interception. Their cameras are reportedly able to discern features as small as 5 cm. What must be the minimum aperture of the camera lens to afford this resolution? (Use $\lambda = 550$ nm.)

* 73. X-rays of wavelength 0.0973 nm are directed at an unknown crystal. The second diffraction maximum is recorded at 23.4°. What is the spacing between crystal planes?

* 74. X-rays of wavelength 0.10 nm fall on a microcrystalline powder sample as in Fig. 36–29. The sample is located 10 cm from the photographic film. The crystal structure of the sample has an atomic spacing of 0.25 nm. Calculate the radii of the diffraction rings corresponding to first- and second-order scattering.

Albert Einstein (1879–1955), one of the great minds of the twentieth century, was the creator of the special and general theories of relativity.

In this chapter we examine the special theory of relativity, which includes a number of non-classical results, including length contraction and time dilation for moving reference frames, and new formulas for momentum and energy.

Special Theory of Relativity

FIGURE 37–1 Albert Einstein and his second wife.

Classical vs. modern physics

Physics at the end of the nineteenth century looked back on a period of great progress. The theories developed over the preceding three centuries had been very successful in explaining a wide range of natural phenomena. Newtonian mechanics beautifully explained the motion of objects on Earth and in the heavens. Furthermore, it formed the basis for successful treatments of fluids, wave motion, and sound. Kinetic theory explained the behavior of gases and other materials. Maxwell's theory of electromagnetism not only brought together and explained electric and magnetic phenomena, but it predicted the existence of electromagnetic (EM) waves that would behave in every way just like light—so light came to be thought of as an electromagnetic wave. Indeed, it seemed that the natural world, as seen through the eyes of physicists, was very well explained. A few puzzles remained, but it was felt that these would soon be explained using already known principles.

But it did not turn out so simply. Instead, these few puzzles were to be solved only by the introduction, in the early part of the twentieth century, of two revolutionary new theories that changed our whole conception of nature: the *theory of relativity* and *quantum theory*.

Physics as it was known at the end of the nineteenth century (what we've covered up to now in this book) is referred to as **classical physics**. The new physics that grew out of the great revolution at the turn of the twentieth century is now called **modern physics**. In this chapter, we present the special theory of relativity, which was first proposed by Albert Einstein (1879–1955; Fig. 37–1) in 1905. In the following chapter, we introduce the equally momentous quantum theory.

37–1 Galilean–Newtonian Relativity

Einstein's special theory of relativity deals with how we observe events, particularly how objects and events are observed from different frames of reference. This subject had, of course, already been explored by Galileo and Newton.

The special theory of relativity deals with events that are observed and measured from so-called **inertial reference frames** which (Sections 4–2 and 11–9) are reference frames in which Newton's first law is valid: if an object experiences no net force the object either remains at rest or continues in motion with constant speed in a straight line. It is easiest to analyze events when they are observed and measured from inertial frames, and the Earth, though not quite an inertial frame (it rotates), is close enough that for most purposes we can consider it an inertial frame. Rotating or otherwise accelerating frames of reference are noninertial frames, and won't concern us in this chapter (they are dealt with in Einstein's general theory of relativity).

A reference frame that moves with constant velocity with respect to an inertial frame is itself also an inertial frame, since Newton's laws hold in it as well. When we say that we observe or make measurements from a certain reference frame, it means that we are at rest in that reference frame.

Both Galileo and Newton were aware of what we now call the **relativity principle** applied to mechanics: that *the basic laws of physics are the same in all inertial reference frames*. You may have recognized its validity in everyday life. For example, objects move in the same way in a smoothly moving (constant-velocity) train or airplane as they do on Earth. (This assumes no vibrations or rocking—for they would make the reference frame noninertial.) When you walk, drink a cup of soup, play Ping-Pong, or drop a pencil on the floor while traveling in a train, airplane, or ship moving at constant velocity, the bodies move just as they do when you are at rest on Earth. Suppose you are in a car traveling rapidly along at constant velocity. If you release a coin from above your head inside the car, how will it fall? It falls straight downward with respect to the car, and hits the floor directly below the point of release, Fig. 37–2a. (If you drop the coin out the car's window, this won't happen because the moving air drags the coin backward relative to the car.) This is just how objects fall on the Earth—straight down—and thus our experiment in the moving car is in accord with the relativity principle.

Note in this example, however, that to an observer on the Earth, the coin follows a curved path, Fig. 37–2b. The actual path followed by the coin is different as viewed from different frames of reference. This does not violate the relativity principle because this principle states that the *laws* of physics are the same in all inertial frames. The same law of gravity, and the same laws of motion, apply in both reference frames. And the acceleration of the coin is the same in both reference frames. The difference in Figs. 37–2a and b is that in the Earth's frame of reference, the coin has an initial velocity (equal to that of the car). The laws of physics therefore predict it will follow a parabolic path like any projectile. In the car's ref-

Relativity principle: the laws of physics are the same in all inertial reference frames

FIGURE 37–2 A coin is dropped by a person in a moving car. (a) In the reference frame of the car, the coin falls straight down. (b) In a reference frame fixed on the Earth, the coin follows a curved (parabolic) path. The upper views show the moment of the coin's release, and the lower views are a short time later.

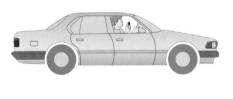

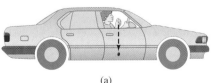

(a)
Reference frame = car

(b)
Reference frame = Earth

SECTION 37–1   **917**

erence frame, there is no initial velocity, and the laws of physics predict that the coin will fall straight down. The laws are the same in both reference frames, although the specific paths are different.

Galilean–Newtonian relativity involves certain unprovable assumptions that make sense from everyday experience. It is assumed that the lengths of objects are the same in one reference frame as in another, and that time passes at the same rate in different reference frames. In classical mechanics, then, space and time are considered to be **absolute**: their measurement doesn't change from one reference frame to another. The mass of an object, as well as all forces, are assumed to be unchanged by a change in inertial reference frame.

The position of an object is, of course, different when specified in different reference frames, and so is velocity. For example, a person may walk inside a bus toward the front with a speed of 5 km/h. But if the bus moves 40 km/h with respect to the Earth, the person is then moving with a speed of 45 km/h with respect to the Earth. The acceleration of a body, however, is the same in any inertial reference frame according to classical mechanics. This is because the change in velocity, and the time interval, will be the same. For example, the person in the bus may accelerate from 0 to 5 km/h in 1.0 seconds, so $a = 5$ km/h/s in the reference frame of the bus. With respect to the Earth, the acceleration is $(45 \text{ km/h} - 40 \text{ km/h})/(1.0 \text{ s}) = 5 \text{ km/h/s}$, which is the same.

Since neither F, m, nor a changes from one inertial frame to another, then Newton's second law, $F = ma$, does not change. Thus Newton's second law satisfies the relativity principle. It is easily shown that the other laws of mechanics also satisfy the relativity principle.

That the laws of mechanics are the same in all inertial reference frames implies that no one inertial frame is special in any sense. We express this important conclusion by saying that **all inertial reference frames are equivalent** for the description of mechanical phenomena. No one inertial reference frame is any better than another. A reference frame fixed to a car or an aircraft traveling at constant velocity is as good as one fixed on the Earth. When you travel smoothly at constant velocity in a car or airplane, it is just as valid to say you are at rest and the Earth is moving as it is to say the reverse. There is no experiment you can do to tell which frame is "really" at rest and which is moving. Thus, there is no way to single out one particular reference frame as being at absolute rest.

All inertial reference frames are equally valid

A complication arose, however, in the last half of the nineteenth century. When Maxwell presented his comprehensive and very successful theory of electromagnetism (Chapter 32), he showed that light can be considered an electromagnetic wave. Maxwell's equations predicted that the velocity of light c would be 3.00×10^8 m/s; and this is just what is measured, within experimental error. The question then arose: in what reference frame does light have precisely the value predicted by Maxwell's theory? For it was assumed that light would have a different speed in different frames of reference. For example, if observers were traveling on a rocket ship at a speed of 1.0×10^8 m/s away from a source of light, we might expect them to measure the speed of the light reaching them to be 3.0×10^8 m/s $- 1.0 \times 10^8$ m/s $= 2.0 \times 10^8$ m/s. But Maxwell's equations have no provision for relative velocity. They predicted the speed of light to be $c = 3.0 \times 10^8$ m/s. This seemed to imply there must be some special reference frame where c would have this value.

We discussed in Chapters 15 and 16 that waves travel on water and along ropes or strings, and sound waves travel in air and other materials. Nineteenth-century physicists viewed the material world in terms of the laws of mechanics, so it was natural for them to assume that light too must travel in some *medium*. They called this transparent medium the **ether** and assumed it permeated all space.[†] It was therefore assumed that the velocity of light given by Maxwell's equations must be with respect to the ether.

The "ether"

However, it appeared that Maxwell's equations did *not* satisfy the relativity principle. They were not the same in all inertial reference frames. They were simplest in the frame where $c = 3.00 \times 10^8$ m/s; that is, in a reference frame at rest in the ether. In any other reference frame, extra terms would have to be added to take into account the relative velocity. Thus, although most of the laws of physics obeyed the relativity principle, the laws of electricity and magnetism apparently did not. Instead, they seemed to single out one reference frame that was better than any other—a reference frame that could be considered absolutely at rest.

Scientists soon set out to determine the speed of the Earth relative to this absolute frame, whatever it might be. A number of clever experiments were designed. The most direct were performed by A. A. Michelson and E. W. Morley in the 1880s. The details of their experiment are discussed in the next Section. Briefly, what they did was measure the difference in the speed of light in different directions. They expected to find a difference depending on the orientation of their apparatus with respect to the ether. For just as a boat has different speeds relative to the land when it moves upstream, downstream, or across the stream, so too light would be expected to have different speeds depending on the velocity of the ether past the Earth.

Strange as it may seem, they detected no difference at all. This was a great puzzle. A number of explanations were put forth over a period of years, but they led to contradictions or were otherwise not generally accepted.

Then in 1905, Albert Einstein proposed a radical new theory that reconciled these many problems in a simple way. But at the same time, as we shall soon see, it completely changed our ideas of space and time.

* 37–2 The Michelson–Morley Experiment

The Michelson–Morley experiment was designed to measure the speed of the *ether*—the medium in which light was assumed to travel—with respect to the Earth. The experimenters thus hoped to find an absolute reference frame, one that could be considered to be at rest.

One of the possibilities nineteenth-century scientists considered was that the ether is fixed relative to the Sun, for even Newton had taken the Sun as the center of the universe. If this were the case (there was no guarantee, of course), the Earth's speed of about 3×10^4 m/s in its orbit around the Sun would produce a change of 1 part in 10^4 in the speed of light $(3.0 \times 10^8$ m/s$)$. Direct measurement of the speed of light to this accuracy was not possible. But A. A. Michelson, later with the help of E. W. Morley, was able to use his interferometer (Section 35–7) to measure the difference in the speed of light in different directions to this accuracy.

[†]The medium for light waves could not be air, since light travels from the Sun to Earth through nearly empty space. Therefore, another medium was postulated, the ether. The ether was not only transparent, but, because of difficulty in detecting it, was assumed to have zero density.

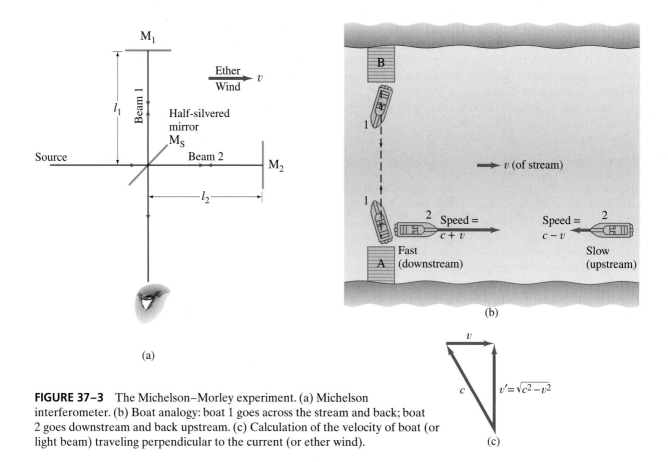

FIGURE 37–3 The Michelson–Morley experiment. (a) Michelson interferometer. (b) Boat analogy: boat 1 goes across the stream and back; boat 2 goes downstream and back upstream. (c) Calculation of the velocity of boat (or light beam) traveling perpendicular to the current (or ether wind).

This famous experiment is based on the principle shown in Fig. 37–3. Part (a) is a diagram of the Michelson interferometer, and it is assumed that the "ether wind" is moving with speed v to the right. (Alternatively, the Earth is assumed to move to the left with respect to the ether at speed v.) The light from the source is split into two beams by the half-silvered mirror M_S. One beam travels to mirror M_1 and the other to mirror M_2. The beams are reflected by M_1 and M_2 and are joined again after passing through M_S. The now superposed beams interfere with each other and the resultant is viewed by the observer's eye as an interference pattern (discussed in Section 35–7).

Whether constructive or destructive interference occurs at the center of the interference pattern depends on the relative phases of the two beams after they have traveled their separate paths. To examine this let us consider an analogy of a boat traveling up and down, and across, a river whose current moves with speed v, as shown in Fig. 37–3b. In still water, the boat can travel with speed c (not the speed of light in this case).

First we consider beam 2 in Fig. 37–3a, which travels parallel to the "ether wind." In its journey from M_S to M_2, we expect the light to travel with speed $c + v$, just as a boat traveling downstream (see Fig. 37–3b) adds the speed of the river to its own speed. Since the beam travels a distance l_2, the time it takes to go from M_S to M_2 is $t = l_2/(c + v)$. To make the return trip from M_2 to M_S, the light must move against the ether wind (like the boat going upstream), so its relative speed is expected to be $c - v$. The time for the return trip is $l_2/(c - v)$. The total time required for beam 2 to travel from M_S to M_2 and back to M_S is

$$t_2 = \frac{l_2}{c + v} + \frac{l_2}{c - v} = \frac{2l_2}{c(1 - v^2/c^2)}.$$

Now let us consider beam 1, which travels crosswise to the ether wind. Here the boat analogy (part b) is especially helpful. The boat is to go from wharf A to wharf B directly across the stream. If it heads directly across, the stream's current will drag it downstream. To reach wharf B, the boat must head at an angle upstream. The precise angle depends on the magnitudes of c and v, but is of no interest to us in itself. Part (c) of Fig. 37–3 shows how to calculate the velocity v' of the boat relative to Earth as it crosses the stream. Since c, v, and v' form a right triangle, we have that $v' = \sqrt{c^2 - v^2}$. The boat has the same speed when it returns. If we now apply these principles to light beam 1 in Fig. 37–3a, we see that the beam travels with a speed $\sqrt{c^2 - v^2}$ in going from M_S to M_1 and back again. The total distance traveled is $2l_1$, so the time required for beam 1 to make the round trip is $2l_1/\sqrt{c^2 - v^2}$, or

$$t_1 = \frac{2l_1}{c\sqrt{1 - v^2/c^2}}.$$

Notice that the denominator in this equation for t_1 involves a square root, whereas that for t_2 does not.

If $l_1 = l_2 = l$, we see that beam 2 will lag behind beam 1 by an amount

$$\Delta t = t_2 - t_1 = \frac{2l}{c}\left(\frac{1}{1 - v^2/c^2} - \frac{1}{\sqrt{1 - v^2/c^2}}\right).$$

If $v = 0$, then $\Delta t = 0$, and the two beams will return in phase since they were initially in phase. But if $v \neq 0$, then $\Delta t \neq 0$, and the two beams will return out of phase. If this change of phase from the condition $v = 0$ to that for $v = v$ could be measured, then v could be determined. But the Earth cannot be stopped. Furthermore, we should not be too quick to assume that lengths are not affected by motion and therefore to assume $l_1 = l_2$.

Michelson and Morley realized that they could detect the difference in phase (assuming that $v \neq 0$) if they rotated their apparatus by 90°, for then the interference pattern between the two beams should change. In the rotated position, beam 1 would now move parallel to the ether and beam 2 perpendicular to it. Thus the roles could be reversed, and in the rotated position the times (designated by primes) would be

$$t_1' = \frac{2l_1}{c(1 - v^2/c^2)} \quad \text{and} \quad t_2' = \frac{2l_2}{c\sqrt{1 - v^2/c^2}}.$$

The time lag between the two beams in the nonrotated position (unprimed) would be

$$\Delta t = t_2 - t_1 = \frac{2l_2}{c(1 - v^2/c^2)} - \frac{2l_1}{c\sqrt{1 - v^2/c^2}}.$$

In the rotated position, the time difference would be

$$\Delta t' = t_2' - t_1' = \frac{2l_2}{c\sqrt{1 - v^2/c^2}} - \frac{2l_1}{c(1 - v^2/c^2)}.$$

When the rotation is made, the fringes of the interference pattern (Section 35–7) will shift an amount determined by the difference:

$$\Delta t - \Delta t' = \frac{2}{c}(l_1 + l_2)\left(\frac{1}{1 - v^2/c^2} - \frac{1}{\sqrt{1 - v^2/c^2}}\right).$$

This expression can be considerably simplified if we assume that $v/c \ll 1$. For in this case we can use the binomial expansion (Appendix A), so

$$\frac{1}{1 - v^2/c^2} \approx 1 + \frac{v^2}{c^2} \quad \text{and} \quad \frac{1}{\sqrt{1 - v^2/c^2}} \approx 1 + \frac{1}{2}\frac{v^2}{c^2}.$$

Then

$$\Delta t - \Delta t' \approx \frac{2}{c}(l_1 + l_2)\left(1 + \frac{v^2}{c^2} - 1 - \frac{1}{2}\frac{v^2}{c^2}\right)$$

$$\approx (l_1 + l_2)\frac{v^2}{c^3}.$$

Now we take $v = 3.0 \times 10^4$ m/s, the speed of the Earth in its orbit around the Sun. In Michelson and Morley's experiments, the arms l_1 and l_2 were about 11 m long. The time difference would then be about

$$\frac{(22\text{ m})(3.0 \times 10^4\text{ m/s})^2}{(3.0 \times 10^8\text{ m/s})^3} \approx 7.0 \times 10^{-16}\text{ s}.$$

For visible light of wavelength $\lambda = 5.5 \times 10^{-7}$ m, say, the frequency would be $f = c/\lambda = (3.0 \times 10^8\text{ m/s})/(5.5 \times 10^{-7}\text{ m}) = 5.5 \times 10^{14}$ Hz, which means that wave crests pass by a point every $1/(5.5 \times 10^{14}\text{ Hz}) = 1.8 \times 10^{-15}$ s. Thus, with a time difference of 7.0×10^{-16} s, Michelson and Morley should have noted a movement in the interference pattern of $(7.0 \times 10^{-16}\text{ s})/(1.8 \times 10^{-15}\text{ s}) = 0.4$ fringe. They could easily have detected this, since their apparatus was capable of observing a fringe shift as small as 0.01 fringe.

The null result   But they found *no significant fringe shift whatever*! They set their apparatus at various orientations. They made observations day and night so that they would be at various orientations with respect to the Sun (due to the Earth's rotation). They tried at different seasons of the year (the Earth at different locations due to its orbit around the Sun). Never did they observe a significant fringe shift.

This **null result** was one of the great puzzles of physics at the end of the nineteenth century. To explain it was a difficult challenge. One possibility to explain the null result was put forth independently by G. F. Fitzgerald and H. A. Lorentz (in the 1890s) in which they proposed that any length (including the arm of an interferometer) contracts by a factor $\sqrt{1 - v^2/c^2}$ in the direction of motion through the ether. According to Lorentz, this could be due to the ether affecting the forces between the molecules of a substance, which were assumed to be electrical in nature. This theory was eventually replaced by the far more comprehensive theory proposed by Albert Einstein in 1905—the special theory of relativity.

37–3 Postulates of the Special Theory of Relativity

The problems that existed at the turn of the century with regard to electromagnetic theory and Newtonian mechanics were beautifully resolved by Einstein's introduction of the theory of relativity in 1905. Einstein, however, was apparently not influenced directly by the null result of the Michelson–Morley experiment. What motivated Einstein were certain questions regarding electromagnetic theory and light waves. For example, he asked himself: "What would I see if I rode a light beam?" The answer was that instead of a traveling electromagnetic wave, he would see alternating electric and magnetic fields at rest whose magnitude changed in space, but did not change in time. Such fields, he realized, had never been detected and indeed were not consistent with Maxwell's electromagnetic theory. He argued, therefore, that it was unreasonable to think that the speed of light relative to any observer could be reduced to zero, or in fact reduced at all. This idea became the second postulate of his theory of relativity.

Einstein concluded that the inconsistencies he found in electromagnetic theory were due to the assumption that an absolute space exists. In his famous 1905 paper, he proposed doing away completely with the idea of the ether and the accompanying assumption of an absolute reference frame at rest. This proposal was embodied in two postulates. The first postulate was an extension of the Newtonian relativity principle to include not only the laws of mechanics but also those of the rest of physics, including electricity and magnetism:

> ***First postulate*** (*the relativity principle*): **The laws of physics have the same form in all inertial reference frames.**

The two postulates of special relativity

The second postulate is consistent with the first:

> ***Second postulate*** (*constancy of the speed of light*): **Light propagates through empty space with a definite speed c independent of the speed of the source or observer.**

These two postulates form the foundation of Einstein's **special theory of relativity**. It is called "special" to distinguish it from his later "general theory of relativity," which deals with noninertial (accelerating) reference frames. The special theory, which is what we discuss here, deals only with inertial frames.

The second postulate may seem hard to accept, for it violates commonsense notions. First of all, we have to think of light traveling through empty space. Giving up the ether is not too hard, however, for after all, it had never been detected. But the second postulate also tells us that the speed of light in vacuum is always the same, 3.00×10^8 m/s, no matter what the speed of the observer or the source. Thus, a person traveling toward or away from a source of light will measure the same speed for that light as someone at rest with respect to the source. This conflicts with our everyday notions, for we would expect to have to add in the velocity of the observer. Part of the problem is that in our everyday experience, we do not measure velocities anywhere near as large as the speed of light. Thus we can't expect our everyday experience to be helpful when dealing with such a high velocity. On the other hand, the Michelson–Morley experiment is fully consistent with the second postulate.[†]

Einstein's proposal has a certain beauty. For by doing away with the idea of an absolute reference frame, it was possible to reconcile classical mechanics with Maxwell's electromagnetic theory. The speed of light predicted by Maxwell's equations *is* the speed of light in vacuum in *any* reference frame.

Einstein's theory required giving up commonsense notions of space and time, and in the following Sections we will examine some strange but interesting consequences of special relativity. Our arguments for the most part will be simple ones. We will use a technique that Einstein himself did: we will imagine very simple experimental situations in which little mathematics is needed. In this way, we can see many of the consequences of relativity theory without getting involved in detailed calculations. Einstein called these "thought" experiments. Starting in Section 37–8 we will look more fully at the mathematics of relativity.

[†]The Michelson–Morley experiment can also be considered as evidence for the first postulate, for it was intended to measure the motion of the Earth relative to an absolute reference frame. Its failure to do so implies the absence of any such preferred frame.

37–4 | Simultaneity

One of the important consequences of the theory of relativity is that we can no longer regard time as an absolute quantity. No one doubts that time flows onward and never turns back. But, as we shall see in this Section and the next, the time interval between two events, and even whether two events are simultaneous, depends on the observer's reference frame.

Two events are said to occur simultaneously if they occur at exactly the same time. But how do we know if two events occur precisely at the same time? If they occur at the same point in space—such as two apples falling on your head at the same time—it is easy. But if the two events occur at widely separated places, it is more difficult to know whether the events are simultaneous since we have to take into account the time it takes for the light from them to reach us. Because light travels at finite speed, a person who sees two events must calculate back to find out when they actually occurred. For example, if two events are *observed* to occur at the same time, but one actually took place farther from the observer than the other, then the more distant one must have occurred earlier, and the two events were not simultaneous.

FIGURE 37–4 A moment after lightning strikes points A and B, the pulses of light are traveling toward the observer O, but O "sees" the lightning only when the light reaches O.

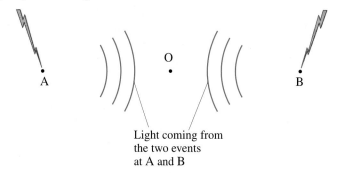

Light coming from the two events at A and B

We will now make use of a simple thought experiment. We assume an observer, called O, is located exactly halfway between points A and B where two events occur, Fig. 37–4. The two events may be lightning that strikes the points A and B, as shown, or any other type of events. For brief events like lightning, only short pulses of light will travel outward from A and B and reach O. O "sees" the events when the pulses of light reach point O. If the two pulses reach O at the same time, then the two events had to be simultaneous. This is because the two light pulses travel at the same speed (postulate 2), and since the distance OA equals OB, the time for the light to travel from A to O and B to O must be the same. Observer O can then definitely state that the two events occurred simultaneously. On the other hand, if O sees the light from one event before that from the other, then it is certain the former event occurred first.

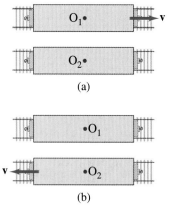

(a)

(b)

FIGURE 37–5 Observers O_1 and O_2, on two different trains (two different reference frames), are moving with relative velocity v. O_2 says that O_1 is moving to the right (a); O_1 says that O_2 is moving to the left (b). Both viewpoints are legitimate—it all depends on your reference frame.

The question we really want to examine is this: if two events are simultaneous to an observer in one reference frame, are they also simultaneous to another observer moving with respect to the first? Let us call the observers O_1 and O_2 and assume they are fixed in reference frames 1 and 2 that move with speed v relative to one another. These two reference frames can be thought of as trains (Fig. 37–5). O_2 says that O_1 is moving to the right with speed v, as in Fig. 37–5a; and O_1 says O_2 is moving to the left with speed v, as in Fig. 37–5b. Both viewpoints are legitimate according to the relativity principle. (There is, of course, no third point of view which will tell us which one is "really" moving.)

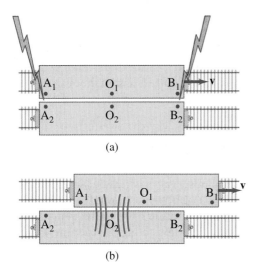

FIGURE 37–6 Thought experiment on simultaneity. To observer O_2, the reference frame of O_1 is moving to the right. In (a), one lightning bolt strikes the two reference frames at A_1 and A_2, and a second lightning bolt strikes at B_1 and B_2. (b) A moment later, the light from the two events reaches O_2 at the same time, so according to observer O_2, the two bolts of lightning strike simultaneously. But in O_1's reference frame, the light from B_1 has already reached O_1, whereas the light from A_1 has not yet reached O_1. So in O_1's reference frame, the event at B_1 must have preceded the event at A_1. Time is not absolute.

(a)

(b)

Now suppose two events occur that are observed and measured by both observers. Let us assume again that the two events are the striking of lightning and that the lightning marks both trains where it struck: at A_1 and B_1 on O_1's train, and at A_2 and B_2 on O_2's train. For simplicity, we assume that O_1 happens to be exactly halfway between A_1 and B_1, and that O_2 is halfway between A_2 and B_2. We now put ourselves in one reference frame or the other, from which we make our observations and measurements. Let us put ourselves in O_2's reference frame, so we observe O_1 moving to the right with speed v. Let us also assume that the two events occur *simultaneously* in O_2's frame, and just at the instant when O_1 and O_2 are opposite each other, Fig. 37–6a. A short time later, Fig. 37–6b, the light from A_2 and B_2 reaches O_2 at the same time (we assumed this). Since O_2 knows (or measures) the distances O_2A_2 and O_2B_2 as equal, O_2 knows the two events are simultaneous in the O_2 reference frame.

But what does observer O_1 observe and measure? From our (O_2) reference frame, we can predict what O_1 will observe. We see that O_1 moves to the right during the time the light is traveling to O_1 from A_1 and B_1. As shown in Fig. 37–6b, we can see from our O_2 reference frame that the light from B_1 has already passed O_1, whereas the light from A_1 has not yet reached O_1. Therefore, it is clear that O_1 will observe the light coming from B_1 before he observes the light coming from A_1. Now O_1's frame is as good as O_2's. Light travels at the same speed c for O_1 as for O_2 (the second postulate)[†]; and in the O_1 reference frame, this speed c is of course the same for light traveling from A_1 to O_1 as it is for light traveling from B_1 to O_1. Furthermore the distance O_1A_1 equals O_1B_1. Hence, since O_1 observes the light from B_1 before he observes the light from A_1 (we established this above, looking from the O_2 reference frame, Fig. 37–6b), then observer O_1 can only conclude that the event at B_1 occurred before the event at A_1. The two events are not simultaneous for O_1, even though they are for O_2.

We thus find that two events which are simultaneous to one observer are not necessarily simultaneous to a second observer.

It may be tempting to ask: "Which observer is right, O_1 or O_2?" The answer, according to relativity, is that they are *both* right. There is no "best" reference frame we can choose to determine which observer is right. Both frames are equally good. We can only conclude that *simultaneity is not an absolute concept*, but is relative. We are not aware of it in everyday life, however, because the effect is noticeable only when the relative speed of the two reference frames is very large (near c), or the distances involved are very large.

Simultaneity is relative

[†]Note that O_1 does not see himself catching up with one light beam and running away from the other (that is O_2's viewpoint of what happens for O_1). O_1 sees both light beams traveling at the same speed, c.

Because of the principle of relativity, the argument we gave for the thought experiment of Fig. 37–6 can be done from O_1's reference frame as well. In this case, O_1 will be at rest and will see event B_1 occur before A_1. But O_1 will recognize (by drawing a diagram equivalent to Fig. 37–6—try it and see!) that O_2, who is moving with speed v to the left, will see the two events as simultaneous.

37–5 | Time Dilation and the Twin Paradox

The fact that two events simultaneous to one observer may not be simultaneous to a second observer suggests that time itself is not absolute. Could it be that time passes differently in one reference frame than in another? This is, indeed, just what Einstein's theory of relativity predicts, as the following thought experiment shows.

Figure 37–7 shows a spaceship traveling past Earth at high speed. The point of view of an observer on the spaceship is shown in part (a), and that of an observer on Earth in part (b). Both observers have accurate clocks. The person on the spaceship (a) flashes a light and measures the time it takes the light to travel across the spaceship and return after reflecting from a mirror. The light travels a distance $2D$ at speed c, so the time required, which we call Δt_0, is

$$\Delta t_0 = \frac{2D}{c}.$$

This is the time interval as measured by the observer on the spaceship.

The observer on Earth, Fig. 37–7b, observes the same process. But to this observer, the spaceship is moving. So the light travels the diagonal path shown going across the spaceship, reflecting off the mirror, and returning to the sender. Although the light travels at the same speed to this observer (the second postulate), it travels a greater distance. Hence the time required, as measured by

FIGURE 37–7 Time dilation can be shown by a thought experiment: the time it takes for light to travel over and back on a spaceship is longer for the observer on Earth (b) than for the observer on the spaceship (a).

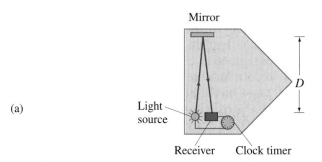

(a)

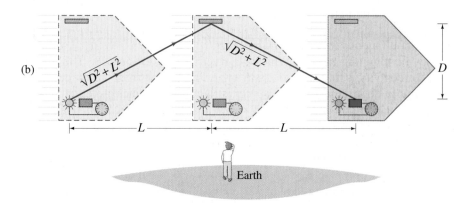

(b)

the observer on Earth, will be *greater* than that measured by the observer on the spaceship. The time interval, Δt, observed by the observer on Earth can be calculated as follows. In the time Δt, the spaceship travels a distance $2L = v \, \Delta t$ where v is the speed of the spaceship (Fig. 37–7b). Thus, the light travels a total distance on its diagonal path of $2\sqrt{D^2 + L^2}$, and therefore

$$c = \frac{2\sqrt{D^2 + L^2}}{\Delta t} = \frac{2\sqrt{D^2 + \dfrac{v^2(\Delta t)^2}{4}}}{\Delta t}.$$

We square both sides, and then solve for Δt, to find

$$c^2 = \frac{4D^2}{(\Delta t)^2} + v^2,$$

$$\Delta t = \frac{2D}{c\sqrt{1 - v^2/c^2}}.$$

We combine this with the formula above for $\Delta t_0 \left(\Delta t_0 = 2D/c \right)$ and find:

$$\Delta t = \frac{\Delta t_0}{\sqrt{1 - v^2/c^2}}. \qquad \textbf{(37–1)} \qquad \textit{Time-dilation formula}$$

Since $\sqrt{1 - v^2/c^2}$ is always less than 1, we see that $\Delta t > \Delta t_0$. That is, the time interval between the two events (the sending of the light, and its reception on the spaceship) is *greater* for the observer on Earth than for the observer on the spaceship. This is a general result of the theory of relativity, and is known as **time dilation**. Stated simply, the time-dilation effect says that

> **clocks moving relative to an observer are measured by that observer to run more slowly (as compared to clocks at rest).**

Time dilation: moving clocks run slow

However, we should not think that the clocks are somehow at fault. Time is actually measured to pass more slowly in any moving reference frame as compared to your own. This remarkable result is an inevitable outcome of the two postulates of the theory of relativity.

The concept of time dilation may be hard to accept, for it violates our commonsense understanding. We can see from Eq. 37–1 that the time dilation effect is negligible unless v is reasonably close to c. If v is much less than c, then the term v^2/c^2 is much smaller than the 1 in the denominator of Eq. 37–1, and then $\Delta t \approx \Delta t_0$ (see Example 37–2). The speeds we experience in everyday life are much smaller than c, so it is little wonder we don't ordinarily notice time dilation. Experiments have tested the time-dilation effect, and have confirmed Einstein's predictions. In 1971, for example, extremely precise atomic clocks were flown around the world in jet planes. The speed of the planes $\left(10^3 \text{ km/h}\right)$ was much less than c, so the clocks had to be accurate to nanoseconds $\left(10^{-9}\text{ s}\right)$ in order to detect any time dilation. They were this accurate, and they confirmed Eq. 37–1 to within experimental error. Time dilation had been confirmed decades earlier, however, by observation on "elementary particles" which have very small masses (typically 10^{-30} to 10^{-27} kg) and so require little energy to be accelerated to speeds close to the speed of light, c. Many of these elementary particles are not stable and decay after a time into lighter particles. One example is the muon, whose mean lifetime is $2.2 \, \mu$s when at rest. Careful experiments showed that when a muon is traveling at high speeds, its lifetime is measured to be longer than when it is at rest, just as predicted by the time-dilation formula.

Why we don't usually notice time dilation

EXAMPLE 37–1 **Lifetime of a moving muon.** (*a*) What will be the mean lifetime of a muon as measured in the laboratory if it is traveling at $v = 0.60c = 1.8 \times 10^8$ m/s with respect to the laboratory? Its mean life at rest is 2.2×10^{-6} s. (*b*) How far does a muon travel in the laboratory, on average, before decaying?

SOLUTION (*a*) If an observer were to move along with the muon (the muon would be at rest to this observer), the muon would have a mean life of 2.2×10^{-6} s. To an observer in the lab, the muon lives longer because of time dilation. From Eq. 37–1 with $v = 0.60c$, we have

$$\Delta t = \frac{\Delta t_0}{\sqrt{1 - \dfrac{v^2}{c^2}}} = \frac{2.2 \times 10^{-6} \text{ s}}{\sqrt{1 - \dfrac{0.36 \, c^2}{c^2}}} = \frac{2.2 \times 10^{-6} \text{ s}}{\sqrt{0.64}} = 2.8 \times 10^{-6} \text{ s}.$$

(*b*) At a speed of 1.8×10^8 m/s, classical physics would tell us that with a mean life of $2.2 \, \mu$s, an average muon would travel $d = vt = \left(1.8 \times 10^8 \text{ m/s}\right)\left(2.2 \times 10^{-6} \text{ s}\right) = 400$ m. But relativity predicts an average distance of $\left(1.8 \times 10^8 \text{ m/s}\right)\left(2.8 \times 10^{-6} \text{ s}\right) = 500$ m, and it is this longer distance that is measured experimentally.

Proper time

We need to make a comment about the use of Eq. 37–1 and the meaning of Δt and Δt_0. The equation is true only when Δt_0 represents the time interval between the two events in a reference frame where the two events occur at *the same point in space* (as in Fig. 37–7a where the two events are the light flash being sent and being received). This time interval, Δt_0, is called the **proper time**. Then Δt in Eq. 37–1 represents the time interval between the two events as measured in a reference frame moving with speed v with respect to the first. In Example 37–1 above, Δt_0 (and not Δt) was set equal to 2.2×10^{-6} s because it is only in the rest frame of the muon that the two events ("birth" and "decay") occur at the same point in space.

EXAMPLE 37–2 **Time dilation at 100 km/h.** Let's check time dilation for everyday speeds. A car traveling 100 km/h covers a certain distance in 10.00 s according to the driver's watch. What does an observer on Earth measure for the time interval?

SOLUTION The car's speed relative to Earth is 100 km/h $= \left(1.00 \times 10^5 \text{ m}\right)/(3600 \text{ s}) = 27.8$ m/s. We set $\Delta t_0 = 10.00$ s in the time-dilation formula (the driver is at rest in the reference frame of the car), and then Δt is

$$\Delta t = \frac{\Delta t_0}{\sqrt{1 - \dfrac{v^2}{c^2}}} = \frac{10.00 \text{ s}}{\sqrt{1 - \left(\dfrac{27.8 \text{ m/s}}{3.00 \times 10^8 \text{ m/s}}\right)^2}} = \frac{10.00 \text{ s}}{\sqrt{1 - 8.59 \times 10^{-15}}}.$$

If you put these numbers into a calculator, you will obtain $\Delta t = 10.00$ s, since the denominator differs from 1 by such a tiny amount. Indeed, the time measured by an observer on Earth would be no different from that measured by the driver, even with the best of today's instruments. A computer that could calculate to a large number of decimal places could reveal a difference between Δt and Δt_0. But we can estimate the difference quite easily using the binomial expansion (Appendix A),

→ **P R O B L E M   S O L V I N G**

Use of the binomial expansion

$$(1 \pm x)^n \approx 1 \pm nx. \qquad\qquad [\text{for } x \ll 1]$$

In our time-dilation formula, we have the factor $\left(1 - v^2/c^2\right)^{-\frac{1}{2}}$. Thus

$$\Delta t = \Delta t_0\left(1 - \frac{v^2}{c^2}\right)^{-\frac{1}{2}} \approx \Delta t_0\left(1 + \frac{1}{2}\frac{v^2}{c^2}\right)$$

$$\approx 10.00\,\text{s}\left[1 + \frac{1}{2}\left(\frac{27.8\,\text{m/s}}{3.00 \times 10^8\,\text{m/s}}\right)^2\right] \approx 10.00\,\text{s} + 4 \times 10^{-15}\,\text{s}.$$

So the difference between Δt and Δt_0 is predicted to be $4 \times 10^{-15}\,\text{s}$, an extremely small amount.

Time dilation has aroused interesting speculation about space travel. According to classical (Newtonian) physics, to reach a star 100 light-years away would not be possible for ordinary mortals (1 light-year is the distance light can travel in 1 year $= 3.0 \times 10^8\,\text{m/s} \times 3.15 \times 10^7\,\text{s} = 9.5 \times 10^{15}\,\text{m}$). Even if a spaceship could travel at close to the speed of light, it would take over 100 years to reach such a star. But time dilation tells us that the time involved would be less for an astronaut. In a spaceship traveling at $v = 0.999c$, the time for such a trip would be only about $\Delta t_0 = \Delta t \sqrt{1 - v^2/c^2} = (100\,\text{yr})\sqrt{1 - (0.999)^2} = 4.5\,\text{yr}$. Thus time dilation allows such a trip, but the enormous practical problems of achieving such speeds will not be overcome in the near future.

Notice, in this example, that whereas 100 years would pass on Earth, only 4.5 years would pass for the astronaut on the trip. Is it just the clocks that would slow down for the astronaut? The answer is no. All processes, including aging and other life processes, run more slowly for the astronaut according to the Earth observer. But to the astronaut, time would pass in a normal way. The astronaut would experience 4.5 years of normal sleeping, eating, reading, and so on. And people on Earth would experience 100 years of ordinary activity.

Not long after Einstein proposed the special theory of relativity, an apparent paradox was pointed out. According to this **twin paradox**, suppose one of a pair of 20-year-old twins takes off in a spaceship traveling at very high speed to a distant star and back again, while the other twin remains on Earth. According to the Earth twin, the traveling twin will age less. Whereas 20 years might pass for the Earth twin, perhaps only 1 year (depending on the spacecraft's speed) would pass for the traveler. Thus, when the traveler returns, the earthbound twin could expect to be 40 years old whereas the traveling twin would be only 21.

This is the viewpoint of the twin on the Earth. But what about the traveling twin? If all inertial reference frames are equally good, won't the traveling twin make all the claims the Earth twin does, only in reverse? Can't the astronaut twin claim that since the Earth is moving away at high speed, time passes more slowly on Earth and the twin on Earth will age less? This is the opposite of what the Earth twin predicts. They cannot both be right, for after all the spacecraft returns to Earth and a direct comparison of ages and clocks can be made.

There is, however, not a paradox at all. The consequences of the special theory of relativity—in this case, time dilation—can be applied only by observers in inertial reference frames. The Earth is such a frame (or nearly so), whereas the spacecraft is not. The spacecraft accelerates at the start and end of its trip and, more importantly, when it turns around at the far point of its journey. During these acceleration periods, the twin on the spacecraft is not maintaining a steady inertial reference frame, so the spacecraft twin's predictions based on special relativity are not valid. The twin on Earth is in an inertial frame and can make valid predictions. Thus, there is no paradox. The traveling twin's point of view expressed above is not correct. The predictions of the Earth twin *are* valid according to the theory of relativity, and the prediction that the traveling twin returns having aged less is the proper one. Einstein's general theory of relativity, which deals with accelerating reference frames, confirms this result.

Twin paradox

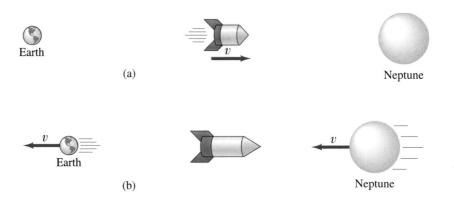

FIGURE 37–8 (a) A spaceship traveling at very high speed from Earth to Neptune, as seen from Earth's frame of reference. (b) As viewed by an observer on the spaceship, Earth and Neptune are moving at the very high velocity v: Earth leaves the spaceship, and a time Δt_0 later planet Neptune arrives at the spaceship. [Note that from the spaceship, part (b), each planet does not look shortened because at high speeds observers see the trailing edge (as in Fig. 37–10), and the net effect is to leave its appearance as a circle.]

37–6 Length Contraction

Not only time intervals are different in different reference frames. Space intervals—lengths and distances—are different as well, according to the special theory of relativity, and we illustrate this with a thought experiment.

Observers on Earth watch a spacecraft traveling at speed v from Earth to, say, Neptune, Fig. 37–8a. The distance between the planets, as measured by the Earth observers, is L_0. The time required for the trip, measured from Earth, is $\Delta t = L_0/v$. In Fig. 37–8b we see the point of view of observers on the spacecraft. In this frame of reference, the spaceship is at rest; Earth and Neptune move with speed v. (We assume v is much greater than the relative speed of Neptune and Earth, so the latter can be ignored.) The time between the departure of Earth and arrival of Neptune (as observed from the spacecraft) is the "proper time" (since the two events occur at the same point in space—i.e., on the spacecraft). Therefore the time interval is less for the spacecraft observers than for the Earth observers, because of time dilation. From Eq. 37–1, the time for the trip as viewed by the spacecraft is $\Delta t_0 = \Delta t\sqrt{1 - v^2/c^2}$. Since the spacecraft observers measure the same speed but less time between these two events, they must also measure the distance as less. If we let L be the distance between the planets as viewed by the spacecraft observers, then $L = v\,\Delta t_0$. We have already seen that $\Delta t_0 = \Delta t\sqrt{1 - v^2/c^2}$ and $\Delta t = L_0/v$, so we have $L = v\,\Delta t_0 = v\,\Delta t\sqrt{1 - v^2/c^2} = L_0\sqrt{1 - v^2/c^2}$. That is,

Length-contraction formula

$$L = L_0\sqrt{1 - \frac{v^2}{c^2}}. \tag{37–2}$$

This is a general result of the special theory of relativity and applies to lengths of objects as well as to distance. The result can be stated most simply in words as:

the length of an object is measured to be shorter when it is moving relative to the observer than when it is at rest.

Length contraction: moving objects are shorter (in the direction of motion)

This is called **length contraction**. The length L_0 in Eq. 37–2 is called the **proper length**. It is the length of the object—or distance between two points whose *positions are measured at the same time*—as determined by observers at rest with respect to it. Equation 37–2 gives the length L that will be measured by observers when the object travels past them at speed v. It is important to note, however, that length contraction occurs *only along the direction of motion*. For example, the moving spaceship in Fig. 37–8a is shortened in length, but its height is the same as when it is at rest.

Length contraction, like time dilation, is not noticeable in everyday life because the factor $\sqrt{1 - v^2/c^2}$ in Eq. 37–2 differs from 1.00 significantly only when v is very large.

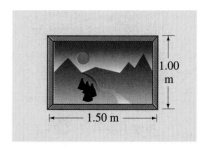

(a)

EXAMPLE 37–3 **Painting's contraction.** A rectangular painting measures 1.00 m tall and 1.50 m wide. It is hung on the side wall of a spaceship which is moving past the Earth at a speed of 0.90 c. See Fig. 37–9a. (a) What are the dimensions of the picture according to the captain of the spaceship? (b) What are the dimensions as seen by an observer on the Earth?

SOLUTION (a) The painting (as well as everything else in the spaceship) looks perfectly normal to everyone on the spaceship, so the captain sees a 1.00 m by 1.50 m painting.
(b) Only the dimension in the direction of motion is shortened, so the height is unchanged at 1.00 m, Fig. 37–9b. The length, however, is contracted to

$$L = L_0 \sqrt{1 - \frac{v^2}{c^2}}$$
$$= (1.50 \text{ m}) \sqrt{1 - (0.90)^2} = 0.65 \text{ m}.$$

So the picture has dimensions 1.00 m × 0.65 m.

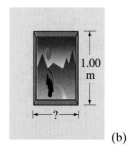

(b)

FIGURE 37–9 Example 37–3.

The appearance *of moving objects*

The Appearance of Objects Moving at Relativistic Speeds

Equation 37–2 tells us what the length of an object will be *measured* to be when traveling at speed v. The *appearance* of the object is another matter. Suppose, for example, you are traveling to the left past a small building at speed $v = 0.85 c$. This is equivalent to the building moving past you to the right at speed v. The building will look narrower (and the same height), but you will also be able to see the side of the building even if you are directly in front of it. This is shown in Fig. 37–10b—part (a) shows the building at rest. The fact that you see the side is not really a relativistic effect, but is due to the finite speed of light. To see how this occurs, we look at Fig. 37–10c which is a top view of the building, looking down. At the instant shown, the observer O is directly in front of the building. Light from points A and B reach O at the same time. If the building were at rest, light from point C could never reach O. But the building is moving at very high speed and does "get out of the way" so that light from C can reach O. Indeed, at the instant shown, light from point C when it was at an earlier location (C′ on the diagram) can reach O because the building has moved. In order to reach the observer at the same time as light from A and B, light from C had to leave at an earlier time since it must travel a greater distance. Thus it is light from C′ that reaches the observer at the same time as light from A and B. This, then, is how an observer might see both the front and side of an object at the same time even when directly in front of it.[†] It can be shown, by the same reasoning, that spherical objects will actually still have a circular outline even at high speeds. That is why the planets in Fig. 37–8b are drawn round rather than contracted.

[†]It would be an error to think that the building in Fig. 37–10b would look rotated. This is not correct since in that case side A would look shorter than side B. In fact, if the observer is directly in front, these sides appear equal in height. Thus the building looks contracted in its front face, but we also see the side, as described above. Also, though not shown in Fig. 37–10b, the walls of the building would appear curved, because of differing distances from the observer's eye of the various points from top to bottom along a vertical wall.

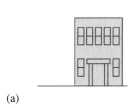

(a)

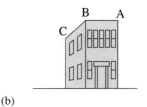

(b)

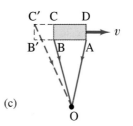

(c)

FIGURE 37–10 Building seen (a) at rest, and (b) moving at high speed. (c) Diagram explains why the side of the building is seen (see the text).

37-7 Four-Dimensional Space–Time

Let us imagine a person is on a train moving at a very high speed, say $0.65\,c$, Fig. 37–11. This person begins a meal at 7:00 and finishes at 7:15, according to a clock on the train. The two events, beginning and ending the meal, take place at the same point on the train. So the proper time between these two events is 15 min. To observers on Earth, the meal will take longer—20 min according to Eq. 37–1. Let us assume that the meal was served on a 20-cm-diameter plate. To observers on the Earth, the plate is only 15 cm wide (length contraction). Thus, to observers on the Earth, the meal looks smaller but lasts longer.

In a sense these two effects, time dilation and length contraction, balance each other. When viewed from the Earth, what the meal seems to lose in size it gains in length of time it lasts. Space, or length, is exchanged for time.

Considerations like this led to the idea of **four-dimensional space–time**: space takes up three dimensions and time is a fourth dimension. Space and time are intimately connected. Just as when we squeeze a balloon we make one dimension larger and another smaller, so when we examine objects and events from different reference frames, a certain amount of space is exchanged for time, or vice versa.

Although the idea of four dimensions may seem strange, it refers to the idea that any object or event is specified by four quantities—three to describe where in space, and one to describe when in time. The really unusual aspect of four-dimensional space–time is that space and time can intermix: a little of one can be exchanged for a little of the other when the reference frame is changed.

It is difficult for most of us to understand the idea of four-dimensional space–time. Somehow we feel, just as physicists did before the advent of relativity, that space and time are completely separate entities. Yet we have found in our thought experiments that they are not completely separate. Our difficulty in accepting this is reminiscent of the situation in the seventeenth century at the time of Galileo and Newton. Before Galileo, the vertical direction, that in which objects fall, was considered to be distinctly different from the two horizontal dimensions. Galileo showed that the vertical dimension differs only in that it happens to be the direction in which gravity acts. Otherwise, all three dimensions are equivalent, a viewpoint we all accept today. Now we are asked to accept one more dimension, time, which we had previously thought of as being somehow different. This is not to say that there is no distinction between space and time. What relativity has shown is that space and time determinations are not independent of one another.

37-8 Galilean and Lorentz Transformations

We now examine in detail the mathematics of relating quantities in one inertial reference frame to the equivalent quantities in another. In particular, we will see how positions and velocities *transform* (that is, change) from one frame to the other.

We begin with the classical or Galilean viewpoint. Consider two reference frames S and S′ which are each characterized by a set of coordinate axes, Fig. 37–12.

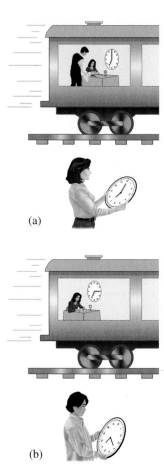

(a)

(b)

FIGURE 37–11 According to an accurate clock on a fast-moving train, a person (a) begins dinner at 7:00 and (b) finishes at 7:15. At the beginning of the meal, observers on Earth set their watches to correspond with the clock on the train. These observers measure the eating time as 20 minutes.

FIGURE 37–12 Inertial reference frame S′ moves to the right at speed v with respect to frame S.

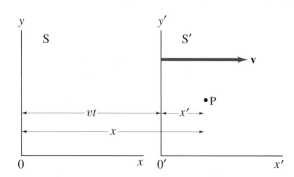

The axes x and y (z is not shown) refer to S and x' and y' to S'. The x' and x axes overlap one another, and we assume that frame S' moves to the right (in the x direction) at speed v with respect to S. And for simplicity let us assume the origins 0 and 0' of the two reference frames are superimposed at time $t = 0$.

Now consider an event that occurs at some point P (Fig. 37–12) represented by the coordinates x', y', z' in reference frame S' at the time t'. What will be the coordinates of P in S? Since S and S' overlap precisely initially, after a time t', S' will have moved a distance vt'. Therefore, at time t', $x = x' + vt'$. The y and z coordinates, on the other hand, are not altered by motion along the x axis; thus $y = y'$ and $z = z'$. Finally, since time is assumed to be absolute in Galilean–Newtonian physics, clocks in the two frames will agree with each other; so $t = t'$. We summarize these in the following **Galilean transformation equations**:

$$
\begin{aligned}
x &= x' + vt' \\
y &= y' \\
z &= z' \\
t &= t'.
\end{aligned}
\qquad \text{[Galilean]} \quad \textbf{(37–3)}
$$

Galilean transformation

These equations give the coordinate of an event in the S frame when those in the S' frame are known. If those in the S frame are known, then the S' coordinates are obtained from

$$
x' = x - vt, \qquad y' = y, \qquad z' = z, \qquad t' = t. \qquad \text{[Galilean]}
$$

These four equations are the "inverse" transformation and are very easily obtained from Eqs. 37–3. Notice that the effect is merely to exchange primed and unprimed quantities and replace v by $-v$. This makes sense because from the S' frame, S moves to the left (negative x direction) with speed v.

Now suppose the point P in Fig. 37–12 represents a particle that is moving. Let the components of its velocity vector in S' be u_x', u_y', u_z'. (We use u to distinguish it from the relative velocity of the two frames, v.) Now $u_x' = dx'/dt'$, $u_y' = dy'/dt'$ and $u_z' = dz'/dt'$. The velocity of P as seen from S will have components u_x, u_y, and u_z. We can show how these are related to the velocity components in S' by differentiating Eqs. 37–3. For u_x we get

$$
u_x = \frac{dx}{dt} = \frac{d(x' + vt')}{dt'} = u_x' + v
$$

since v is assumed constant. For the other components, $u_y' = u_y$ and $u_z' = u_z$, so we have

$$
\begin{aligned}
u_x &= u_x' + v \\
u_y &= u_y' \\
u_z &= u_z'.
\end{aligned}
\qquad \text{[Galilean]} \quad \textbf{(37–4)}
$$

Galilean velocity transformation

These are known as the **Galilean velocity transformation equations**. We see that the y and z components of velocity are unchanged, but the x components differ by v: $u_x = u_x' + v$. This is just what we have used before when dealing with relative velocity.

The Galilean transformations, Eqs. 37–3 and 37–4, are valid only when the velocities involved are much less than c. We can see, for example, that the first of Eqs. 37–4 will not work for the speed of light; for light traveling in S' with speed $u_x' = c$ will have speed $c + v$ in S, whereas the theory of relativity insists it must be c in S. Clearly, then, a new set of transformation equations is needed to deal with relativistic velocities.

We will derive the required equations in a simple way, again looking at Fig. 37–12.

We assume that the transformation is linear and of the form

$$x = \gamma(x' + vt'), \qquad y = y', \qquad z = z'. \tag{i}$$

That is, we modify the first of Eqs. 37–3 by multiplying by a constant γ which is yet to be determined ($\gamma = 1$ non-relativistically). But we assume the y and z equations are unchanged since there is no length contraction in these directions. We will not assume a form for t, but will derive it. The inverse equations must have the same form with v replaced by $-v$. (The principle of relativity demands it, since S′ moving to the right with respect to S is equivalent to S moving to the left with respect to S′.) Therefore

$$x' = \gamma(x - vt). \tag{ii}$$

Now if a light pulse leaves the common origin of S and S′ at time $t = t' = 0$, after a time t it will have traveled a distance $x = ct$ or $x' = ct'$ along the x axis. Therefore, from Eqs. (i) and (ii) above,

$$ct = \gamma(ct' + vt') = \gamma(c + v)t', \tag{iii}$$

$$ct' = \gamma(ct - vt) = \gamma(c - v)t. \tag{iv}$$

We substitute t' from Eq. (iv) into Eq. (iii) and find $ct = \gamma(c + v)\gamma(c - v)(t/c) = \gamma^2(c^2 - v^2)t/c$. We cancel out the t on each side and solve for γ to find

$$\gamma = \frac{1}{\sqrt{1 - v^2/c^2}}.$$

Now that we have found γ, we need only find the relation between t and t'. To do so, we combine $x' = \gamma(x - vt)$ with $x = \gamma(x' + vt')$:

$$x' = \gamma(x - vt) = \gamma(\gamma[x' + vt'] - vt).$$

We solve for t and find $t = \gamma(t' + vx'/c^2)$. In summary,

Lorentz

transformation

equations

$$x = \gamma(x' + vt') \quad = \frac{1}{\sqrt{1 - v^2/c^2}}(x' + vt')$$

$$y = y'$$

$$z = z' \tag{37–5}$$

$$t = \gamma\left(t' + \frac{vx'}{c^2}\right) \quad = \frac{1}{\sqrt{1 - v^2/c^2}}\left(t' + \frac{vx'}{c^2}\right).$$

These are called the **Lorentz transformation equations**. They were first proposed, in a slightly different form, by Lorentz in 1904 to explain the null result of the Michelson–Morley experiment and to make Maxwell's equations take the same form in all inertial reference frames. A year later Einstein derived them independently based on his theory of relativity. Notice that not only is the x equation modified as compared to the Galilean transformation, but so is the t equation; indeed, we see directly in this last equation how the space and time coordinates mix.

EXAMPLE 37–4 **Deriving length contraction.** Derive the length contraction formula, Eq. 37–2, from the Lorentz transformation equations.

SOLUTION Let an object of length L_0 be at rest on the x axis in S. The coordinates of its two end points are x_1 and x_2, so that $x_2 - x_1 = L_0$. At any instant in S′, the end points will be at x_1' and x_2' as given by the Lorentz transformation equations. The length measured in S′ is $L = x_2' - x_1'$. An observer in S′ measures this length by measuring x_2' and x_1' at the same time (in the S′ reference frame), so $t_2' = t_1'$. Then, from the first of Eqs. 37–5,

$$L_0 = x_2 - x_1 = \frac{1}{\sqrt{1 - v^2/c^2}}(x_2' + vt_2' - x_1' - vt_1').$$

Since $t_2' = t_1'$, we have

$$L_0 = \frac{1}{\sqrt{1 - v^2/c^2}}(x_2' - x_1') = \frac{L}{\sqrt{1 - v^2/c^2}},$$

or

$$L = L_0\sqrt{1 - v^2/c^2},$$

which is Eq. 37–2.

EXAMPLE 37-5 **Deriving time dilation.** Derive the time-dilation formula, Eq. 37–1, using the Lorentz transformation equations.

SOLUTION The time Δt_0 between two events that occur at the same place $(x_2' = x_1')$ in S' is measured to be $\Delta t_0 = t_2' - t_1'$. Since $x_2' = x_1'$, then from the last of Eqs. 37–5, the time Δt between the events as measured in S is

$$\Delta t = t_2 - t_1 = \frac{1}{\sqrt{1 - v^2/c^2}}\left(t_2' + \frac{vx_2'}{c^2} - t_1' - \frac{vx_1'}{c^2}\right)$$

$$= \frac{1}{\sqrt{1 - v^2/c^2}}(t_2' - t_1') = \frac{\Delta t_0}{\sqrt{1 - v^2/c^2}},$$

which is Eq. 37–1. Notice that we chose S' to be the frame in which the two events occur at the same place, so that $x_1' = x_2'$ and the terms containing x_1' and x_2' cancel out.

The relativistically correct velocity equations are readily obtained by differentiating Eqs. 37–5 with respect to time. For example (using $\gamma = 1/\sqrt{1 - v^2/c^2}$ and the chain rule for derivatives):

$$u_x = \frac{dx}{dt} = \frac{d}{dt}\left[\gamma(x' + vt')\right]$$

$$= \frac{d}{dt'}\left[\gamma(x' + vt')\right]\frac{dt'}{dt} = \gamma\left[\frac{dx'}{dt'} + v\right]\frac{dt'}{dt}.$$

But $dx'/dt' = u_x'$ and $dt'/dt = 1/(dt/dt') = 1/\left[\gamma(1 + vu_x'/c^2)\right]$ where we have differentiated the last of Eqs. 37–5 with respect to time. Therefore

$$u_x = \frac{\left[\gamma(u_x' + v)\right]}{\left[\gamma(1 + vu_x'/c^2)\right]} = \frac{u_x' + v}{1 + vu_x'/c^2}.$$

The others are obtained in the same way and we collect them here:

$$u_x = \frac{u_x' + v}{1 + vu_x'/c^2} \qquad \text{(37–6a)}$$

$$u_y = \frac{u_y'\sqrt{1 - v^2/c^2}}{1 + vu_x'/c^2} \qquad \text{(37–6b)}$$

$$u_z = \frac{u_z'\sqrt{1 - v^2/c^2}}{1 + vu_x'/c^2}. \qquad \text{(37–6c)}$$

Relativistic

velocity

transformation

Note that even though the relative velocity **v** is in the x direction, the transformation of all the components of a particle's velocity are affected by v and the x component of the particle's velocity; this was not true for the Galilean transformation, Eqs. 37–4.

FIGURE 37-13 Rocket 2 is fired from rocket 1 with speed $u' = 0.60c$. What is the speed of rocket 2 with respect to the Earth?

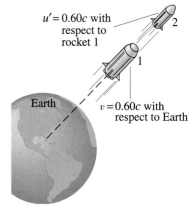

EXAMPLE 37-6 **Adding velocities.** Calculate the speed of rocket 2 in Fig. 37–13 with respect to Earth.

SOLUTION Rocket 2 moves with speed $u' = 0.60c$ with respect to rocket 1. Rocket 1 has speed $v = 0.60c$ with respect to Earth. The velocities are along the same straight line which we take to be the x (and x') axis. We need use only the first of Eqs. 37–6. Then the speed of rocket 2 with respect to Earth is

$$u = \frac{u' + v}{1 + \frac{vu'}{c^2}} = \frac{0.60c + 0.60c}{1 + \frac{(0.60c)(0.60c)}{c^2}} = \frac{1.20c}{1.36} = 0.88c.$$

(The Galilean transformation would have given $u = 1.20c$.)

Notice that Eqs. 37–6 reduce to the classical (Galilean) forms for velocities small compared to the speed of light, since $1 + vu'/c^2 \approx 1$ for v and $u' \ll c$. At the other extreme, let rocket 1 in Fig. 37–13 send out a beam of light, so that $u' = c$. Then Eq. 37–6a tells us the speed of light relative to Earth is

$$u = \frac{0.60\,c + c}{1 + \dfrac{(0.60\,c)(c)}{c^2}} = c,$$

which is consistent with the second postulate of relativity.

37–9 | Relativistic Momentum and Mass

So far in this chapter, we have seen that two basic mechanical quantities, length and time intervals, need modification because they are relative—their value depends on the reference frame from which they are measured. We might expect that other physical quantities might need some modification according to the theory of relativity, such as momentum, energy, and mass.

Let us first examine momentum, which classically is defined as mass times velocity, $\mathbf{p} = m\mathbf{v}$. Classically, momentum is a conserved quantity. It would be a great benefit if the law of conservation of momentum were still valid in the relativistic domain, and we will try to insist that it does as we investigate any possible modification to the definition of momentum. A good guess as to how momentum might be altered would be the use of the factor $\gamma = 1/\sqrt{1 - v^2/c^2}$, as in Eqs. 37–1, 2, and 37–5. But let us be a little more general and let momentum be defined by $p = fmv$ where f is some function of v, $f(v)$. Now we consider a hypothetical collision between two objects—a thought experiment—and see what form $f(v)$ must take if momentum is to be conserved.

Our thought experiment involves the elastic collision of two *identical* balls. If the two balls travel at the same speed, v, we can safely say that they will have the same magnitude of momentum.

The collision in this thought experiment takes place as follows. We consider two inertial reference frames, S and S′, moving along the x axis with a speed v with respect to each other. In reference frame S, a ball (call it ball A) is thrown with speed u along the y axis. In reference frame S′, a second ball (B) is thrown with speed u along the negative y' axis. The two balls are thrown at just the right time so that they collide. We assume that they rebound elastically and, from symmetry, that each moves with the same speed u back along the y axis in its thrower's reference frame. Figure 37–14a shows the collision as seen by the person in reference frame S; and Fig. 37–14b shows the collision as seen from reference frame S′. In reference frame S ball A has velocity $+u$ along the y axis before the collision and $-u$ along the y axis after the collision. In frame S′, ball A has, both before and after the collision, an x component of velocity equal to v, and a y component (see Eq. 37–6b with $u'_x = 0$) of magnitude

$$u\sqrt{1 - v^2/c^2}.$$

The same holds true for ball B, except in reverse. The velocity components are indicated in Fig. 37–14. Let us further assume that $u \ll v$, so that the speed of ball A is essentially v as seen in reference frame S′, and thus A's momentum can be written $f(v)mv$. Similarly, the momentum of B in reference frame S is $f(v)mv$. We now apply the law of conservation of momentum, which we hope remains valid in relativity, even if momentum has to be redefined. That is, we assume that the total

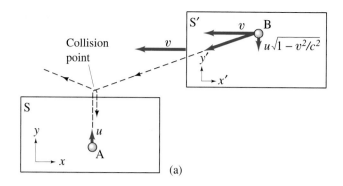

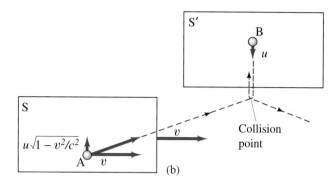

FIGURE 37–14 Deriving the momentum formula. Collision as seen (a) from reference frame S, (b) from reference frame S′.

momentum before the collision is equal to the total after the collision. We apply this to the y component of momentum in reference frame S (Fig. 37–14a):

$$f(u)mu - f(v)mu \sqrt{1 - v^2/c^2} = -f(u)mu + f(v)mu \sqrt{1 - v^2/c^2},$$

where we use $f(u)$ for ball A because its speed in S is only u.

We solve this for $f(v)$ and obtain

$$f(v) = \frac{f(u)}{\sqrt{1 - v^2/c^2}}.$$

This relation is valid for any u and v, (as long as $u \ll v$), and to simplify it so we can solve for f let us consider what happens if we let u become very small so that it approaches zero (this corresponds to a glancing collision with one of the balls essentially at rest and the other moving with speed v). Then the momentum terms $f(u)mu$ are in the nonrelativistic realm and take on the classical form, simply mu, meaning that $f(u) = 1$. So the previous equation becomes

$$f(v) = \frac{1}{\sqrt{1 - v^2/c^2}}.$$

We see that $f(v)$ comes out to be the factor we used before and called γ, and here has been shown to be valid for ball A. Using Fig. 37–14b we can derive the same relation for ball B. Thus we can conclude that we need to define the relativistic momentum of a particle as

Relativistic momentum

$$\mathbf{p} = \frac{m\mathbf{v}}{\sqrt{1 - v^2/c^2}} = \gamma m\mathbf{v}. \tag{37–7}$$

With this definition, as we saw in our thought experiment, the law of conservation of momentum will remain valid even in the relativistic realm. This relativistic momentum formula (Eq. 37–7) has been tested countless times on tiny elementary particles and been found valid.

This relativistic definition of momentum, Eq. 37–7, is sometimes interpreted as an increase in the mass of an object. That is, we can keep the form of our classical definition of momentum as

$$\mathbf{p} = m_{\text{rel}}\mathbf{v},$$

but only if we interpret m_{rel} to be the **relativistic mass**, which increases with speed according to

$$m_{\text{rel}} = \frac{m}{\sqrt{1 - v^2/c^2}}. \qquad (37\text{–}8)$$

Rest mass In this "mass-increase" formula, m is referred to as the **rest mass** of the object— the mass it has as measured in a reference frame in which it is at rest; and m_{rel} is the object's relativistic mass in a reference frame in which it moves at speed v. With this interpretation, *the mass of an object appears to increase as its speed increases.* But we must be careful in the use of relativistic mass. We cannot just plug it into formulas like $F = ma$ or $K = \frac{1}{2}mv^2$ (as we can here for momentum, obtaining Eq. 37–7). For example, if we substitute Eq. 37–8 into $F = ma$, we obtain a formula that does not agree with experiment. If however, we write Newton's second law in its more general form, $\mathbf{F} = d\mathbf{p}/dt$, we do get a correct result. That is, Newton's second law, stated in its most general form is

$$\mathbf{F} = \frac{d\mathbf{p}}{dt} = \frac{d}{dt}(\gamma m\mathbf{v}) = \frac{d}{dt}\left(\frac{m\mathbf{v}}{\sqrt{1 - v^2/c^2}}\right) \qquad (37\text{–}9)$$

and is valid relativistically.[†]

Whenever we talk about the mass of an object, we will always mean its rest mass (a fixed value). On the rare occasion when we want to refer to an object's relativistic mass, we will say so explicitly.

37–10 | The Ultimate Speed

A basic result of the special theory of relativity is that the speed of an object cannot equal or exceed the speed of light. That the speed of light is a natural speed limit in the universe can be seen from any of Eqs. 37–1, 37–2, 37–7, or the addition of velocities formula. It is perhaps easiest to see from Eq. 37–7. As an object is accelerated to greater and greater speeds, its momentum becomes larger and larger. Indeed, if v were to equal c, the denominator in this equation would be zero, and the momentum would be infinite. To accelerate an object up to $v = c$ would thus require infinite energy, and so is not possible.

37–11 | Energy and Mass; $E = mc^2$

When a steady net force is applied to an object, the object increases in speed. Since the force is acting through a distance, work is done on the object and its energy increases. As the speed of the object approaches c, the speed cannot increase indefinitely since it cannot exceed c. On the other hand, the relativistic mass of the object increases with increasing speed. That is, the work done on an object not only increases its speed but also contributes to increasing its inertia. Normally, the work done on an object increases its energy. This new twist from the theory of relativity leads to the idea that mass is a form of energy, a crucial part of Einstein's theory of relativity.

[†]Clearly, Newton's second law written as $\mathbf{F} = m\mathbf{a}$ is not valid, even if we use the relativistic mass. There is an extra term:

$$\mathbf{F} = \frac{d\mathbf{p}}{dt} = \frac{d(\gamma m\mathbf{v})}{dt} = \frac{d(m_{\text{rel}}\mathbf{v})}{dt} = m_{\text{rel}}\mathbf{a} + \frac{dm_{\text{rel}}}{dt}\mathbf{v}.$$

To find the mathematical relationship between mass and energy, we assume that the work-energy theorem is still valid in relativity, and we take the motion to be along the x axis. The work done to increase a particle's speed from zero to v is

$$W = \int_i^f F\, dx = \int_i^f \frac{dp}{dt}\, dx = \int_i^f \frac{dp}{dt}\, v\, dt = \int_i^f v\, dp$$

where i and f refer to the initial $(v=0)$ and final $(v=v)$ states. Since $d(pv) = p\, dv + v\, dp$ we can write

$$v\, dp = d(pv) - p\, dv$$

so

$$W = \int_i^f d(pv) - \int_i^f p\, dv.$$

Since integration is the exact inverse of differentiation, the first term on the right becomes

$$\int_i^f d(pv) = pv\Big|_i^f = (\gamma m v)v = \frac{mv^2}{\sqrt{1 - v^2/c^2}}.$$

The second term in our equation for W above is easily integrated since

$$\frac{d}{dv}\left(\sqrt{1 - v^2/c^2}\right) = -(v/c^2)/\sqrt{1 - v^2/c^2},$$

and so becomes

$$-\int_i^f p\, dv = -\int_0^v \frac{mv}{\sqrt{1 - v^2/c^2}}\, dv = mc^2\sqrt{1 - v^2/c^2}\,\Big|_0^v$$
$$= mc^2\sqrt{1 - v^2/c^2} - mc^2.$$

Finally, we have for W:

$$W = \frac{mv^2}{\sqrt{1 - v^2/c^2}} + mc^2\sqrt{1 - v^2/c^2} - mc^2.$$

We multiply the second term by $\sqrt{1 - v^2/c^2}/\sqrt{1 - v^2/c^2} = 1$, and obtain

$$W = \frac{mc^2}{\sqrt{1 - v^2/c^2}} - mc^2.$$

By the work-energy theorem, the work done must equal the final kinetic energy K since the particle started from rest. Therefore

$$K = \frac{mc^2}{\sqrt{1 - v^2/c^2}} - mc^2 \qquad \text{(37–10a)}$$

$$= \gamma mc^2 - mc^2 = (\gamma - 1)mc^2. \qquad \text{(37–10b)}$$

Relativistic kinetic energy

Clearly, the kinetic energy K is not $\frac{1}{2}mv^2$ at high speeds. $\left[\frac{1}{2}mv^2$ does not work for m as rest mass, nor for relativistic mass.$\right]$

Equations 37–10 require some interpretation. First of all, what does the second term in Eq. 37–10a mean, the mc^2? Consistent with the idea that mass is a form of energy, Einstein called mc^2 the **rest energy** of the object. We can rearrange Eq. 37–10a to get

Rest energy, mc^2

$$E = K + mc^2, \qquad \text{(37–11a)}$$

where

$$E = \gamma mc^2 = \frac{mc^2}{\sqrt{1 - v^2/c^2}} \qquad \text{(37–11b)}$$

Total energy (defined)

is called the *total energy* E of the particle (assuming no potential energy), and equals the rest energy plus the kinetic energy.

For a particle at rest in a given reference frame, K is zero in Eq. 37–11a, so its total energy is its rest energy:

Mass related to energy

$$E = mc^2. \qquad (37\text{–}12)$$

Here we have Einstein's famous formula, $E = mc^2$. This formula mathematically relates the concepts of energy and mass. But if this idea is to have any meaning from a practical point of view, then mass ought to be convertible to energy and vice versa. That is, if mass is just one form of energy, then it should be convertible to

Mass and energy interchangeable

other forms of energy just as other types of energy are interconvertible. Einstein suggested that this might be possible, and indeed changes of mass to other forms of energy, and vice versa, have been experimentally confirmed countless times. The interconversion of mass and energy is most easily detected in nuclear and elementary particle physics. For example, the neutral pion (π^0) of rest mass 2.4×10^{-28} kg is observed to decay into pure electromagnetic radiation (photons). The π^0 completely disappears in the process. The amount of electromagnetic energy produced is found to be exactly equal to that predicted by Einstein's formula, $E = mc^2$. The reverse process is also commonly observed in the laboratory: electromagnetic radiation under certain conditions can be converted into material particles such as electrons (see Section 38–4 on pair production). On a larger scale, the energy produced in nuclear power plants is a result of the loss in rest mass of the uranium fuel as it undergoes the process called fission. Even the radiant energy we receive from the Sun is an example of $E = mc^2$; the Sun's mass is continually decreasing as it radiates electromagnetic energy outward.

The relation $E = mc^2$ is now believed to apply to all processes, although the changes are often too small to measure. That is, when the energy of a system changes by an amount ΔE, the mass of the system changes by an amount Δm given by

$$\Delta E = (\Delta m)(c^2).$$

In a chemical reaction where heat is gained or lost, the masses of the reactants and the products will be different. Even when water is heated on a stove, the mass of the water increases very slightly.[†]

EXAMPLE 37–7 **Pion's kinetic energy.** A π^0 meson $(m = 2.4 \times 10^{-28}\,\text{kg})$ travels at a speed $v = 0.80\,c = 2.4 \times 10^8\,\text{m/s}$. What is its kinetic energy? Compare to a classical calculation.

SOLUTION We substitute values into Eq. 37–10b

$$K = (\gamma - 1)mc^2$$

where

$$\gamma = \frac{1}{\sqrt{1 - v^2/c^2}} = \frac{1}{\sqrt{1 - (0.80)^2}} = 1.67.$$

Then

$$K = (1.67 - 1)(2.4 \times 10^{-28}\,\text{kg})(3.0 \times 10^8\,\text{m/s})^2$$
$$= 1.4 \times 10^{-11}\,\text{J}.$$

PROBLEM SOLVING

Relativistic kinetic energy

Notice that the units of mc^2 are kg·m²/s², which is the joule. A classical calculation would give $K = \frac{1}{2}mv^2 = \frac{1}{2}(2.4 \times 10^{-28}\,\text{kg})(2.4 \times 10^8\,\text{m/s})^2 = 6.9 \times 10^{-12}$ J, about half as much, but this is not a correct result.

EXAMPLE 37–8 **Energy from nuclear decay.** The energy required or released in nuclear reactions and decays comes from a change in mass between the initial and final particles. In one type of radioactive decay, an atom of uranium $(m = 232.03714\,\text{u})$ decays to an atom of thorium $(m = 228.02873\,\text{u})$ plus an atom of helium $(m = 4.00260\,\text{u})$ where the masses (always rest masses) given are in atomic mass units $(1\,\text{u} = 1.6605 \times 10^{-27}\,\text{kg})$. Calculate the energy released in this decay.

[†]This last example is also easy to understand from the point of view of kinetic theory (Chapter 18): as heat is added, the temperature and therefore the average speed of the molecules increases, and Eq. 37–8 tells us that the relativistic mass of the molecules also increases.

940 CHAPTER 37

SOLUTION The initial mass is 232.03714 u, and after the decay it is 228.02873 u + 4.00260 u = 232.03133 u, so there is a decrease in mass of 0.00581 u. This mass decrease, which equals $(0.00581 \text{ u})(1.66 \times 10^{-27} \text{ kg}) = 9.64 \times 10^{-30} \text{ kg}$, is changed into kinetic energy. Thus

Energy released in nuclear process

$$E = \Delta m c^2 = (9.64 \times 10^{-30} \text{ kg})(3.0 \times 10^8 \text{ m/s})^2 = 8.68 \times 10^{-13} \text{ J}.$$

Since $1 \text{ MeV} = 1.60 \times 10^{-13} \text{ J}$, the energy released is 5.4 MeV.

EXAMPLE 37–9 **Mass change in a chemical reaction.** When two moles of hydrogen and one mole of oxygen react to form two moles of water, the energy released is 484 kJ. How much does the mass decrease in this reaction?

SOLUTION Using Eq. 37–12 we have for the change in mass Δm:

$$\Delta m = \frac{\Delta E}{c^2} = \frac{(-484 \times 10^3 \text{ J})}{(3.00 \times 10^8 \text{ m/s})^2} = -5.38 \times 10^{-12} \text{ kg}.$$

The initial mass of the system is $0.002 \text{ kg} + 0.016 \text{ kg} = 0.018 \text{ kg}$. Thus the change in mass is relatively very tiny and can normally be neglected. [Conservation of mass is usually a reasonable principle to apply to chemical reactions.]

Equation 37–10a for the kinetic energy is

$$K = mc^2\left(\frac{1}{\sqrt{1 - v^2/c^2}} - 1\right).$$

At low speeds, $v \ll c$, we can expand the square root in the denominator using the binomial expansion, $(1 \pm x)^n = 1 \pm nx + n(n-1)x^2/2! + \cdots$. With $n = -\frac{1}{2}$, we get

$$K \approx mc^2\left(1 + \frac{1}{2}\frac{v^2}{c^2} + \cdots - 1\right)$$

$$\approx \tfrac{1}{2}mv^2,$$

where the dots in the first expression represent very small terms in the expansion which we have neglected since we assumed that $v \ll c$. Thus at low speeds, the relativistic form for kinetic energy reduces to the classical form, $K = \frac{1}{2}mv^2$. This is, of course, what we would like. It makes relativity a more viable theory in that it can predict accurate results at low speed as well as at high. Indeed, the other equations of special relativity also reduce to their classical equivalents at ordinary speeds: length contraction, time dilation, and modifications to momentum as well as kinetic energy, all disappear for $v \ll c$ since $\sqrt{1 - v^2/c^2} \approx 1$.

A useful relation between the total energy E of a particle and its momentum p can also be derived. The momentum of a particle of mass m and speed v is given by Eq. 37–7

$$p = \gamma m v = \frac{mv}{\sqrt{1 - v^2/c^2}}.$$

Relativistic momentum

The total energy is

$$E = K + mc^2$$

or

$$E = \gamma mc^2 = \frac{mc^2}{\sqrt{1 - v^2/c^2}}.$$

We square this equation (and we insert "$v^2 - v^2$" which is zero, but will help us):

$$E^2 = \frac{m^2c^2(v^2 - v^2 + c^2)}{1 - v^2/c^2}$$

$$= p^2c^2 + \frac{m^2c^4(1 - v^2/c^2)}{1 - v^2/c^2}$$

or

$$E^2 = p^2c^2 + m^2c^4. \tag{37–13}$$

Energy related to momentum

Thus, the total energy can be written in terms of the momentum p, or in terms of the kinetic energy (Eq. 37–11), where we have assumed there is no potential energy.

We can rewrite Eq. 37–13 as $E^2 - p^2c^2 = m^2c^4$. Since the rest mass m of a given particle is the same in any reference frame, we see that the quantity $E^2 - p^2c^2$ must also be the same in any reference frame. Thus, at any given moment the total energy E and momentum p of a particle will be different in different reference frames, but the quantity $E^2 - p^2c^2$ will have the same value in all inertial reference frames. We say that the quantity $E^2 - p^2c^2$ is **invariant** under a Lorentz transformation.

$E^2 - p^2c^2$
is invariant

*37–12 | Doppler Shift for Light

In Section 16–7 we discussed how the frequency and wavelength of sound are altered if the source of the sound and the observer are moving toward or away from each other. When a source is moving toward us, the frequency is higher than when the source is at rest. If the source moves away from us, the frequency is lower. We obtained four different equations for the Doppler shift (Eqs. 16–9a and b, Eqs. 16–10a and b), depending on the direction of the relative motion and whether the source or the observer is moving. The Doppler effect occurs also for light; but the shifted frequency or wavelength is given by slightly different equations, and there are only two of them, because for light—according to special relativity—we can make no distinction between motion of the source and motion of the observer. (Recall that sound travels in a medium such as air, whereas light does not—there is no evidence for an ether.)

To derive the Doppler shift for light, let us consider a light source and an observer that move toward each other, and let their relative velocity be v as measured in the reference frame of either the source or the observer. Figure 37–15a shows a source at rest emitting light waves of frequency f_0 and wavelength $\lambda_0 = c/f_0$. Two wavecrests are shown, a distance λ_0 apart, the second crest just having been emitted. In Fig. 37–15b, the source is shown moving at speed v toward a stationary observer who will see the wavelength λ being somewhat less than λ_0. (This is much like Fig. 16–20 for sound.) Let Δt represent the time between crests as detected by the observer, whose reference frame is shown in Fig. 37–15b. From Fig. 37–15b we see that

$$\lambda = c\,\Delta t - v\,\Delta t$$

where $c\,\Delta t$ is the distance crest 1 has moved in the time Δt after it was emitted, and $v\,\Delta t$ is the distance the source has moved in time Δt. So far our derivation has not differed from that for sound (Section 16–7). Now we invoke the theory of relativity. The time between emission of wavecrests has undergone time dilation:

$$\Delta t = \Delta t_0 / \sqrt{1 - v^2/c^2}$$

where Δt_0 is the time between emissions of wavecrests in the reference frame where the source is at rest (the "proper" time). In the source's reference frame (Fig. 37–15a), we have

$$\Delta t_0 = \frac{1}{f_0} = \frac{\lambda_0}{c}$$

(Eqs. 3–15 and 32–14). Thus

$$\lambda = (c - v)\,\Delta t = (c - v)\frac{\Delta t_0}{\sqrt{1 - v^2/c^2}} = \frac{(c - v)}{\sqrt{c^2 - v^2}}\lambda_0$$

or

$$\lambda = \lambda_0\sqrt{\frac{c - v}{c + v}}. \qquad \begin{bmatrix} \text{source and observer} \\ \text{moving toward} \\ \text{each other} \end{bmatrix} \qquad \textbf{(37–14a)}$$

The frequency f is (recall $\lambda_0 = c/f_0$)

$$f = \frac{c}{\lambda} = f_0\sqrt{\frac{c + v}{c - v}}. \qquad \begin{bmatrix} \text{source and observer} \\ \text{moving toward} \\ \text{each other} \end{bmatrix} \qquad \textbf{(37–14b)}$$

Doppler shift for light

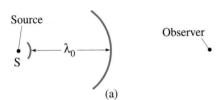

Source
S
λ_0

Observer

(a)

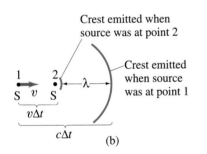

Crest emitted when source was at point 2

Crest emitted when source was at point 1

1   2
S v S
λ
$v\Delta t$
$c\Delta t$

(b)

FIGURE 37–15 Doppler shift for light. (a) Reference frame of light source. (b) Reference frame of observer, toward whom source is moving.

Here f_0 is the frequency of the light as seen in the source's reference frame, and f is the frequency as measured by an observer moving toward the source or toward whom the source is moving. Equations 37–14 depend only on the relative velocity v. For relative motion *away* from each other we set $v < 0$ in Eqs. 37–14, and obtain

$$\lambda = \lambda_0 \sqrt{\frac{c+v}{c-v}} \qquad \begin{bmatrix} \text{source and observer} \\ \text{moving away from} \\ \text{each other} \end{bmatrix} \quad \textbf{(37–15a)}$$

$$f = f_0 \sqrt{\frac{c-v}{c+v}}. \qquad\qquad\qquad\qquad\qquad \textbf{(37–15b)}$$

From Eqs. 37–14 and 37–15 we see that light from a source moving toward us will have a higher frequency and shorter wavelength, whereas if a light source moves away from us, we will see a lower frequency and a longer wavelength. In the latter case, visible light will have its wavelength lengthened toward the red end of the visible spectrum (Fig. 33–24), an effect called a **redshift**. As we will see in the next chapter, all atoms have their own distinctive signature in terms of the frequencies of the light they emit. In 1929 the American astronomer Edwin Hubble (1889–1953) found that radiation from atoms in many galaxies is redshifted. That is, the frequencies of light emitted are lower than those emitted by stationary atoms on Earth, suggesting that the galaxies are receding from us. This is the origin of the idea that the universe is expanding.

37–13 | The Impact of Special Relativity

A great many experiments have been performed to test the predictions of the special theory of relativity. Within experimental error, no contradictions have been found. Scientists have therefore accepted relativity as an accurate description of nature.

At speeds much less than the speed of light, the relativistic formulas reduce to the old classical ones, as we have discussed. We would, of course, hope—or rather, insist—that this be true since Newtonian mechanics works so well for objects moving with speeds $v \ll c$. This insistence that a more general theory (such as relativity) give the same results as a more restricted theory (such as classical mechanics which works for $v \ll c$) is called the **correspondence principle**. The two theories must correspond where their realms of validity overlap. Relativity thus does not contradict classical mechanics. Rather, it is a more general theory, of which classical mechanics is now considered to be a limiting case.

Correspondence principle

The importance of relativity is not simply that it gives more accurate results, especially at very high speeds. Much more than that, it has changed the way we view the world. The concepts of space and time are now seen to be relative, and intertwined with one another, whereas before they were considered absolute and separate. Even our concepts of matter and energy have changed: either can be converted to the other. The impact of relativity extends far beyond physics. It has influenced the other sciences, and even the world of art and literature; it has, indeed, entered the general culture.

From a practical point of view, we do not have much opportunity in our daily lives to use the mathematics of relativity. For example, the factor $\sqrt{1 - v^2/c^2}$, which appears in many relativistic formulas, has a value of 0.995 when $v = 0.10c$. Thus, for speeds even as high as $0.10c = 3.0 \times 10^7$ m/s, the factor $\sqrt{1 - v^2/c^2}$ in relativistic formulas gives a numerical correction of less than 1 percent. For speeds less than $0.10c$, or unless mass and energy are interchanged, we don't usually need to use the more complicated relativistic formulas, and can use the simpler classical formulas.

Summary

An **inertial reference frame** is one in which Newton's law of inertia holds. Inertial reference frames can move at constant velocity relative to one another; accelerating reference frames are noninertial.

The **special theory of relativity** is based on two principles: the **relativity principle**, which states that the laws of physics are the same in all inertial reference frames, and the principle of the **constancy of the speed of light**, which states that the speed of light in empty space has the same value in all inertial reference frames.

One consequence of relativity theory is that two events that are simultaneous in one reference frame may not be simultaneous in another. Other effects are **time dilation**: moving clocks are measured to run slow; and **length contraction**: the length of a moving object is measured to be shorter (in its direction of motion) than when it is at rest. Quantitatively,

$$L = L_0 \sqrt{1 - v^2/c^2}$$

$$\Delta t = \frac{\Delta t_0}{\sqrt{1 - v^2/c^2}}$$

where L and Δt are the length and time interval of objects (or events) observed as they move by at the speed v; L_0 and Δt_0 are the **proper length** and **proper time**—that is, the same quantities as measured in the rest frame of the objects or events.

The **Lorentz transformations** relate the positions and times of events in one inertial reference frame to their positions and times in a second inertial reference frame.

$$x = \gamma(x' + vt')$$
$$y = y'$$
$$z = z'$$
$$t = \gamma\left(t' + \frac{vx'}{c^2}\right)$$

where $\gamma = 1/\sqrt{1 - v^2/c^2}$.

Velocity addition also must be done in a special way. All these relativistic effects are significant only at high speeds, close to the speed of light, which itself is the ultimate speed in the universe.

The theory of relativity has changed our notions of space and time, and of momentum, energy, and mass. Space and time are seen to be intimately connected, with time being the fourth dimension in addition to the three dimensions of space.

The **momentum** of an object is given by

$$\mathbf{p} = \gamma m \mathbf{v} = \frac{m \mathbf{v}}{\sqrt{1 - v^2/c^2}}.$$

This formula can be interpreted as a *mass increase*, where the relativistic mass is $m_{rel} = \gamma m$ and m is the **rest mass** of the object ($v = 0$).

Mass and energy are interconvertible. The equation

$$E = mc^2$$

tells how much energy E is needed to create a mass m, or vice versa. Said another way, $E = mc^2$ is the amount of energy an object has because of its mass m. The law of conservation of energy must include mass as a form of energy.

The kinetic energy K of an object moving at speed v is given by

$$K = (\gamma - 1)mc^2 = \frac{mc^2}{\sqrt{1 - v^2/c^2}} - mc^2$$

where m is the rest mass of the object. The total energy E is

$$E = K + mc^2 = \gamma mc^2.$$

The momentum p of an object is related to its total energy E (assuming no potential energy) by

$$E^2 = p^2c^2 + m^2c^4.$$

Questions

1. You are in a windowless car in an exceptionally smooth train. Is there any physical experiment you can do in the train car to determine whether you are moving?

2. You might have had the experience of being at a red light when, out of the corner of your eye, you see the car beside you creep forward. Instinctively you stomp on the brake pedal, thinking that you are rolling backward. What does this say about absolute and relative motion?

3. A worker stands on top of a moving railroad car, and throws a heavy ball straight up (from his point of view). Ignoring air resistance, will the ball land on the car or behind it?

4. Does the Earth really go around the Sun? Or is it also valid to say that the Sun goes around the Earth? Discuss in view of the first principle of relativity (that there is no best reference frame).

5. If you were on a spaceship traveling at $0.5c$ away from a star, at what speed would the starlight pass you?

6. Will two events that occur at the same place and same time for one observer be simultaneous to a second observer moving with respect to the first?

7. Analyze the thought experiment of Section 37–4 from O_1's point of view. (Make a diagram analogous to Fig. 37–6.)

8. The time-dilation effect is sometimes expressed as "moving clocks run slowly." Actually, this effect has nothing to do with motion affecting the functioning of clocks. What then does it deal with?

9. Does time dilation mean that time actually passes more slowly in moving reference frames or that it only *seems* to pass more slowly?

10. A young-looking woman astronaut has just arrived home from a long trip. She rushes up to an old gray-haired man and in the ensuing conversation refers to him as her son. How might this be possible?

11. If you were traveling away from Earth at speed $0.5c$, would you notice a change in your heartbeat? Would your mass, height, or waistline change? What would observers on Earth using telescopes say about you?

12. Do time dilation and length contraction occur at ordinary speeds, say 90 km/h?

13. Suppose the speed of light were infinite. What would happen to the relativistic predictions of length contraction and time dilation?

14. Explain how the length-contraction and time-dilation formulas might be used to indicate that c is the limiting speed in the universe.

15. Does the equation $E = mc^2$ conflict with the conservation of energy principle? Explain.

16. If mass is a form of energy, does this mean that a spring has more mass when compressed than when relaxed?

17. It is not correct to say that "matter can neither be created nor destroyed." What must we say instead?

18. Is our intuitive notion that velocities simply add, as we did in Section 3–10, completely wrong?

Problems

Sections 37–4 to 37–6

1. (I) Lengths and time intervals depend on the factor
$$\sqrt{1 - v^2/c^2}$$
according to the theory of relativity (Eqs. 37–1 and 37–2). Evaluate this factor for speeds of: (*a*) $v = 20,000$ m/s (typical speed of a satellite); (*b*) $v = 0.0100c$; (*c*) $v = 0.100c$; (*d*) $v = 0.900c$; (*e*) $v = 0.990c$; (*f*) $v = 0.999c$.

2. (I) A spaceship passes you at a speed of $0.750c$. You measure its length to be 28.2 m. How long would it be when at rest?

3. (I) A certain type of elementary particle travels at a speed of 2.70×10^8 m/s. At this speed, the average lifetime is measured to be 4.76×10^{-6} s. What is the particle's lifetime at rest?

4. (I) If you were to travel to a star 100 light-years from Earth at a speed of 2.50×10^8 m/s, what would you measure this distance to be?

5. (II) What is the speed of a pion if its average lifetime is measured to be 4.10×10^{-8} s? At rest, its mean lifetime is 2.60×10^{-8} s.

6. (II) Suppose you decide to travel to a star 75 light-years away. How fast would you have to travel so the distance would be only 25 light-years?

7. (II) At what speed do the relativistic formulas for length and time intervals differ from classical values by 1.00 percent? (This is a reasonable way to estimate when to do relativistic calculations rather than classical.)

8. (II) Suppose a news report stated that starship *Enterprise* had just returned from a 5-year voyage while traveling at $0.84c$. (*a*) If the report meant 5.0 years of *Earth time*, how much time elapsed on the ship? (*b*) If the report meant 5.0 years of *ship time*, how much time passed on Earth?

9. (II) A certain star is 95.0 light-years away. How long would it take a spacecraft traveling $0.960c$ to reach that star from Earth, as measured by observers: (*a*) on Earth, (*b*) on the spacecraft? (*c*) What is the distance traveled according to observers on the spacecraft? (*d*) What will the spacecraft occupants compute their speed to be from the results of (*b*) and (*c*)?

10. (II) A friend of yours travels by you in her fast sports car at a speed of $0.660c$. It is measured in your frame to be 4.80 m long and 1.25 m high. (*a*) What will be its length and height at rest? (*b*) How many seconds would you say elapsed on your friend's watch when 20.0 s passed on yours? (*c*) How fast did you appear to be traveling according to your friend? (*d*) How many seconds would she say elapsed on your watch when she saw 20.0 s pass on hers?

11. (II) How fast must an average pion be moving to travel 15 m before it decays? The average lifetime, at rest, is 2.6×10^{-8} s.

Section 37–8

12. (I) Suppose in Fig. 37–12 that the origins of S and S' overlap at $t = t' = 0$ and that S' moves at speed $v = 30$ m/s with respect to S. In S', a person is resting at a point whose coordinates are $x' = 25$ m, $y' = 20$ m, and $z' = 0$. Calculate this person's coordinates in S (x, y, z) at (*a*) $t = 2.5$ s, (*b*) $t = 10.0$ s. Use the Galilean transformation.

13. (I) Repeat Problem 12 using the Lorentz transformation and a relative speed $v = 1.80 \times 10^8$ m/s, but choose the time to be (*a*) $2.5 \,\mu$s and (*b*) $10.0 \,\mu$s.

14. (I) A person on a rocket traveling at $0.50c$ (with respect to the Earth) observes a meteor come from behind and pass her at a speed she measures as $0.50c$. How fast is the meteor moving with respect to the Earth?

15. (II) Two spaceships leave Earth in opposite directions, each with a speed of $0.50c$ with respect to Earth. (*a*) What is the velocity of spaceship 1 relative to spaceship 2? (*b*) What is the velocity of spaceship 2 relative to spaceship 1?

16. (II) A spaceship leaves Earth traveling at $0.71c$. A second spaceship leaves the first at a speed of $0.87c$ with respect to the first. Calculate the speed of the second ship with respect to Earth if it is fired (*a*) in the same direction the first spaceship is already moving, (*b*) directly backward toward Earth.

17. (II) In Problem 12, suppose that the person moves with a velocity whose components are $u'_x = u'_y = 25.0$ m/s, What will be her velocity with respect to S? (Give magnitude and direction.)

18. (II) In Problem 13, suppose that the person moves with a velocity (with a rocket) whose components are $u'_x = u'_y = 2.0 \times 10^8$ m/s. What will be her velocity (magnitude and direction) with respect to S?

19. (II) A spaceship traveling at $0.66c$ away from Earth fires a module with a speed of $0.82c$ at right angles to its own direction of travel (as seen by the spaceship). What is the speed of the module, and its direction of travel (relative to the spaceship's direction), as seen by an observer on Earth?

20. (II) If a particle moves in the xy plane of system S (Fig. 37–12) with speed u in a direction that makes an angle θ with the x axis, show that it makes an angle θ' in S' given by $\tan \theta' = (\sin \theta)\sqrt{1 - v^2/c^2}/(\cos \theta - v/u)$.

21. (II) A stick of length L_0, at rest in reference frame S, makes an angle θ with the x axis. In reference frame S', which moves to the right with velocity $\mathbf{v} = v\mathbf{i}$ with respect to S, determine (*a*) the length L of the stick, and (*b*) the angle θ' it makes with the x' axis.

22. (III) In the old West, a marshal riding on a train traveling 50 m/s sees a duel between two men standing on the Earth 50 m apart parallel to the train. The marshal's instruments indicate that in his reference frame the two men fired simultaneously. (*a*) Which of the two men, the first one the train passes (A) or the second one (B) should be arrested for firing the first shot? That is, in the gunfighter's frame of reference, who fired first? (*b*) How much earlier did he fire? (*c*) Who was struck first?

23. (III) A farm boy studying physics believes that he can fit a 13.0-m long pole into a 10.0-m long barn if he runs fast enough, carrying the pole. Can he do it? Explain in detail. How does this fit with the idea that when he is running the barn looks even shorter than 10.0 m?

Section 37–9

24. (I) What is the momentum of a proton traveling at $v = 0.85c$?

25. (I) At what speed will an object's relativistic mass be twice its rest mass?

26. (II) A particle of rest mass m travels at a speed $v = 0.20c$. At what speed will its momentum be doubled?

27. (II) (*a*) A particle travels at $v = 0.10c$. By what percentage will a calculation of its momentum be wrong if you use the classical formula? (*b*) Repeat for $v = 0.50c$.

28. (II) What is the percent change in momentum of a proton that accelerates (*a*) from $0.45c$ to $0.90c$, (*b*) from $0.90c$ to $0.98c$?

Section 37–11

29. (I) A certain chemical reaction requires 4.82×10^4 J of energy input for it to go. What is the increase in mass of the products over the reactants?

30. (I) When a uranium nucleus at rest breaks apart in the process known as fission in a nuclear reactor, the resulting fragments have a total kinetic energy of about 200 MeV. How much mass was lost in the process?

31. (I) Calculate the rest energy of an electron in joules and in MeV $(1 \text{ MeV} = 1.60 \times 10^{-13}$ J$)$.

32. (I) Calculate the rest mass of a proton in MeV/c^2.

33. (I) The total annual energy consumption in the United States is about 8×10^{19} J. How much mass would have to be converted to energy to fuel this need?

34. (II) How much energy can be obtained from conversion of 1.0 gram of mass? How much mass could this energy raise to a height of 100 m?

35. (II) Show that when the kinetic energy of a particle equals its rest energy, the speed of the particle is about $0.866c$.

36. (II) At what speed will an object's kinetic energy be 25 percent of its rest energy?

37. (II) (*a*) How much work is required to accelerate a proton from rest up to a speed of $0.997c$? (*b*) What would be the momentum of this proton?

38. (II) Calculate the kinetic energy and momentum of a proton traveling 2.60×10^8 m/s.

39. (II) What is the momentum of a 750-MeV proton (that is, its kinetic energy is 750 MeV)?

40. (II) What is the speed of a proton accelerated by a potential difference of 95 MV?

41. (II) What is the speed of an electron whose kinetic energy is 1.00 MeV?

42. (II) What is the speed of an electron when it hits a television screen after being accelerated by the 25,000 V of the picture tube?

43. (II) Two identical particles of rest mass m approach each other at equal and opposite speeds, v. The collision is completely inelastic and results in a single particle at rest. What is the rest mass of the new particle? How much energy was lost in the collision? How much kinetic energy is lost in this collision?

44. (II) Calculate the speed of a proton $(m = 1.67 \times 10^{-27}$ kg$)$ whose kinetic energy is exactly half its total energy.

45. (II) What is the speed and momentum of an electron $(m = 9.11 \times 10^{-31}$ kg$)$ whose kinetic energy equals its rest energy?

46. (II) Suppose a spacecraft of mass 27,000 kg is accelerated to $0.21c$. (*a*) How much kinetic energy would it have? (*b*) If you used the classical formula for kinetic energy, by what percentage would you be in error?

47. (II) Calculate the kinetic energy and momentum of a proton $(m = 1.67 \times 10^{-27}$ kg$)$ traveling 8.4×10^7 m/s. By what percentages would your calculations have been in error if you had used classical formulas?

48. (II) The americium nucleus, $^{241}_{95}$Am, decays to a neptunium nucleus, $^{237}_{93}$Np, by emitting an alpha particle of mass 4.00260 u and kinetic energy 5.5 MeV. Estimate the mass of the neptunium nucleus, ignoring its recoil, given that the americium mass is 241.05682 u.

49. (II) An electron $(m = 9.11 \times 10^{-31}$ kg$)$ is accelerated from rest to speed v by a conservative force. In this process, its potential energy decreases by 5.60×10^{-14} J. Determine the electron's speed, v.

50. (II) Make a graph of the kinetic energy versus momentum for (a) a particle of nonzero rest mass, and (b) a particle with zero rest mass.

51. (II) What magnetic field intensity is needed to keep 900-GeV protons revolving in a circle of radius 1.0 km (at, say, the Fermilab synchrotron)? Use the relativistic mass. The proton's rest mass is 0.938 GeV/c^2. $(1$ GeV $= 10^9$ eV.$)$

52. (II) A negative muon traveling at 33 percent the speed of light collides head on with a positive muon traveling at 50 percent the speed of light. The two muons $($each of mass 105.7 MeV/$c^2)$ annihilate, and produce how much electromagnetic energy?

53. (II) Show that the energy of a particle of charge e revolving in a circle of radius r in a magnetic field B is given by $E(\text{in eV}) = Brc$ in the relativistic limit $(v \approx c)$.

54. (II) Show that the kinetic energy K of a particle of rest mass m is related to its momentum p by the equation

$$p = \sqrt{K^2 + 2Kmc^2}/c.$$

55. (III) (a) In reference frame S, a particle has momentum $\mathbf{p} = p_x\mathbf{i}$ along the positive x axis. Show that in frame S', which moves with speed v as in Fig. 37–12, the momentum has components

$$p'_x = \frac{p_x - vE/c^2}{\sqrt{1 - v^2/c^2}}$$

$$p'_y = p_y$$

$$p'_z = p_z$$

$$E' = \frac{E - p_x v}{\sqrt{1 - v^2/c^2}}.$$

(These transformation equations hold, actually, for any direction of $\mathbf{p}$.) (b) Show that p_x, p_y, p_z, E/c transform according to the Lorentz transformation in the same way as x, y, z, ct.

* Section 37–12

* 56. (II) A certain galaxy has a Doppler shift given by $f_0 - f = 0.797 f_0$. How fast is it moving away from us?

* 57. (II) A quasar emits familiar hydrogen lines whose wavelengths are 3.0 times longer than what we measure in the laboratory. (a) What is the speed of this quasar? (b) What result would you obtain if you used the "classical" Doppler shift discussed in Chapter 16?

* 58. (II) A spaceship moving toward Earth at 0.80 c transmits radio signals at 95.0 MHz. At what frequency should Earth receivers be tuned?

* 59. (II) Starting from Eq. 37–15a, show that the Doppler shift in wavelength is

$$\frac{\Delta\lambda}{\lambda} = \frac{v}{c}$$

if $v \ll c$.

General Problems

60. As a rule of thumb, anything traveling faster than about 0.1 c is called *relativistic*—i.e., for which the correction using special relativity is a significant effect. Is the electron in a hydrogen atom (radius 0.5×10^{-10} m) relativistic? (Treat the electron as though it were in a circular orbit around the proton.)

61. An atomic clock is taken to the North Pole, while another stays at the Equator. How far will they be out of synchronization after a year has elapsed?

62. The nearest star to Earth is Proxima Centauri, 4.3 light-years away. (a) At what constant velocity must a spacecraft travel from Earth if it is to reach the star in 4.0 years, as measured by travelers on the spacecraft? (b) How long does the trip take according to Earth observers?

63. Derive a formula showing how the apparent density of an object changes with speed v relative to an observer. [Hint: Use relativistic mass.]

64. An airplane travels 1500 km/h around the world, returning to the same place, in a circle of radius essentially equal to that of the Earth. Estimate the difference in time to make the trip as seen by Earth and airplane observers. [Hint: Use the binomial expansion, Appendix A.]

65. (a) What is the speed of an electron whose kinetic energy is 10,000 times its rest energy? Such speeds are reached in the Stanford Linear Accelerator, SLAC. (b) If the electrons travel in the lab through a tube 3.0 km long (as at SLAC), how long is this tube in the electrons' reference frame?

66. How many grams of matter would have to be totally destroyed to run a 100-W lightbulb for 1 year?

67. What minimum amount of electromagnetic energy is needed to produce an electron and a positron together? A positron is a particle with the same mass as an electron, but has the opposite charge. (Note that electric charge is conserved in this process. See Section 38–4.)

68. A 1.68-kg mass oscillates on the end of a spring whose spring constant is $k = 48.7 \, \text{N/m}$. If this system is in a spaceship moving past Earth at $0.900 c$, what is its period of oscillation according to (a) observers on the ship, and (b) observers on Earth?

69. An electron $(m = 9.11 \times 10^{-31} \, \text{kg})$ enters a uniform magnetic field $B = 1.8 \, \text{T}$, and moves perpendicular to the field lines with a speed $v = 0.92 c$. What is the radius of curvature of its path?

70. An observer on Earth sees an alien vessel approach at a speed of $0.60 c$. The *Enterprise* comes to the rescue (Fig. 37–16), overtaking the aliens while moving directly toward Earth at a speed of $0.90 c$ relative to Earth. What is the relative speed of one vessel as seen by the other?

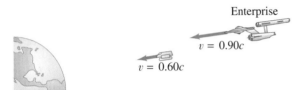

FIGURE 37–16 Problem 70.

71. A free neutron can decay into a proton, an electron, and a neutrino. Assume the neutrino's rest mass is zero, and the other masses can be found in the table inside the front cover. Determine the total kinetic energy shared among the three particles when a neutron decays at rest.

72. The Sun radiates energy at a rate of about $4 \times 10^{26} \, \text{W}$. (a) At what rate is the Sun's mass decreasing? (b) How long does it take for the Sun to lose a mass equal to that of Earth? (c) Estimate how long the Sun could last if it radiated constantly at this rate.

73. An unknown particle is measured to have a negative charge and a speed of $2.24 \times 10^8 \, \text{m/s}$. Its momentum is determined to be $3.07 \times 10^{-22} \, \text{kg·m/s}$. Identify the particle by finding its mass.

74. How much energy would be required to break a helium nucleus into its constituents, two protons and two neutrons? The rest masses of a proton (including an electron), a neutron, and helium are, respectively, 1.00783 u, 1.00867 u, and 4.00260 u. (This is called the *total binding energy* of the ^4_2He nucleus.)

75. What is the percentage increase in the (relativistic) mass of a car traveling 110 km/h as compared to at rest?

76. Two protons, each having a speed of $0.935 c$ in the laboratory, are moving toward each other. Determine (a) the momentum of each proton in the laboratory, (b) the total momentum of the two protons in the laboratory, and (c) the momentum of one proton as seen by the other proton.

77. Show analytically that a particle with momentum p and energy E has a speed given by
$$v = \frac{pc^2}{E} = \frac{pc}{\sqrt{m^2c^2 + p^2}}.$$

78. A slab of glass moves to the right with speed v. A flash of light is emitted at point A (Fig. 37–17) and passes through the glass arriving at point B a distance L away. The glass has thickness d in the reference frame where it is at rest, and the speed of light in the glass is c/n. How long does it take the light to go from point A to point B according to an observer at rest with respect to points A and B? Check your answer for the cases $v = 0$, $n = 1$, and for $v = c$.

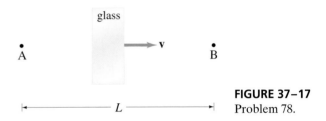

FIGURE 37–17
Problem 78.

79. Show that the space-time "distance" $(c\Delta t)^2 - (\Delta x)^2$ is invariant, meaning that all observers in all inertial reference frames calculate the same number for this quantity for any pair of events.

80. The fictional starship *Enterprise* obtains its power by combining matter and antimatter, achieving complete conversion of mass into energy. If the mass of the *Enterprise* is approximately $5 \times 10^9 \, \text{kg}$, how much mass must be converted into kinetic energy to accelerate it from rest to one-tenth the speed of light?

81. A spacecraft (reference frame S′) moves past Earth (reference frame S) at velocity **v**, which points along the x and x' axes. The spacecraft emits light along its y' axis as shown in Fig. 37–18. (a) What angle does this light make with the x axis in the Earth's reference frame? (b) Show that the light moves with speed c also in the Earth's reference frame (i.e., given c in frame S′). (c) Compare these relativistic results to what you would have obtained classically (Galilean transformations).

FIGURE 37–18 Problem 81.

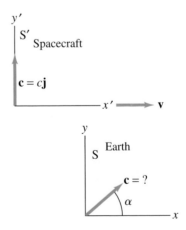

APPENDIX

Mathematical Formulas

A–1 Quadratic Formula

If

$$ax^2 + bx + c = 0$$

then

$$x = \frac{-b \pm \sqrt{b^2 - 4ac}}{2a}$$

A–2 Binomial Expansion

$$(1 \pm x)^n = 1 \pm nx + \frac{n(n-1)}{2!}x^2 \pm \frac{n(n-1)(n-2)}{3!}x^3 + \cdots$$

$$(x + y)^n = x^n\left(1 + \frac{y}{x}\right)^n = x^n\left(1 + n\frac{y}{x} + \frac{n(n-1)}{2!}\frac{y^2}{x^2} + \cdots\right)$$

A–3 Other Expansions

$$e^x = 1 + x + \frac{x^2}{2!} + \frac{x^3}{3!} + \cdots$$

$$\ln(1 + x) = x - \frac{x^2}{2} + \frac{x^3}{3} - \frac{x^4}{4} + \cdots$$

$$\sin\theta = \theta - \frac{\theta^3}{3!} + \frac{\theta^5}{5!} - \cdots$$

$$\cos\theta = 1 - \frac{\theta^2}{2!} + \frac{\theta^4}{4!} - \cdots$$

$$\tan\theta = \theta + \frac{\theta^3}{3} + \frac{2}{15}\theta^5 + \cdots \qquad |\theta| < \frac{\pi}{2}$$

In general:

$$f(x) = f(0) + \left(\frac{df}{dx}\right)_0 x + \left(\frac{d^2f}{dx^2}\right)_0 \frac{x^2}{2!} + \cdots$$

A–4 Areas and Volumes

Object	Surface area	Volume
Circle, radius r	πr^2	—
Sphere, radius r	$4\pi r^2$	$\frac{4}{3}\pi r^3$
Right circular cylinder, radius r, height h	$2\pi r^2 + 2\pi rh$	$\pi r^2 h$
Right circular cone, radius r, height h	$\pi r^2 + \pi r\sqrt{r^2 + h^2}$	$\frac{1}{3}\pi r^2 h$

A–5 Plane Geometry

1.

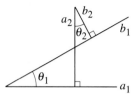

FIGURE A–1

If line a_1 is parallel to line a_2, then $\theta_1 = \theta_2$.

2.

FIGURE A–2

If $a_1 \perp a_2$ and $b_1 \perp b_2$, then $\theta_1 = \theta_2$.

3. The sum of the angles in any plane triangle is 180°.

4. *Pythagorean theorem:*

In any right triangle (one angle = 90°) of sides a, b, and c:
$$a^2 + b^2 = c^2$$
where c is the length of the hypotenuse (opposite the 90° angle).

FIGURE A–3

A–6 Trigonometric Functions and Identities

(See Fig. A–4.)

$$\sin \theta = \frac{o}{h} \qquad\qquad \csc \theta = \frac{1}{\sin \theta} = \frac{h}{o}$$

$$\cos \theta = \frac{a}{h} \qquad\qquad \sec \theta = \frac{1}{\cos \theta} = \frac{h}{a}$$

$$\tan \theta = \frac{o}{a} = \frac{\sin \theta}{\cos \theta} \qquad\qquad \cot \theta = \frac{1}{\tan \theta} = \frac{a}{o}$$

$$a^2 + o^2 = h^2 \qquad\qquad \text{[Pythagorean theorem]}.$$

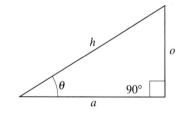

FIGURE A–4

Figure A–5 shows the signs (+ or −) that cosine, sine, and tangent take on for angles θ in the four quadrants (0° to 360°). Note that angles are measured counterclockwise from the x axis as shown; negative angles are measured from *below* the x axis, clockwise: for example, $-30° = +330°$, and so on.

FIGURE A–5

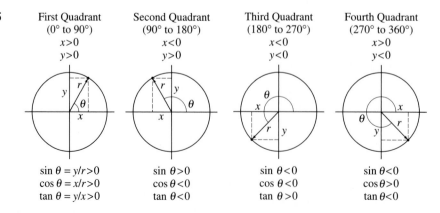

First Quadrant (0° to 90°)	Second Quadrant (90° to 180°)	Third Quadrant (180° to 270°)	Fourth Quadrant (270° to 360°)
$x>0$ $y>0$	$x<0$ $y>0$	$x<0$ $y<0$	$x>0$ $y<0$
$\sin \theta = y/r > 0$ $\cos \theta = x/r > 0$ $\tan \theta = y/x > 0$	$\sin \theta > 0$ $\cos \theta < 0$ $\tan \theta < 0$	$\sin \theta < 0$ $\cos \theta < 0$ $\tan \theta > 0$	$\sin \theta < 0$ $\cos \theta > 0$ $\tan \theta < 0$

The following are some useful identities among the trigonometric functions:

$$\sin^2\theta + \cos^2\theta = 1, \quad \sec^2\theta - \tan^2\theta = 1, \quad \csc^2\theta - \cot^2\theta = 1$$

$$\sin 2\theta = 2\sin\theta\cos\theta$$

$$\cos 2\theta = \cos^2\theta - \sin^2\theta = 2\cos^2\theta - 1 = 1 - 2\sin^2\theta$$

$$\tan 2\theta = \frac{2\tan\theta}{1 - \tan^2\theta}$$

$$\sin(A \pm B) = \sin A \cos B \pm \cos A \sin B$$

$$\cos(A \pm B) = \cos A \cos B \mp \sin A \sin B$$

$$\tan(A \pm B) = \frac{\tan A \pm \tan B}{1 \mp \tan A \tan B}$$

$$\sin(180° - \theta) = \sin\theta$$

$$\cos(180° - \theta) = -\cos\theta$$

$$\sin(90° - \theta) = \cos\theta$$

$$\cos(90° - \theta) = \sin\theta$$

$$\cos(-\theta) = \cos\theta$$

$$\sin(-\theta) = -\sin\theta$$

$$\tan(-\theta) = -\tan\theta$$

$$\sin\tfrac{1}{2}\theta = \sqrt{\frac{1 - \cos\theta}{2}}, \quad \cos\tfrac{1}{2}\theta = \sqrt{\frac{1 + \cos\theta}{2}}, \quad \tan\tfrac{1}{2}\theta = \sqrt{\frac{1 - \cos\theta}{1 + \cos\theta}}$$

$$\sin A \pm \sin B = 2\sin\left(\frac{A \pm B}{2}\right)\cos\left(\frac{A \mp B}{2}\right).$$

For any triangle (see Fig. A–6):

$$\frac{\sin\alpha}{a} = \frac{\sin\beta}{b} = \frac{\sin\gamma}{c} \qquad \text{[Law of sines]}$$

$$c^2 = a^2 + b^2 - 2ab\cos\gamma. \qquad \text{[Law of cosines]}$$

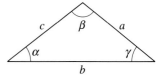

FIGURE A–6

A–7 Logarithms

The following identities apply to common logs (base 10), natural logs (base e) which are often abbreviated ln, or logs to any other base.

$$\log(ab) = \log a + \log b$$

$$\log\left(\frac{a}{b}\right) = \log a - \log b$$

$$\log a^n = n\log a.$$

A–8 Vectors

Vector addition is covered in Sections 3–2 to 3–5.
Vector multiplication is covered in Sections 3–3, 7–2 and 11–1.

Derivatives and Integrals

B–1 Derivatives: General Rules

(See also Section 2–3.)

$$\frac{dx}{dx} = 1$$

$$\frac{d}{dx}\left[af(x)\right] = a\frac{df}{dx} \qquad (a = \text{constant})$$

$$\frac{d}{dx}\left[f(x) + g(x)\right] = \frac{df}{dx} + \frac{dg}{dx}$$

$$\frac{d}{dx}\left[f(x)g(x)\right] = \frac{df}{dx}g + f\frac{dg}{dx}$$

$$\frac{d}{dx}\left[f(y)\right] = \frac{df}{dy}\frac{dy}{dx} \qquad \text{[chain rule]}$$

$$\frac{dx}{dy} = \frac{1}{\left(\dfrac{dy}{dx}\right)} \qquad \text{if } \frac{dy}{dx} \neq 0.$$

B–2 Derivatives: Particular Functions

$$\frac{da}{dx} = 0 \qquad (a = \text{constant})$$

$$\frac{d}{dx}x^n = nx^{n-1}$$

$$\frac{d}{dx}\sin ax = a\cos ax$$

$$\frac{d}{dx}\cos ax = -a\sin ax$$

$$\frac{d}{dx}\tan ax = a\sec^2 ax$$

$$\frac{d}{dx}\ln ax = \frac{1}{x}$$

$$\frac{d}{dx}e^{ax} = ae^{ax}$$

B–3 Indefinite Integrals: General Rules

(See also Section 7–3.)

$$\int dx = x$$

$$\int af(x)\,dx = a\int f(x)\,dx \qquad (a = \text{constant})$$

$$\int [f(x) + g(x)]\,dx = \int f(x)\,dx + \int g(x)\,dx$$

$$\int u\,dv = uv - \int v\,du \qquad \text{[integration by parts]}$$

B–4 Indefinite Integrals: Particular Functions

(An arbitrary constant can be added to the right side of each equation.)

$$\int a\,dx = ax \qquad (a = \text{constant})$$

$$\int x^m\,dx = \frac{1}{m+1}x^{m+1} \qquad (m \neq -1)$$

$$\int \sin ax\,dx = -\frac{1}{a}\cos ax$$

$$\int \cos ax\,dx = \frac{1}{a}\sin ax$$

$$\int \tan ax\,dx = \frac{1}{a}\ln|\sec ax|$$

$$\int \frac{1}{x}\,dx = \ln x$$

$$\int e^{ax}\,dx = \frac{1}{a}e^{ax}$$

$$\int \frac{dx}{x^2 + a^2} = \frac{1}{a}\tan^{-1}\frac{x}{a}$$

$$\int \frac{dx}{x^2 - a^2} = \frac{1}{2a}\ln\left(\frac{x-a}{x+a}\right) \qquad (x^2 > a^2)$$

$$= -\frac{1}{2a}\ln\left(\frac{a+x}{a-x}\right) \qquad (x^2 < a^2)$$

$$\int \frac{dx}{\sqrt{x^2 \pm a^2}} = \ln(x + \sqrt{x^2 \pm a^2})$$

$$\int \frac{dx}{(x^2 \pm a^2)^{\frac{3}{2}}} = \frac{\pm x}{a^2\sqrt{x^2 \pm a^2}}$$

$$\int \frac{x\,dx}{(x^2 \pm a^2)^{\frac{3}{2}}} = \frac{-1}{\sqrt{x^2 \pm a^2}}$$

$$\int \sin^2 ax\,dx = \frac{x}{2} - \frac{\sin 2ax}{4a}$$

$$\int xe^{-ax}\,dx = -\frac{e^{-ax}}{a^2}(ax + 1)$$

$$\int x^2 e^{-ax}\,dx = -\frac{e^{-ax}}{a^3}(a^2x^2 + 2ax + 2)$$

B–5 A few Definite Integrals

$$\int_0^\infty x^n e^{-ax}\,dx = \frac{n!}{a^{n+1}}$$

$$\int_0^\infty e^{-ax^2}\,dx = \sqrt{\frac{\pi}{4a}}$$

$$\int_0^\infty xe^{-ax^2}\,dx = \frac{1}{2a}$$

$$\int_0^\infty x^2 e^{-ax^2}\,dx = \sqrt{\frac{\pi}{16a^3}}$$

$$\int_0^\infty x^3 e^{-ax^2}\,dx = \frac{1}{2a^2}$$

$$\int_0^\infty x^{2n} e^{-ax^2}\,dx = \frac{1 \cdot 3 \cdot 5 \cdots (2n-1)}{2^{n+1}a^n}\sqrt{\frac{\pi}{a}}$$

Gravitational Force due to a Spherical Mass Distribution

In Chapter 6 (Section 6–1), we stated that the gravitational force exerted by or on a uniform sphere acts as if all the mass of the sphere were concentrated at its center, if the other mass is outside the sphere. In other words, the gravitational force that a uniform sphere exerts on a particle outside it is

$$F = G\frac{mM}{r^2}, \qquad\qquad [m \text{ outside sphere of mass } M]$$

where m is the mass of the particle, M the mass of the sphere, and r the distance of m from the center of the sphere. Now we will derive this result. We will use the concepts of infinitesimally small quantities and integration.

First we consider a very thin, uniform spherical shell (like a thin-walled basketball) of mass M whose thickness t is small compared to its radius R (Fig. C–1). The force on a particle of mass m at a distance r from the center of the shell can be calculated as the vector sum of the forces due to all the particles of the shell. We imagine the shell divided up into thin (infinitesimal) circular strips so that all points on a strip are equidistant from our particle m. One of these circular strips, labeled AB, is shown in Fig. C–1. It is $R\,d\theta$ wide, t thick, and has a radius $R\sin\theta$. The force on our particle m due to a tiny piece of the strip at point A is represented by the vector $\mathbf{F}_A$ shown. The force due to a tiny piece of the strip at point B, which is diametrically opposite A, is the force $\mathbf{F}_B$. We take the two pieces at A and B to be of equal mass, so $F_A = F_B$. The horizontal components of $\mathbf{F}_A$ and $\mathbf{F}_B$ are each equal to

$$F_A \cos\phi$$

and point toward the center of the shell. The vertical components of $\mathbf{F}_A$ and $\mathbf{F}_B$ are of equal magnitude and point in opposite directions, and so cancel. Since for every point on the strip there is a corresponding point diametrically opposite (as with A and B), we see that the net force due to the entire strip points toward the center of the shell. Its magnitude will be

$$dF = G\frac{m\,dM}{l^2}\cos\phi,$$

where dM is the mass of the entire circular strip and l is the distance from all

FIGURE C–1 Calculating the gravitational force on a particle of mass m due to a uniform spherical shell of radius R and mass M.

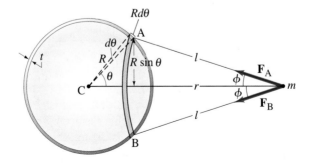

points on the strip to m, as shown. We write dM in terms of the density ρ; by density we mean the mass per unit volume (Section 13–1). Hence, $dM = \rho \, dV$, where dV is the volume of the strip and equals $(2\pi R \sin\theta)(t)(R \, d\theta)$. Then the force dF due to the circular strip shown is

$$dF = G\frac{m\rho 2\pi R^2 t \sin\theta \, d\theta}{l^2} \cos\phi. \qquad (C\text{–}1)$$

To get the total force F that the entire shell exerts on the particle m, we must integrate over all the circular strips: that is, from $\theta = 0°$ to $\theta = 180°$. But our expression for dF contains l and ϕ, which are functions of θ. From Fig. C–1 we can see that

$$l \cos\phi = r - R\cos\theta.$$

Furthermore, we can write the law of cosines for triangle CmA:

$$\cos\theta = \frac{r^2 + R^2 - l^2}{2rR}. \qquad (C\text{–}2)$$

With these two expressions we can reduce our three variables (l, θ, ϕ) to only one, which we take to be l. We do two things with Eq. C–2: (1) We put it into the equation for $l\cos\phi$ above:

$$\cos\phi = \frac{1}{l}(r - R\cos\theta) = \frac{r^2 + l^2 - R^2}{2rl};$$

and (2) we take the differential of both sides of Eq. C–2 (because $\sin\theta \, d\theta$ appears in the expression for dF, Eq. C–1):

$$-\sin\theta \, d\theta = -\frac{2l \, dl}{2rR} \qquad \text{or} \qquad \sin\theta \, d\theta = \frac{l \, dl}{rR},$$

since r and R are considered constants when summing over the strips. Now we insert these into Eq. C–1 for dF and find

$$dF = Gm\rho\pi t\frac{R}{r^2}\left(1 + \frac{r^2 - R^2}{l^2}\right) dl.$$

Now we integrate to get the net force on our thin shell of radius R. To integrate over all the strips ($\theta = 0°$ to $180°$), we must go from $l = r - R$ to $l = r + R$ (see Fig. C–1). Thus,

$$F = Gm\rho\pi t\frac{R}{r^2}\left[l - \frac{r^2 - R^2}{l}\right]_{l=r-R}^{l=r+R}$$

$$= Gm\rho\pi t\frac{R}{r^2}(4R).$$

The volume V of the spherical shell is its area $(4\pi R^2)$ times the thickness t. Hence the mass $M = \rho V = \rho 4\pi R^2 t$, and finally

$$F = G\frac{mM}{r^2}. \qquad \left[\begin{array}{c}\text{particle of mass } m \text{ outside a}\\ \text{thin uniform spherical shell of mass } M\end{array}\right]$$

This result gives us the force a thin shell exerts on a particle of mass m a distance r from the center of the shell, and *outside* the shell. We see that the force is the same as that between m and a particle of mass M at the center of the shell. In other words, for purposes of calculating the gravitational force exerted on or by a uniform spherical shell, we can consider all its mass concentrated at its center.

What we have derived for a shell holds also for a solid sphere, since a solid sphere can be considered as made up of many concentric shells, from $R = 0$ to $R = R_0$, where R_0 is the radius of the solid sphere. Why? Because if each shell has mass dM, we write for each shell, $dF = Gm\,dM/r^2$, where r is the distance from the center C to mass m and is the same for all shells. Then the total force equals the sum or integral over dM, which gives the total mass M. Thus the result

$$F = G\frac{mM}{r^2} \qquad \left[\begin{array}{l}\text{particle of mass } m \text{ outside} \\ \text{solid sphere of mass } M\end{array}\right] \quad \text{(C–3)}$$

is valid for a solid sphere of mass M even if the density varies with distance from the center. (It is not valid if the density varies within each shell—that is, depends not only on R.) Thus the gravitational force exerted on or by spherical objects, including nearly spherical objects like the Earth, Sun, and Moon, can be considered to act as if the objects were point particles.

This result, Eq. C–3, is true only if the mass m is outside the sphere. Let us next consider a point mass m that is located inside the spherical shell of Fig. C–1. Here, r would be less than R, and the integration over l would be from $l = R - r$ to $l = R + r$, so

$$\left[l - \frac{r^2 - R^2}{l}\right]_{R-r}^{R+r} = 0.$$

Thus the force on any mass inside the shell would be zero. This result has particular importance for the electrostatic force, which is also an inverse square law. For the gravitational situation, we see that at points within a solid sphere, say 1000 km below the earth's surface, only the mass up to that radius contributes to the net force. The outer shells beyond the point in question contribute no net gravitational effect.

The results we have obtained here can also be reached using the gravitational analog of Gauss's law for electrostatics (Chapter 22).

Selected Isotopes

(1) Atomic Number Z	(2) Element	(3) Symbol	(4) Mass Number A	(5) Atomic Mass†	(6) % Abundance (or Radioactive Decay Mode)	(7) Half-life (if radioactive)
0	(Neutron)	n	1	1.008665	β^-	10.4 min
1	Hydrogen	H	1	1.007825	99.985%	
	Deuterium	D	2	2.014102	0.015%	
	Tritium	T	3	3.016049	β^-	12.33 yr
2	Helium	He	3	3.016029	0.000137%	
			4	4.002603	99.999863%	
3	Lithium	Li	6	6.015122	7.5%	
			7	7.016004	92.5%	
4	Beryllium	Be	7	7.016929	EC, γ	53.12 days
			9	9.012182	100%	
5	Boron	B	10	10.012937	19.9%	
			11	11.009305	80.1%	
6	Carbon	C	11	11.011434	β^+, EC	20.39 min
			12	12.000000	98.90%	
			13	13.003355	1.10%	
			14	14.003242	β^-	5730 yr
7	Nitrogen	N	13	13.005739	β^+	9.965 min
			14	14.003074	99.63%	
			15	15.000108	0.37%	
8	Oxygen	O	15	15.003065	β^+, EC	122.24 s
			16	15.994915	99.76%	
			18	17.999160	0.20%	
9	Fluorine	F	19	18.998403	100%	
10	Neon	Ne	20	19.992440	90.48%	
			22	21.991386	9.25%	
11	Sodium	Na	22	21.994437	β^+, EC, γ	2.6019 yr
			23	22.989770	100%	
			24	23.990963	β^-, γ	14.9590 h
12	Magnesium	Mg	24	23.985042	78.99%	
13	Aluminum	Al	27	26.981538	100%	

†The masses given in column (5) are those for the neutral atom, including the Z electrons.

(1) Atomic Number Z	(2) Element	(3) Symbol	(4) Mass Number A	(5) Atomic Mass[†]	(6) % Abundance (or Radioactive Decay Mode)	(7) Half-life (if radioactive)
14	Silicon	Si	28	27.976927	92.23%	
			31	30.975363	β^-, γ	157.3 min
15	Phosphorus	P	31	30.973762	100%	
			32	31.973907	β^-	14.262 days
16	Sulfur	S	32	31.972071	95.02%	
			35	34.969032	β^-	87.32 days
17	Chlorine	Cl	35	34.968853	75.77%	
			37	36.965903	24.23%	
18	Argon	Ar	40	39.962383	99.600%	
19	Potassium	K	39	38.963707	93.2581%	
			40	39.963999	0.0117%	
					$\beta^-, EC, \gamma, \beta^+$	1.28×10^9 yr
20	Calcium	Ca	40	39.962591	96.941%	
21	Scandium	Sc	45	44.955910	100%	
22	Titanium	Ti	48	47.947947	73.8%	
23	Vanadium	V	51	50.943964	99.750%	
24	Chromium	Cr	52	51.940512	83.79%	
25	Manganese	Mn	55	54.938049	100%	
26	Iron	Fe	56	55.934942	91.72%	
27	Cobalt	Co	59	58.933200	100%	
			60	59.933822	β^-, γ	5.2714 yr
28	Nickel	Ni	58	57.935348	68.077%	
			60	59.930791	26.233%	
29	Copper	Cu	63	62.929601	69.17%	
			65	64.927794	30.83%	
30	Zinc	Zn	64	63.929147	48.6%	
			66	65.926037	27.9%	
31	Gallium	Ga	69	68.925581	60.108%	
32	Germanium	Ge	72	71.922076	27.66%	
			74	73.921178	35.94%	
33	Arsenic	As	75	74.921596	100%	
34	Selenium	Se	80	79.916522	49.61%	
35	Bromine	Br	79	78.918338	50.69%	
36	Krypton	Kr	84	83.911507	57.0%	
37	Rubidium	Rb	85	84.911789	72.17%	
38	Strontium	Sr	86	85.909262	9.86%	
			88	87.905614	82.58%	
			90	89.907737	β^-	28.79 yr
39	Yttrium	Y	89	88.905848	100%	
40	Zirconium	Zr	90	89.904704	51.45%	
41	Niobium	Nb	93	92.906377	100%	
42	Molybdenum	Mo	98	97.905408	24.13%	

[†] The masses given in column (5) are those for the neutral atom, including the Z electrons.

(1) Atomic Number Z	(2) Element	(3) Symbol	(4) Mass Number A	(5) Atomic Mass†	(6) % Abundance (or Radioactive Decay Mode)	(7) Half-life (if radioactive)
43	Technetium	Tc	98	97.907216	β^-, γ	4.2×10^6 yr
44	Ruthenium	Ru	102	101.904349	31.6%	
45	Rhodium	Rh	103	102.905504	100%	
46	Palladium	Pd	106	105.903483	27.33%	
47	Silver	Ag	107	106.905093	51.839%	
			109	108.904756	48.161%	
48	Cadmium	Cd	114	113.903358	28.73%	
49	Indium	In	115	114.903878	95.7%; β^-, γ	4.41×10^{14} yr
50	Tin	Sn	120	119.902197	32.59%	
51	Antimony	Sb	121	120.903818	57.36%	
52	Tellurium	Te	130	129.906223	33.80%	7.9×10^{20} yr
53	Iodine	I	127	126.904468	100%	
			131	130.906124	β^-, γ	8.0207 days
54	Xenon	Xe	132	131.904154	26.9%	
			136	135.907220	8.9%	
55	Cesium	Cs	133	132.905446	100%	
56	Barium	Ba	137	136.905821	11.23%	
			138	137.905241	71.70%	
57	Lanthanum	La	139	138.906348	99.9098%	
58	Cerium	Ce	140	139.905434	88.48%	
59	Praseodymium	Pr	141	140.907647	100%	
60	Neodymium	Nd	142	141.907718	27.13%	
61	Promethium	Pm	145	144.912744	EC, γ, α	17.7 yr
62	Samarium	Sm	152	151.919728	26.7%	
63	Europium	Eu	153	152.921226	52.2%	
64	Gadolinium	Gd	158	157.924101	24.84%	
65	Terbium	Tb	159	158.925343	100%	
66	Dysprosium	Dy	164	163.929171	28.2%	
67	Holmium	Ho	165	164.930319	100%	
68	Erbium	Er	166	165.930290	33.6%	
69	Thulium	Tm	169	168.934211	100%	
70	Ytterbium	Yb	174	173.938858	31.8%	
71	Lutecium	Lu	175	174.940767	97.4%	
72	Hafnium	Hf	180	179.946549	35.100%	
73	Tantalum	Ta	181	180.947996	99.988%	
74	Tungsten (wolfram)	W	184	183.950933	30.67%	
75	Rhenium	Re	187	186.955751	62.60%; β^-	4.35×10^{10} yr
76	Osmium	Os	191	190.960927	β^-, γ	15.4 days
			192	191.961479	41.0%	
77	Iridium	Ir	191	190.960591	37.3%	
			193	192.962923	62.7%	
78	Platinum	Pt	195	194.964774	33.8%	

†The masses given in column (5) are those for the neutral atom, including the Z electrons.

(1) Atomic Number Z	(2) Element	(3) Symbol	(4) Mass Number A	(5) Atomic Mass†	(6) % Abundance (or Radioactive Decay Mode)	(7) Half-life (if radioactive)
79	Gold	Au	197	196.966551	100%	
80	Mercury	Hg	199	198.968262	16.87%	
			202	201.970625	29.86%	
81	Thallium	Tl	205	204.974412	70.476%	
82	Lead	Pb	206	205.974449	24.1%	
			207	206.975880	22.1%	
			208	207.976635	52.4%	
			210	209.984173	β^-, γ, α	22.3 yr
			211	210.988731	β^-, γ	36.1 min
			212	211.991887	β^-, γ	10.64 h
			214	213.999798	β^-, γ	26.8 min
83	Bismuth	Bi	209	208.980383	100%	
			211	210.987258	α, γ, β^-	2.14 min
84	Polonium	Po	210	209.982857	α, γ	138.376 days
			214	213.995185	α, γ	164.3 μs
85	Astatine	At	218	218.008681	α, β^-	1.5 s
86	Radon	Rn	222	222.017570	α, γ	3.8235 days
87	Francium	Fr	223	223.019731	β^-, γ, α	21.8 min
88	Radium	Ra	226	226.025402	α, γ	1600 yr
89	Actinium	Ac	227	227.027746	β^-, γ, α	21.773 yr
90	Thorium	Th	228	228.028731	α, γ	1.9116 yr
			232	232.038050	100%; α, γ	1.405×10^{10} yr
91	Protactinium	Pa	231	231.035878	α, γ	3.276×10^4 yr
92	Uranium	U	232	232.037146	α, γ	68.9 yr
			233	233.039628	α, γ	1.592×10^5 yr
			235	235.043923	0.720%, α, γ	7.038×10^8 yr
			236	236.045561	α, γ	2.342×10^7 yr
			238	238.050782	99.2745%; α, γ	4.468×10^9 yr
			239	239.054287	β^-, γ	23.45 min
93	Neptunium	Np	237	237.048166	α, γ	2.144×10^6 yr
			239	239.052931	β^-, γ	2.3565 d
94	Plutonium	Pu	239	239.052157	α, γ	24,110 yr
			244	244.064197	α	8.08×10^7 yr
95	Americium	Am	243	243.061373	α, γ	7370 yr
96	Curium	Cm	247	247.070346	α, γ	1.56×10^7 yr
97	Berkelium	Bk	247	247.070298	α, γ	1380 yr
98	Californium	Cf	251	251.079580	α, γ	898 yr
99	Einsteinium	Es	252	252.082972	α, EC, γ	472 d
100	Fermium	Fm	257	257.095099	α, γ	101 d
101	Mendelevium	Md	258	258.098425	α, γ	51.5 d
102	Nobelium	No	259	259.10102	α, EC	58 min
103	Lawrencium	Lr	262	262.10969	α, EC, fission	216 min

†The masses given in column (5) are those for the neutral atom, including the Z electrons.

(1) Atomic Number Z	(2) Element	(3) Symbol	(4) Mass Number A	(5) Atomic Mass[†]	(6) % Abundance (or Radioactive Decay Mode)	(7) Half-life (if radioactive)
104	Rutherfordium	Rf	261	261.10875	α	65 s
105	Dubnium	Db	262	262.11415	α, fission, EC	34 s
106	Seaborgium	Sg	266	266.12193	α, fission	21 s
107	Bohrium	Bh	264	264.12473	α	0.44 s
108	Hassium	Hs	269	269.13411	α	9 s
109	Meitnerium	Mt	268	268.13882	α	0.07 s
110			271	271.14608	α	0.06 s
111			272	272.15348	α	1.5 ms
112			277	277	α	0.24 ms
114			289	289	α	20 s
116			289	289	α	0.6 ms
118			293	293	α	0.1 ms

[†] The masses given in column (5) are those for the neutral atom, including the Z electrons.

Answers to Odd-Numbered Problems

CHAPTER 1

1. (a) 1×10^{10} yr; (b) 3×10^{17} s.
3. (a) 1.156×10^3; (b) 2.18×10^1;
   (c) 6.8×10^{-3}; (d) 2.7635×10^1;
   (e) 2.19×10^{-1}; (f) 2.2×10^1.
5. 7.7%.
7. (a) 4%; (b) 0.4%; (c) 0.07%.
9. 1.0×10^5 s.
11. 9%.
13. (a) 0.286 6 m; (b) 0.000 085 V;
    (c) 0.000 760 kg;
    (d) 0.000 000 000 060 0 s;
    (e) 0.000 000 000 000 022 5 m;
    (f) 2,500,000,000 volts.
15. 1.8 m.
17. (a) 0.111 yd^2; (b) 10.76 ft^2.
19. (a) 3.9×10^{-9} in;
    (b) 1.0×10^8 atoms.
21. (a) 0.621 mi/h;
    (b) 1 m/s = 3.28 ft/s; (c) 0.278 m/s.
23. (a) 9.46×10^{15} m;
    (b) 6.31×10^4 AU; (c) 7.20 AU/h.
25. (a) 10^3; (b) 10^4; (c) 10^{-2}; (d) 10^9.
27. ≈20%.
29. 1×10^5 cm^3.
31. (a) ≈600 dentists.
33. ≈3×10^8 kg/yr.
35. 51 km.
37. $A = [L/T^4] = $ m/s^4,
    $B = [L/T^2] = $ m/s^2.
39. (a) 0.10 nm; (b) 1.0×10^5 fm;
    (c) 1.0×10^{10} Å; (d) 9.5×10^{25} Å.
41. (a) 3.16×10^7 s; (b) 3.16×10^{16} ns;
    (c) 3.17×10^{-8} yr.
43. (a) 1,000 drivers.
45. 1×10^{11} gal/yr.
47. 9 cm.
49. 4×10^5 t.
51. ≈4 yr.
53. 1.9×10^2 m.
55. (a) 3%, 3%; (b) 0.7%, 0.2%.

CHAPTER 2

1. 5.0 h.
3. 61 m.
5. 0.78 cm/s (toward $+x$).
7. ≈300 m/s.
9. (a) 10.1 m/s;
   (b) +3.4 m/s, away from trainer.
11. (a) 0.28 m/s; (b) 1.2 m/s;
    (c) 0.28 m/s; (d) 1.6 m/s;
    (e) −1.0 m/s.
13. (a) 13.4 m/s;
    (b) +4.5 m/s, away from master.
15. 24 s.
17. 55 km/h, 0.
19. 6.73 m/s.
21. 5.2 s
23. -7.0 m/s^2, 0.72.
25. (a) 4.7 m/s^2; (b) 2.2 m/s^2;
    (c) 0.3 m/s^2; (d) 1.6 m/s^2.
27. $v = (6.0$ m/s$) + (17$ m/s$^2)t$,
    $a = 17$ m/s^2.
29. 1.5 m/s^2, 99 m.
31. 1.7×10^2 m.
33. 4.41 m/s^2, $t = 2.61$ s.
35. 55.0 m.
37. (a) 2.3×10^2 m; (b) 31 s;
    (c) 15 m, 13 m.
39. (a) 103 m; (b) 64 m.
41. 31 m/s.
43. (b) 3.45 s.
45. 32 m/s (110 km/h).
47. 2.83 s.
49. (a) 8.81 s; (b) 86.3 m/s.
51. 1.44 s.
53. 15 m/s.
55. 5.44 s.
59. 0.035 s.
61. 1.8 m above the top of the window.
63. 52 m.
65. 19.8 m/s, 20.0 m.
67. (a) $v = (g/k)(1 - e^{-kt})$;
    (b) $v_{term} = g/k$.
69. $6h_{Earth}$.
71. 1.3 m.
73. (b) $H_{50} = 9.8$ m; (c) $H_{100} = 39$ m.
75. (a) 1.3 m; (b) 6.1 m/s; (c) 1.2 s.
77. (a) 3.88 s; (b) 73.9 m; (c) 38.0 m/s,
    48.4 m/s.
79. (a) 52 min; (b) 31 min.
81. (a) $v_0 = 26$ m/s; (b) 35 m; (c) 1.2 s;
    (d) 4.1 s.
83. (a) 4.80 s; (b) 37.0 m/s; (c) 75.2 m.
85. She should decide to stop!

87. $\Delta v_{0down} = 0.8$ m/s, $\Delta v_{0up} = 0.9$ m/s.
89. 29.0 m.

CHAPTER 3

1. 263 km, 13° S of W.
3. $\mathbf{V}_{wrong} = \mathbf{V}_2 - \mathbf{V}_1$.
5. 13.6 m, 18° N of E,

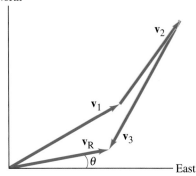

7. (a)

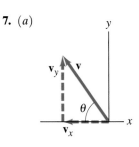

   (b) $V_x = -11.7$, $V_y = 8.16$;
   (c) 14.3, 34.9° above $-x$-axis.
9. (a) $V_N = 476$ km/h,
   $V_W = 421$ km/h;
   (b) $d_N = 1.43 \times 10^3$ km,
   $d_W = 1.26 \times 10^3$ km.
11. (a) 4.2, 45° below $+x$-axis;
    (b) 5.1, 79° below $+x$-axis.
13. (a) 53.7, 1.40° above $-x$-axis;
    (b) 53.7, 1.40° below $+x$-axis.
15. (a) 94.5, 11.8° below $-x$-axis;
    (b) 150, 35.3° below $+x$-axis.
17. (a) $A_x = \pm 82.9$; (b) 166.6, 12.1°
    above $-x$-axis.
19. $(7.60$ m/s$)\mathbf{i} - (4.00$ m/s$)\mathbf{k}$; 8.59 m/s.
21. (a) Unknown; (b) 4.11 m/s^2, 33.2°
    north of east; (c) unknown.

23. (a) $\mathbf{v} = (4.0 \text{ m/s}^2)t\mathbf{i} + (3.0 \text{ m/s}^2)t\mathbf{j}$;
   (b) $(5.0 \text{ m/s}^2)t$;
   (c) $\mathbf{r} = (2.0 \text{ m/s}^2)t^2\mathbf{i} + (1.5 \text{ m/s}^2)t^2\mathbf{j}$;
   (d) $\mathbf{v} = (8.0 \text{ m/s})\mathbf{i} + (6.0 \text{ m/s})\mathbf{j}$,
      $|\mathbf{v}| = 10.0 \text{ m/s}$,
      $\mathbf{r} = (8.0 \text{ m})\mathbf{i} + (6.0 \text{ m})\mathbf{j}$.

25. (a) $-(18.0 \text{ m/s})\sin(3.0 \text{ s}^{-1})t\mathbf{i}$
      $+ (18.0 \text{ m/s})\cos(3.0 \text{ s}^{-1})t\mathbf{j}$;
   (b) $-(54.0 \text{ m/s}^2)\cos(3.0 \text{ s}^{-1})t\mathbf{i}$
      $- (54.0 \text{ m/s}^2)\sin(3.0 \text{ s}^{-1})t\mathbf{j}$;
   (c) circle; (d) $a = (9.0 \text{ s}^{-2})r$, 180°.

27. 44 m, 6.3 m.

29. 38° and 52°.

31. 1.95 s.

33. 22 m.

35.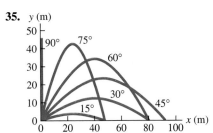

37. 5.71 s.

39. (a) 65.7 m; (b) 7.32 s; (c) 267 m;
   (d) 42.2 m/s, 30.1° above the
   horizontal.

43. Unsuccessful, 34.7 m.

45. (a) $\mathbf{v}_0 = 3.42 \text{ m/s}$, 47.5° above the
   horizontal; (b) 5.32 m above the
   water; (c) $\mathbf{v}_f = 10.5 \text{ m/s}$, 77° below
   the horizontal.

47. $\theta = \tan^{-1}(gt/v_0)$ below the
   horizontal.

49. $\theta = \frac{1}{2}\tan^{-1}(-\cot\phi)$.

51. $7.29g$ up.

53. $5.9 \times 10^{-3} \text{ m/s}^2$ toward the Sun.

55. $0.94g$.

59. 2.7 m/s, 22° from the river bank.

61. 23.1 s.

63. 1.41 m/s.

65. (a) 1.82 m/s; (b) 3.22 m/s.

67. (a) 60 m; (b) 75 s.

69. 58 km/h, 31°, 58 km/h opposite
   to $\mathbf{v}_{12}$.

71. 0.0889 m/s^2.

73. $D_x = 60 \text{ m}, D_y = -35 \text{ m}$,
   $D_z = -12 \text{ m}$; 70 m; $\theta_h = 30°$ from
   the x-axis toward the $-y$-axis,
   $\theta_v = 9.8°$ below the horizontal.

75. 7.0 m/s.

77. $\pm28.5°, \pm25.2$.

79. 170 km/h, 41.5° N of E.

81. 1.6 m/s^2.

83. 2.7 s, 1.9 m/s.

85. (a) $Dv/(v^2 - u^2)$;
   (b) $D/(v^2 - u^2)^{1/2}$.

87. 54.6° below the horizontal.

89. (a) 464 m/s; (b) 355 m/s.

91. Row at an angle of 23° upstream
   and run 243 m in a total time of
   20.7 min.

93. 1.8×10^3 rev/day.

CHAPTER 4

3. 6.9×10^2 N.

5. (a) 5.7×10^2 N; (b) 99 N;
   (c) 2.1×10^2 N; (d) 0.

7. 107 N.

9. -9.3×10^5 N, 25% of the weight of
   the train.

11. $m > 1.9$ kg.

13. 2.1×10^2 N.

15. -1.40 m/s^2 (down).

17. a (downward) $\geq 1.2 \text{ m/s}^2$.

19. $a_{max} = 0.557 \text{ m/s}^2$.

21. (a) 2.2 m/s^2; (b) 18 m/s; (c) 93 s.

23. 3.0×10^3 N downward.

25. (a) 1.4×10^2 N; (b) 14.5 m/s.

27. Southwesterly direction.

29.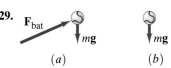
   (a) (b)

31. (a) 1.13 m/s^2, 52.2° below $-x$-axis;
   (b) 0.814 m/s^2, 42.3° above $+x$-axis.

33. (a) $m_1g - F_T = m_1a$, $F_T - m_2g$
      $= m_2a$.

35. (a) lower bucket $= 34$ N, upper
   bucket $= 68$N; (b) lower bucket
   $= 40$ N, upper bucket $= 80$ N.

37. 1.4×10^3 N.

39. $F_B = 6890 \text{ N}, \mathbf{F}_A + \mathbf{F}_B = 8860 \text{ N}$.

41. (a) 2.2 m up the plane; (b) 2.2 s.

43. $\frac{5}{2}(F_0/m)t_0^2$.

47. (a)

(b) $a = m_2g/(m_1 + m_2)$,
   $F_T = m_1m_2g/(m_1 + m_2)$.

49. $a = [m_2 + m_C(\ell_2/\ell)]g/$
   $(m_1 + m_2 + m_C)$.

51. $1.74 \text{ m/s}^2, F_{T1} = 22.6$ N,
   $F_{T2} = 20.9$ N.

53. $(m + M)g\tan\theta$.

55. $F_{T1} = [4m_1m_2m_3/$
      $(m_1m_3 + m_2m_3 + 4m_1m_2)]g$,
   $F_{T3} = [8m_1m_2m_3/$
      $(m_1m_3 + m_2m_3 + 4m_1m_2)]g$,
   $a_1 = [(m_1m_3 - 3m_2m_3 + 4m_1m_2)/$
      $(m_1m_3 + m_2m_3 + 4m_1m_2)]g$,
   $a_2 = [(-3m_1m_3 + m_2m_3 + 4m_1m_2)/$
      $(m_1m_3 + m_2m_3 + 4m_1m_2)]g$,
   $a_3 = [(m_1m_3 + m_2m_3 - 4m_1m_2)/$
      $(m_1m_3 + m_2m_3 + 4m_1m_2)]g$.

57. $v = \{[2m_2\ell_2 + m_C(\ell_2^2/\ell)]g/$
   $(m_1 + m_2 + m_C)\}^{1/2}$.

59. 2.0×10^{-2} N.

61. 4.3 N.

63. 1.5×10^4 N.

65. 1.2 s, no change.

67. (a) 2.45 m/s^2 (up the incline);
   (b) 0.50 kg; (c) 7.35 N, 4.9 N.

69. 1.3×10^2 N.

71. 8.8°.

73. 82 m/s (300 km/h).

75. (a) $F = \frac{1}{2}Mg$;
   (b) $F_{T1} = F_{T2} = \frac{1}{2}Mg, F_{T3} = \frac{3}{2}Mg$,
   $F_{T4} = Mg$.

77. -8.3×10^2 N.

79. (a) 0.606 m/s^2; (b) 150 kN.

CHAPTER 5

1. 35 N, no force.

3. (a)

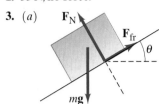

(b)

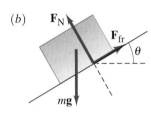

(c)

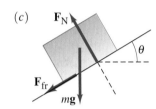

5. 0.20.

7. 69 N, $\mu_k = 0.54$.

9. 8.0 kg.

11. 1.3 m.

13. 1.3×10^3 N.

15. (a) 0.58; (b) 5.7 m/s; (c) 15 m/s.

17. (a) 1.4 m/s²; (b) 5.4×10^2 N;
(c) 1.41 m/s², 2.1×10^2 N.

19. (a) 86 cm up the plane; (b) 1.5 s.

21. (a) 2.8 m/s²; (b) 2.1 N.

23. $a = \{\sin\theta - [(\mu_1 m_1 + \mu_2 m_2)/$
$(m_1 + m_2)]\cos\theta\}g$,
$F_T = [m_1 m_2(\mu_2 - \mu_1)/$
$(m_1 + m_2)]g\cos\theta$.

25. (a) $\mu_k = (v_0^2/2gd\cos\theta) - \tan\theta$;
(b) $\mu_s \geq \tan\theta$.

27. (a) 2.0 m/s² up the plane;
(b) 5.4 m/s² up the plane.

29. $\mu_k = 0.40$.

31. (a) c = 14 kg/m; (b) 5.7×10^2 N.

33. $F_{min} = (m + M)g(\sin\theta + \mu\cos\theta)/$
$(\cos\theta - \mu_s\sin\theta)$.

35. $v_{max} = 21$ m/s, independent of the mass.

37. (a) 0.25 m/s² toward the center;
(b) 6.3 N toward the center.

39. Yes, $v_{top, min} = (gR)^{1/2}$.

41. 0.34.

43. 2.1×10^2 N.

45. 5.91°, 14.3 N.

47. (a) 5.8×10^3 N; (b) 4.1×10^2 N;
(c) 31 m/s.

49. $F_T = 2\pi mRf^2$.

51. 66 km/h < v < 123 km/h.

53. (a) $(1.6 \text{ m/s}^2)\mathbf{i}$;
(b) $(0.98 \text{ m/s}^2)\mathbf{i} - (1.7 \text{ m/s}^2)\mathbf{j}$;
(c) $-(4.9 \text{ m/s}^2)\mathbf{i} - (1.6 \text{ m/s}^2)\mathbf{j}$.

55. (a) 9.0 m/s²; (b) 15 m/s².

57. $\tau = m/b$.

59. (a) $v = (mg/b)$
$+ [v_0 - (mg/b)]e^{-bt/m}$;
(b) $v = -(mg/b)$
$+ [v_0 + (mg/b)]e^{-bt/m}$,
$v \geq 0$.

61. (b) 1.8°.

63. 10 m.

65. $\mu_s = 0.41$.

67. 2.3.

69. 101 N, $\mu_k = 0.719$.

71. (b) Will slide.

73. Emerges with a speed of 13 m/s.

75. 27.6 m/s, 0.439 rev/s.

77. $\Sigma F_{tan} = 3.3 \times 10^3$ N,
$\Sigma F_R = 2.0 \times 10^3$ N.

79. (a) $F_{NC} > F_{NB} > F_{NA}$;
(b) heaviest at C, lightest at A;
(c) $v_{Amax} = (gR)^{1/2}$.

81. (a) 1.23 m/s; (b) 3.01 m/s.

83. $\phi = 31°$.

85. (a) $r = v^2/g\cos\theta$; (b) 92 m.

87. (a) 59 s; (b) greater normal force.

89. 29.2 m/s.

91. 302 m, 735 m.

93. $g(1 - \mu_s\tan\phi)/4\pi^2 f^2(\tan\phi + \mu_s)$
$< r < g(1 + \mu_s\tan\phi)/$
$4\pi^2 f^2(\tan\phi - \mu_s)$.

CHAPTER 6

1. 1.52×10^3 N.

3. 1.6 m/s².

5. $g_h = 0.91 g_{surface}$.

7. 1.9×10^{-8} N toward center of square.

9. $Gm^2\{(2/x_0^2) + [3x_0/(x_0^2 + y_0^2)^{3/2}]\}\mathbf{i}$
$+ Gm^2\{[3y_0/(x_0^2 + y_0^2)^{3/2}]$
$+ (4/y_0^2)\}\mathbf{j}$.

11. 1.26.

13. 3.46×10^8 m from Earth's center.

15. (b) g decreases with an increase in height; (c) 9.493 m/s².

19. 7.56×10^3 m/s.

21. 2.0 h.

23. (a) 56 kg; (b) 56 kg; (c) 75 kg;
(d) 38 kg; (e) 0.

25. (a) 22 N (toward the Moon);
(b) -1.7×10^2 N (away from the Moon).

27. (a) Gravitational force provides required centripetal acceleration;
(b) 9.6×10^{26} kg.

29. 7.9×10^3 m/s.

31. $v = (Gm/L)^{1/2}$.

33. 0.0587 days (1.41 h).

35. 1.6×10^2 yr.

37. 2×10^8 yr.

39. $r_{Europa} = 6.71 \times 10^5$ km,
$r_{Ganymede} = 1.07 \times 10^6$ km,
$r_{Callisto} = 1.88 \times 10^6$ km.

41. 9.0 Earth-days.

43. (a) 2.1×10^2 A.U. $(3.1 \times 10^{13}$ m);
(b) 4.2×10^2 A.U.; (c) 4.2×10^2.

45. (a) 5.9×10^{-3} N/kg;
(b) not significant.

47. 2.7×10^3 km.

49. 6.7×10^{12} m/s².

51. 4.4×10^7 m/s².

53. $G' = 1 \times 10^{-4}$ N·m²/kg² $\approx 10^6 G$.

55. 5 h 35 min, 19 h 50 min.

57. (a) 10 h; (b) 6.5 km;
(c) 4.2×10^{-3} m/s².

59. 5.4×10^{12} m, in the Solar System, Pluto.

61. $2.3 g_{Earth}$.

63. $m_P = g_P r^2/G$.

67. 7.9×10^3 m/s.

CHAPTER 7

1. 6.86×10^3 J.

3. 1.27×10^4 J.

5. 8.1×10^3 J.

7. 1 J = 1×10^7 erg = 0.738 ft·lb.

9. 1.0×10^4 J.

13. (a) 3.6×10^2 N; (b) -1.3×10^3 J;
(c) -4.6×10^3 J; (d) 5.9×10^3 J;
(e) 0.

15. $W_{FN} = W_{mg} = 0$,
$W_{FP} = -W_{fr} = 2.0 \times 10^2$ J.

21. (a) −16.1; (b) −238; (c) −3.9.

23. $\mathbf{C} = -1.3\mathbf{i} + 1.8\mathbf{j}$.

25. $\theta_x = 42.7°, \theta_y = 63.8°, \theta_z = 121°$.

27. 95°, −35° from x-axis.

31. 0.089 J.

33. 2.3×10^3 J.

35. 2.7×10^3 J.

37. $(kX^2/2) + (aX^4/4) + (bX^5/5)$.

39. (a) 5.0×10^{10} J.

41. (a) $\sqrt{3}$; (b) $\frac{1}{4}$.

43. -5.02×10^{5} J.

45. 3.0×10^{2} N in the direction of the motion of the ball.

47. 24 m/s (87 km/h or 54 mi/h), the mass cancels.

49. (a) 72 J; (b) -35 J; (c) 37 J.

51. 10.2 m/s.

53. $\mu_k = F/2mg$.

55. (a) 6.5×10^{2} J; (b) -4.9×10^{2} J; (c) 0; (d) 4.0 m/s.

57. (a) 1.66×10^{5} J; (b) 21.0 m/s; (c) 2.13 m.

59. $v_p = 2.0 \times 10^{7}$ m/s, $v_{pc} = 2.0 \times 10^{7}$ m/s; $v_e = 2.9 \times 10^{8}$ m/s, $v_{ec} = 8.4 \times 10^{8}$ m/s.

61. 1.74×10^{3} J.

63. (a) 15 J; (b) 4.2×10^{2} J; (c) -1.8×10^{2} J; (d) -2.5×10^{2} J; (e) 0; (f) 10 J.

65. (a) 12 J; (b) 10 J; (c) -2.1 J.

67. 86 kJ, $\theta = 42°$.

69. $(A/k)e^{-(0.10\,\mathrm{m})k}$.

71. 1.5 N.

73. 5.0×10^{3} N/m.

75. (a) 6.6°; (b) 10.3°.

CHAPTER 8

1. 0.924 m.

3. 2.2×10^{3} J.

5. (a) 51.7 J; (b) 15.1 J; (c) 51.7 J.

7. (a) Conservative; (b) $\frac{1}{2}kx^2 - \frac{1}{4}ax^4 - \frac{1}{5}bx^5$ + constant.

9. (a) $\frac{1}{2}k(x^2 - x_0^2)$; (b) same.

11. 45.4 m/s.

13. 6.5 m/s.

15. (a) 1.0×10^{2} N/m; (b) 22 m/s².

17. (a) 8.03 m/s; (b) 3.44 m.

19. (a) $v_{\max} = [v_0^2 + (kx_0^2/m)]^{1/2}$; (b) $x_{\max} = [x_0^2 + (mv_0^2/k)]^{1/2}$.

21. (a) 2.29 m/s; (b) 1.98 m/s; (c) 1.98 m/s; (d) $F_{Ta} = 0.87$ N, $F_{Tb} = 0.80$ N, $F_{Tc} = 0.80$ N; (e) $v_a = 2.59$ m/s, $v_b = 2.31$ m/s, $v_c = 2.31$ m/s.

23. $k = 12Mg/h$.

25. 4.5×10^{6} J.

27. (a) 22 m/s; (b) 2.9×10^{2} m.

29. 13 m/s.

31. 0.23.

33. 0.40.

35. (a) 0.13 m; (b) 0.77; (c) 0.46 m/s.

37. (a) $K = GM_E m_S/2r_S$; (b) $U = -GM_E m_S/r_S$; (c) $-\frac{1}{2}$.

39. (a) 6.2×10^{5} m/s; (b) 4.2×10^{4} m/s, $v_{esc}/v_{orbit} = \sqrt{2}$.

45. (a) 1.07×10^{4} m/s; (b) 1.17×10^{4} m/s; (c) 1.12×10^{4} m/s.

47. (a) $dv_{esc}/dr = -\frac{1}{2}(2GM_E/r^3)^{1/2}$ $= -v_{esc}/2r$; (b) 1.09×10^{4} m/s.

49. $GmM_E/12r_E$.

51. 1.1×10^{4} m/s.

55. 5.4×10^{2} N.

57. (a) 1.0×10^{3} J; (b) 1.0×10^{3} W.

59. 2.1×10^{4} W, 28 hp.

61. 4.8×10^{2} W.

63. 1.2×10^{3} W.

65. 1.8×10^{6} W.

67. (a) -25 W; (b) $+4.3 \times 10^{3}$ W; (c) $+1.5 \times 10^{3}$ W.

69. (a) 80 J; (b) 60 J; (c) 80 J; (d) 5.7 m/s at $x = 0$; (e) 32 m/s² at $x = \pm x_0$.

71. $a^2/4b$.

73. 8.0 m/s.

75. 32.5 hp.

77. (a) 28 m/s; (b) 1.2×10^{2} m.

79. (a) $(2gL)^{1/2}$; (b) $(1.2gL)^{1/2}$.

81. (a) 1.1×10^{6} J; (b) 60 W (0.081 hp); (c) 4.0×10^{2} W (0.54 hp).

83. (a) 40 m/s; (b) 2.6×10^{5} W.

87. (a) 29°; (b) 6.4×10^{2} N; (c) 9.2×10^{2} N.

89. (a) $-\dfrac{U_0}{r}\left(\dfrac{r_0}{r} + 1\right)e^{-r/r_0}$; (b) 0.030; (c) $F(r) = -C/r^2$, 0.11.

91. 6.7 hp.

93. (a) 2.8 m; (b) 1.5 m; (c) 1.5 m.

95. 76 hp.

97. (a) 5.00×10^{3} m/s; (b) 2.89×10^{3} m/s.

CHAPTER 9

1. 6.0×10^{7} N, up.

3. (a) 0.36 kg·m/s; (b) 0.12 kg·m/s.

5. $(26 \,\mathrm{N\cdot s})\mathbf{i} - (28 \,\mathrm{N\cdot s})\mathbf{j}$.

7. (a) $(8h/g)^{1/2}$; (b) $(2gh)^{1/2}$; (c) $-(8m^2 gh)^{1/2}$ (up); (d) mg (down), a surprising result.

9. 3.4×10^{4} kg.

11. 4.4×10^{3} m/s.

13. -0.667 m/s (opposite to the direction of the package).

15. 2, lesser kinetic energy has greater mass.

17. $\frac{3}{2}v_0\mathbf{i} - v_0\mathbf{j}$.

19. 1.1×10^{-22} kg·m/s, 36° from the direction opposite to the electron's.

21. (a) $(100 \,\mathrm{m/s})\mathbf{i} + (50 \,\mathrm{m/s})\mathbf{j}$; (b) 3.3×10^{5} J.

23. 130 N, not large enough.

25. 1.1×10^{3} N.

27. (a) $2mv/\Delta t$; (b) $2mv/t$.

29. (a)

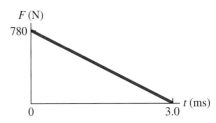

(b) 1.2 N·s; (c) 1.2 N·s; (d) 3.9 g.

31. (a) $(0.84\,\mathrm{N}) + (1.2\,\mathrm{N/s})t$; (b) 18.5 N; (c) $(0.12 \,\mathrm{kg/s})\{[(49 \,\mathrm{m^2/s^2}) - (1.18 \,\mathrm{m^2/s^3})t]^{1/2} + (9.80 \,\mathrm{m/s^2})t\}$, 18.3 N.

33. $v_1' = -1.40$ m/s (rebound), $v_2' = 2.80$ m/s.

35. (a) 2.7 m/s; (b) 0.84 kg.

37. 3.2×10^{3} m/s.

39. (a) 1.00; (b) 0.89; (c) 0.29; (d) 0.019.

41. (a) 0.32 m; (b) -3.1 m/s (rebound), 4.9 m/s; (c) Yes.

43. (a) $+M/(m + M)$; (b) 0.964.

45. 141°.

47. (b) $e = (h'/h)^{1/2}$.

49. (a) $v_1' = v_2' = 1.9$ m/s; (b) $v_1' = -1.6$ m/s, $v_2' = 7.9$ m/s; (c) $v_1' = 0$, $v_2' = 5.2$ m/s; (d) $v_1' = 3.1$ m/s, $v_2' = 0$; (e) $v_1' = -4.0$ m/s, $v_2' = 12$ m/s; result for (c) is reasonable, result for (d) is not reasonable, result for (e) is not reasonable.

51. 61° from first eagles's direction, 6.8 m/s.

53. (a) 30°; (b) $v_2' = v/\sqrt{3}$, $v_1' = v/\sqrt{3}$; (c) $\frac{2}{3}$.

55. $\theta'_1 = 76°$, $v'_n = 5.1 \times 10^5$ m/s,
$v'_{He} = 1.8 \times 10^5$ m/s.

59. 6.5×10^{-11} m from the carbon atom.

61. 0.030 nm above center of H triangle.

63. $x_{CM} = 1.10$ m (East),
$y_{CM} = -1.10$ m (South).

65. $x_{CM} = 0$, $y_{CM} = 2r/\pi$.

67. $x_{CM} = 0$, $y_{CM} = 0$,
$z_{CM} = 3h/4$ above the point.

69. (a) 4.66×10^6 m.

71. (a) $x_{CM} = 4.6$ m; (b) 4.3 m;
(c) 4.6 m.

73. $mv/(M + m)$ up, balloon will also stop.

75. 55 m.

77. 0.899 hp.

79. (a) 2.3×10^3 N; (b) 2.8×10^4 N;
(c) 1.1×10^4 hp.

81. A "scratch shot".

83. 1.4×10^4 N, 43.3°.

85. 5.1×10^2 m/s.

87. $m_2 = 4.00m$.

89. 50%.

91. (a) No; (b) $v_1/v_2 = -m_2/m_1$;
(c) m_2/m_1; (d) does not move;
(e) center of mass will move.

93. 8.29 m/s.

95. (a) 2.5×10^{-13} m/s; (b) 1.7×10^{-17};
(c) 0.19 J.

97. $m \le M/3$.

99. 29.6 km/s.

101. (a) 2.3 N·s; (b) 4.5×10^2 N.

103. (a) Inelastic collision; (b) 0.10 s;
(c) -1.4×10^5 N.

105. 0.28 m, 1.1 m.

CHAPTER 10

1. (a) $\pi/6$ rad $= 0.524$ rad;
(b) $19\pi/60 = 0.995$ rad;
(c) $\pi/2 = 1.571$ rad;
(d) $2\pi = 6.283$ rad;
(e) $7\pi/3 = 7.330$ rad.

3. 2.3×10^3 m.

5. (a) 0.105 rad/s;
(b) 1.75×10^{-3} rad/s;
(c) 1.45×10^{-4} rad/s; (d) zero.

7. (a) 464 m/s; (b) 185 m/s;
(c) 355 m/s.

9. (a) 262 rad/s;
(b) 46 m/s, 1.2×10^4 m/s² radial.

11. 7.4 cm.

13. (a) 1.75×10^{-4} rad/s²;
(b) $a_R = 1.17 \times 10^{-2}$ m/s²,
$a_{tan} = 7.44 \times 10^{-4}$ m/s².

15. (a) 0.58 rad/s2; (b) 12 s.

17. (a) $(1.67 \text{ rad/s}^4)t^3 - (1.75 \text{ rad/s}^3)t^2$;
(b) $(0.418 \text{ rad/s}^4)t^4 - (0.583 \text{ rad/s}^3)t^3$;
(c) 6.4 rad/s, 2.0 rad.

19. (a) ω_1 is in the $-x$-direction,
ω_2 is in the $+z$-direction;
(b) $\omega = 61.0$ rad/s, 35.0° above
$-x$-axis;
(c) $-(1.75 \times 10^3 \text{ rad/s}^2)\mathbf{j}$.

21. (a) 35 m·N; (b) 30 m·N.

23. 1.2 m·N (clockwise).

25. 3.5×10^2 N, 2.0×10^3 N.

27. 53 m·N.

29. (a) 3.5 kg·m²; (b) 0.024 m·N.

31. 2.25×10^3 kg·m², 8.8×10^3 m·N.

33. 9.5×10^4 m·N.

35. 10 m/s.

37. (a)

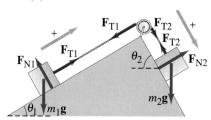

(b) $F_{T1} = 47$ N, $F_{T2} = 75$ N;
(c) 7.0 m·N, 1.7 kg·m².

39. Thin hoop (through center):
$k = R_0$;

Thin hoop (through diameter):
$k = [(R_0^2/2) + (w^2/12)]^{1/2}$;

Solid cylinder (through center):
$k = R/\sqrt{2}$;

Hollow cylinder (through center):
$k = [(R_1^2 + R_2^2)/2]^{1/2}$;

Uniform sphere (through center):
$k = (2r_0^2/5)^{1/2}$;

Rod (through center): $k = \ell/\sqrt{12}$;
Rod (through end): $k = \ell/\sqrt{3}$;

Plate (through center):
$k = [(\ell^2 + w^2)/12]^{1/2}$.

41. (a) 4.18 rad/s²; (b) 8.37 m/s²;
(c) 421 m/s²; (d) 3.07×10^3 N;
(e) 1.14°.

43. (a) $I_a = Ms^2/12$; (b) $I_b = Ms^2/12$.

45. (a) $5.30MR_0^2$; (b) -15%.

47. (a) $9MR_0^2/16$; (b) $MR_0^2/4$;
(c) $5MR_0^2/4$.

51. (b) $M\ell^2/12$, $Mw^2/12$.

53. 0.38 rev/s.

55. (a) As moment of inertia increases,
angular velocity must decrease;
(b) 1.6.

57. (a) 7.1×10^{33} kg·m²/s;
(b) 2.7×10^{40} kg·m²/s.

59. 0.45 rad/s, 0.80 rad/s.

61. 2.33×10^4 J.

63. 5×10^9, loss of gravitational
potential energy.

65. 1.4 m/s.

67. (a) 2.5 kg·m²; (b) 0.58 kg·m²;
(c) 0.35 s; (d) -72 J; (e) rotating.

69. 12.4 m/s.

71. 1.4×10^2 J.

73. (a) 4.48 m/s; (b) 1.21 J;
(c) $\mu_s \ge 0.197$.

75. $v = [10g(R_0 - r_0)/7]^{1/2}$.

77. (a) 4.5×10^5 J; (b) 0.18 (18%);
(c) 1.71 m/s²; (d) 6.4%.

79. (a) $12v_0^2/49\mu_k g$;
(b) $v = 5v_0/7$, $\omega = 5v_0/7R$.

81. (a) 4.5 m/s², 19 rad/s²; (b) 5.8 m/s;
(c) 15.3 J; (d) 1.4 J;
(e) $K = 16.7$ J, $\Delta E = 0$;
(f) $a = 4.5$ m/s², $v = 5.8$ m/s, 14.1 J.

83. $\theta_{Sun} = 9.30 \times 10^{-3}$ rad (0.53°),
$\theta_{Moon} = 9.06 \times 10^{-3}$ rad (0.52°).

85. $\omega_1/\omega_2 = R_2/R_1$.

87. $\ell/2$, $\ell/2$.

89. (a) $-(I_W/I_P)\omega_W$ (down);
(b) $-(I_W/2I_P)\omega_W$ (down);
(c) $(I_W/I_P)\omega_W$ (up); (d) 0.

91. (a) $\omega_R/\omega_F = N_F/N_R$; (b) 4.0;
(c) 1.5.

93. (a) 1.5×10^2 rad/s²;
(b) 1.2×10^3 N.

95. (a) 0.070 rad/s²; (b) 40 rpm.

97. 7.9 N.

99. (b) 2.2×10^3 rad/s; (c) 24 min.

101. (a) 2.9 m; (b) 3.6 m.

103. (a) 1.2 rad/s; (b) 2.0×10^3 J,
1.2×10^3 J, loss of 8.0×10^2 J,
decrease of 40%.

105. (a) 1.7 m/s; (b) 0.84 m/s.

107. (a) $h_{min} = 2.7R_0$;
(b) $h_{min} = 2.7R_0 - 1.7r_0$.

109. (a) 0.84 m/s; (b) 0.96.

CHAPTER 11

7. (a) $-7.0\mathbf{i} - 14.0\mathbf{j} + 19.3\mathbf{k}$; (b) 164°.

11. $-(30.3 \text{ m·N})\mathbf{k}$ (in $-z$-direction).

13. $(18 \text{ m·kN})\mathbf{i} \pm (14 \text{ m·kN})\mathbf{j}$
$\mp (19 \text{ m·kN})\mathbf{k}$.

19. $(55\mathbf{i} - 90\mathbf{j} + 42\mathbf{k})\,\text{kg}\cdot\text{m}^2/\text{s}$.

21. (a) $\left[(7m/9) + (M/6)\right]\ell^2\omega^2$;
(b) $\left[(14m/9) + (M/3)\right]\ell^2\omega$.

23. $2.30\,\text{m/s}^2$.

25. (a) $L = \left[R_0M_1 + R_0M_2 + (I/R_0)\right]v$;
(b) $a = M_2g/\left[M_1 + M_2 + (I/R_0^2)\right]$.

27. Rod rotates at 7.8 rad/s about the center of mass, which moves with constant velocity of 0.21 m/s.

31. $F_1 = \left[(d + r\cos\phi)/2d\right]m_1r\omega^2\sin\phi$,
$F_2 = \left[(d - r\cos\phi)/2d\right]m_1r\omega^2\sin\phi$.

33. $16\,\text{N}, -7.5\,\text{N}$.

35. $3m^2v^2/g(3m + 4M)(m + M)$.

37. $(1 - 4.7 \times 10^{-13})\omega_E$.

39. (a) 14 m/s; (b) 6.8 rad/s.

41. $1.02 \times 10^{-3}\,\text{kg}\cdot\text{m}^2$.

43. 2.2 rad/s (0.35 rev/s).

45. $\tan^{-1}(r\omega^2/g)$.

47. (a) g, along a radial line; (b) $0.998g$, 0.0988° south from a radial line; (c) $0.997g$, along a radial line.

49. North or south direction.

51. (a) South; (b) $\omega D^2\sin\lambda/v_0$; (c) 0.46 m.

53. (a) $(-9.0\mathbf{i} + 12\mathbf{j} - 8.0\mathbf{k})\,\text{kg}\cdot\text{m}^2/\text{s}$;
(b) $(9.0\mathbf{j} - 6.0\mathbf{k})\,\text{m}\cdot\text{N}$.

55. (a) Turn in the direction of the lean;
(b) $\Delta L = 0.98\,\text{kg}\cdot\text{m}^2/\text{s}$,
$\Delta L = 0.18L_0$.

57. (a) $1.8 \times 10^3\,\text{kg}\cdot\text{m}^2/\text{s}^2$;
(b) $1.8 \times 10^3\,\text{m}\cdot\text{N}$; (c) $2.1 \times 10^3\,\text{W}$.

59. $v_{CM} = (3g\ell/4)^{1/2}$.

61. $(19\,\text{m/s})(1 - \cos\theta)^{1/2}$.

63. (a) $2.3 \times 10^4\,\text{rev/s}$;
(b) $5.7 \times 10^3\,\text{rev/s}$.

CHAPTER 12

1. 379 N, 141°.

3. $1.6 \times 10^3\,\text{m}\cdot\text{N}$.

5. 6.52 kg.

7. 2.84 m from the adult.

9. 0.32 m.

11. $F_{T1} = 3.4 \times 10^3\,\text{N}$,
$F_{T2} = 3.9 \times 10^3\,\text{N}$.

13. $F_1 = -2.94 \times 10^3\,\text{N}$ (down),
$F_2 = 1.47 \times 10^4\,\text{N}$.

15. Top hinge: $F_{Ax} = 55.2\,\text{N}$,
$F_{Ay} = 63.7\,\text{N}$; bottom hinge:
$F_{Bx} = -55.2\,\text{N}, F_{By} = 63.7\,\text{N}$.

17. (a)

(b) $1.5 \times 10^4\,\text{N}$; (c) $6.7 \times 10^3\,\text{N}$.

19. $F_T = 1.4 \times 10^3\,\text{N}$ (up),
$F_{bone} = 2.1 \times 10^3\,\text{N}$ (down).

21. $2.7 \times 10^3\,\text{N}$.

23. 89.5 cm from the feet.

25. $F_1 = 5.8 \times 10^3\,\text{N}, F_2 = 5.6 \times 10^3\,\text{N}$.

27. (a) $2.1 \times 10^2\,\text{N}$; (b) $2.0 \times 10^3\,\text{N}$.

29. $7.1 \times 10^2\,\text{N}$.

31. $F_T = 2.5 \times 10^2\,\text{N}$,
$F_{AH} = 2.5 \times 10^2\,\text{N}$,
$F_{AV} = 2.0 \times 10^2\,\text{N}$.

33. (a) 1.00 N; (b) 1.25 N.

35. $\theta_{max} = 40°$, same.

37. (a) $F_T = 182\,\text{N}$; (b) $F_{N1} = 352\,\text{N}$,
$F_{N2} = 236\,\text{N}$; (c) $F_B = 298\,\text{N}, 52.4°$.

39. $1.0 \times 10^2\,\text{N}$.

41. (a) $1.2 \times 10^5\,\text{N/m}^2$; (b) 2.4×10^{-6}.

43. (a) $1.3 \times 10^5\,\text{N/m}^2$; (b) 6.5×10^{-7};
(c) 0.0062 mm.

45. $9.6 \times 10^6\,\text{N/m}^2$.

47. $9.0 \times 10^7\,\text{N/m}^2, 9.0 \times 10^2\,\text{atm}$.

49. $2.2 \times 10^7\,\text{N}$.

51. (a) $1.1 \times 10^2\,\text{m}\cdot\text{N}$; (b) wall;
(c) all three.

53. $3.9 \times 10^2\,\text{N}$, thicker strings, maximum strength is exceeded.

55. (a) $4.4 \times 10^{-5}\,\text{m}^2$; (b) 2.7 mm.

57. 1.2 cm.

61. (a) $F_T = 129\,\text{kN}$;
$F_A = 141\,\text{kN}, 23.5°$;
(b) $F_{DE} = 64.7\,\text{kN}$ (tension),
$F_{CE} = 32.3\,\text{kN}$ (compression),
$F_{CD} = 64.7\,\text{kN}$ (compression),
$F_{BD} = 64.7\,\text{kN}$ (tension),
$F_{BC} = 64.7\,\text{kN}$ (tension),
$F_{AC} = 97.0\,\text{kN}$ (compression),
$F_{AB} = 64.7\,\text{kN}$ (compression).

63. (a) $4.8 \times 10^{-2}\,\text{m}^2$; (b) $6.8 \times 10^{-2}\,\text{m}^2$.

65. $F_{AB} = 5.44 \times 10^4\,\text{N}$ (compression),
$F_{ACx} = 2.72 \times 10^4\,\text{N}$ (tension),
$F_{BC} = 5.44 \times 10^4\,\text{N}$ (tension),
$F_{BD} = 5.44 \times 10^4\,\text{N}$ (compression),
$F_{CD} = 5.44 \times 10^4\,\text{N}$ (tension),
$F_{CE} = 2.72 \times 10^4\,\text{N}$ (tension),
$F_{DE} = 5.44 \times 10^4\,\text{N}$ (compression).

67. 12 m.

69. $M_C = 0.191\,\text{kg}, M_D = 0.0544\,\text{kg}$,
$M_A = 0.245\,\text{kg}$.

71. (a) $Mg\left[h/(2R - h)\right]^{1/2}$;
(b) $Mg\left[h(2R - h)\right]^{1/2}/(R - h)$.

73. $\theta_{max} = 29°$.

75. 6, 2.0 m apart.

77. 3.8.

79. $5.0 \times 10^5\,\text{N}, 3.2\,\text{m}$.

81. (a) 600 N; (b) $F_A = 0, F_B = 1200\,\text{N}$;
(c) $F_A = 150\,\text{N}, F_B = 1050\,\text{N}$;
(d) $F_A = 750\,\text{N}, F_B = 450\,\text{N}$.

83. $6.5 \times 10^2\,\text{N}$.

85. 0.67 m.

87. Right end is safe, left end is not safe, 0.10 m.

89. (a) $F_L = 3.3 \times 10^2\,\text{N}$ up,
$F_R = 2.3 \times 10^2\,\text{N}$ down;
(b) 65 cm from right hand;
(c) 123 cm from right hand.

91. $\theta \geq 40°$.

93. (b) beyond the table;
(c) $D = L\sum_{i=1}^{n}\frac{1}{2i}$; (d) 32 bricks.

95. $F_{TB} = 134\,\text{N}, F_{TA} = 300\,\text{N}$.

97. $2.6w, 31°$ above horizontal.

CHAPTER 13

1. $3 \times 10^{11}\,\text{kg}$.

3. $4.3 \times 10^2\,\text{kg}$.

5. 0.8477.

7. (a) $3 \times 10^7\,\text{N/m}^2$;
(b) $2 \times 10^5\,\text{N/m}^2$.

9. 1.1 m.

11. $8.28 \times 10^3\,\text{kg}$.

13. $1.2 \times 10^5\,\text{N/m}^2$,
$2.3 \times 10^7\,\text{N}$ (down),
$1.2 \times 10^5\,\text{N/m}^2$.

15. $6.54 \times 10^2\,\text{kg/m}^3$.

17. $3.36 \times 10^4\,\text{N/m}^2$ (0.331 atm).

19. (a) $1.41 \times 10^5\,\text{Pa}$; (b) $9.8 \times 10^4\,\text{Pa}$.

21. (a) 0.34 kg; (b) $1.5 \times 10^4\,\text{N}$ (up).

23. (c) $\geq 0.38h$, no.

27. $4.70 \times 10^3\,\text{kg/m}^3$.

29. $8.5 \times 10^2\,\text{kg}$.

31. Copper.

33. (a) $1.14 \times 10^6\,\text{N}$; (b) $4.0 \times 10^5\,\text{N}$.

35. (b) Above the center of gravity.

37. 0.88.

39. $7.9 \times 10^2\,\text{kg}$.

43. 4.1 m/s.

45. 9.5 m/s.

47. $1.5 \times 10^5\,\text{N/m}^2 = 1.5\,\text{atm}$.

49. $4.11 \times 10^{-3}\,\text{m}^3/\text{s}$.

51. $1.7 \times 10^6\,\text{N}$.

59. (a) $2[h_1(h_2 - h_1)]^{1/2}$;
(b) $h_1' = h_2 - h_1$.

61. 0.072 Pa·s.

63. 4.0×10^3 Pa.

65. 11 cm.

67. (a) Laminar; (b) 3200, turbulent.

69. 1.9 m.

71. 9.1×10^{-3} N.

73. (a) $\gamma = F/4\pi r$; (b) 0.024 N/m.

75. (a) 0.88 m; (b) 0.55 m; (c) 0.24 m.

77. $1.5 \times 10^2\,\text{N} \le F \le 2.2 \times 10^2\,\text{N}$.

79. 0.051 atm.

81. 0.63 N.

83. 5 km.

85. 5.3×10^{18} kg.

87. 2.6 m.

89. 39 people.

91. 37 N, not float.

93. $d = D[v_0^2/(v_0^2 + 2gy)]^{1/4}$.

95. (a) 3.2 m/s; (b) 19 s.

97. 1.9×10^2 m/s.

CHAPTER 14

1. 0.60 m.

3. 1.15 Hz.

5. (a) 2.4 N/m; (b) 12 Hz.

7. (a) $0.866\,x_{max}$; (b) $0.500\,x_{max}$.

9. $0.866\,A$.

11. $[(k_1 + k_2)/m]^{1/2}/2\pi$.

13. (a) 8/7 s, 0.875 Hz; (b) 3.3 m, -10.4 m/s; (c) $+18$ m/s, -57 m/s^2.

15. 3.6 Hz.

19. (a) $y = -(0.220\,\text{m}) \sin[(37.1\,\text{s}^{-1})t]$;
(b) maximum extensions at 0.0423 s, 0.211 s, 0.381 s,…; minimum extensions at 0.127 s, 0.296 s, 0.465 s,….

21. $f = (3k/M)^{1/2}/2\pi$.

25. (a) $x = (12.0\,\text{cm}) \cos[(25.6\,\text{s}^{-1})t + 1.89\,\text{rad}]$;
(b) $t_{max} = 0.294\,\text{s}, 0.539\,\text{s}, 0.784\,\text{s},…$;
$t_{min} = 0.171\,\text{s}, 0.416\,\text{s}, 0.661\,\text{s},…$;
(c) -3.77 cm; (d) $+13.1$ N (up);
(e) 3.07 m/s, 0.110 s.

27. (a) 0.650 m; (b) 1.34 Hz; (c) 29.8 J;
(d) $K = 25.0\,\text{J}, U = 4.8\,\text{J}$.

29. 9.37 m/s.

31. $A_1 = 2.24A_2$.

33. (a) 4.2×10^2 N/m; (b) 3.3 kg.

35. 352.6 m/s.

39. 0.9929 m.

41. (a)0.248 m; (b)2.01 s.

43. (a) $-12°$; (b) $+1.9°$; (c) $-13°$.

45. $\frac{1}{3}$.

47. 1.08 s.

49. 0.31 g.

51. (a) 1.6 s.

53. 3.5 s.

55. (a) 0.727 s; (b) 0.0755;
(c) $x = (0.189\,\text{m})e^{-(0.108/\text{s})t}$
$\sin[(8.64\,\text{s}^{-1})t]$.

57. (a) $8.3 \times 10^{-4}\%$; (b) 39 periods.

59. (a) 5.03 Hz; (b) 0.0634 s^{-1};
(c) 110 oscillations.

61. A_0k/F_0

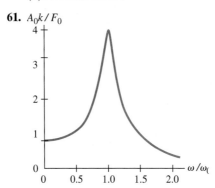

65. (a) 198 s; (b) 8.7×10^{-6} W;
(c) 8.8×10^{-4} Hz on either side of f_0.

69. (a) 0.63 Hz; (b) 0.65 m/s; (c) 0.077 J.

71. 151 N/m, 20.3 m.

73. 0.11 m.

75. 3.6 Hz.

77. (a) 1.1 Hz; (b) 13 J.

79. (a) 90 N/m; (b) 8.9 cm.

81. $k = \rho_{water}\,gA$.

83. Water will oscillate with SHM, $k = 2\rho gA$, the density and the cross section.

85. $T = 2\pi(ma/2k\,\Delta a)^{1/2}$.

87. (a) 1.64 s; (b) 0.67 m.

CHAPTER 15

1. 2.3 m/s.

3. 1.26 m.

5. 0.72 m.

7. 2.7 N.

9. (a) 75 m/s; (b) 7.8×10^3 N.

11. (a) 1.3×10^3 km;
(b) cannot be determined.

13. (a) 0.25; (b) 0.50.

17. (a) 0.30 W; (b) 0.25 cm.

19. $D = D_M \sin[2\pi(x/\lambda + t/T) + \phi]$.

21. (a) 41 m/s; (b) 6.4×10^4 m/s^2;
(c) 41 m/s, 8.2×10^3 m/s^2.

23. (a, c)

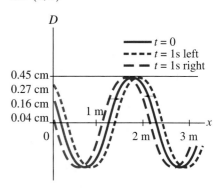

(b) $D = (0.45\,\text{m}) \cos[(3.0\,\text{m}^{-1})x - (6.0\,\text{s}^{-1})t + 1.2]$.
(d) $D = (0.45\,\text{m}) \cos[(3.0\,\text{m}^{-1})x + (6.0\,\text{s}^{-1})t + 1.2]$.

25. $D = -(0.020\,\text{cm}) \cos[(8.01\,\text{m}^{-1})x - (2.76 \times 10^3\,\text{s}^{-1})t]$.

27. The function is a solution.

31. (a) $v_2/v_1 = (\mu_1/\mu_2)^{1/2}$;
(b) $\lambda_2/\lambda_1 = v_2/v_1 = (\mu_1/\mu_2)^{1/2}$;
(c) lighter cord.

33. (c) $A_T = [2k_1/(k_2 + k_1)]A = [2v_2/(v_1 + v_2)]A$.

35. (b) $2D_M \cos(\frac{1}{2}\phi)$, purely sinusoidal;
(d) $D = \sqrt{2}\,D_M \sin(kx - \omega t + \pi/4)$.

37. 440 Hz, 880 Hz, 1320 Hz, 1760 Hz.

39. $f_n = n(0.50\,\text{Hz}), n = 1, 2, 3,…$;
$T_n = (2.0\,\text{s})/n, n = 1, 2, 3,…$.

41. 70 Hz.

45. 4.

47. (a) $D_2 = (4.2\,\text{cm}) \sin[(0.71\,\text{cm}^{-1})x + (47\,\text{s}^{-1})t + 2.1]$;
(b) $D_{resultant} = (8.4\,\text{cm})$
$\sin[(0.71\,\text{cm}^{-1})x + 2.1]$
$\cos[(47\,\text{s}^{-1})t]$.

49. 308 Hz.

51. (a)

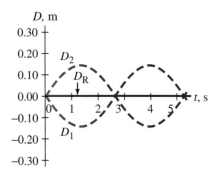

(b)

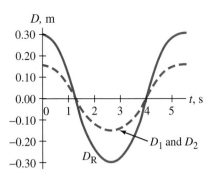

53. 5.4 km/s.

55. 29°.

57. 24°.

59. Speed will be greater in the less dense rod by a factor of $\sqrt{2}$.

61. (a) 0.050 m; (b) 2.3.

63. 0.99 m.

65. (a) solid curves,

(c) dashed curves;

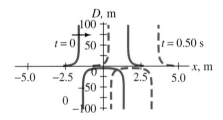

(b)
$D = (4.0 \, \text{m}^3)/\{[x - (3.0 \, \text{m/s})t]^2 - 2.0 \, \text{m}^2\};$

(d)

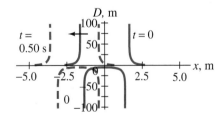

$D = (4.0 \, \text{m}^3)/\{[x + (3.0 \, \text{m/s})t]^2 - 2.0 \, \text{m}^2\}.$

67. (a) 784 Hz, 1176 Hz, 880 Hz,

1320 Hz; (b) 1.26; (c) 1.12; (d) 0.794.

69. $\lambda_n = 4L/(2n - 1), n = 1, 2, 3, \ldots$

71. $y = (3.5 \, \text{cm}) \cos[(1.05 \, \text{cm}^{-1})x$

$- (1.39 \, \text{s}^{-1})t].$

73.

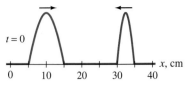

$t = 0$

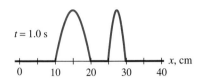

$t = 1.0 \, \text{s}$

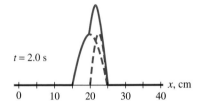

$t = 2.0 \, \text{s}$

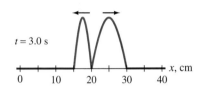

$t = 3.0 \, \text{s}$

CHAPTER 16

1. $2.6 \times 10^2 \, \text{m}.$

3. $5.4 \times 10^2 \, \text{m}.$

5. 1200 m, 300 m.

7. (a) $1.1 \times 10^{-8} \, \text{m}$; (b) $1.1 \times 10^{-10} \, \text{m}.$

9. (a) $\Delta P = (4 \times 10^{-5} \, \text{Pa})$

$\sin[(0.949 \, \text{m}^{-1})x$

$- (315 \, \text{s}^{-1})t];$

(b) $\Delta P = (4 \times 10^{-3} \, \text{Pa})$

$\sin[(94.9 \, \text{m}^{-1})x$

$- (3.15 \times 10^4 \, \text{s}^{-1})t].$

11. (a) 49 dB; (b) $3.2 \times 10^{-10} \, \text{W/m}^2.$

13. 150 Hz to 20,000 Hz.

15. (a) 9; (b) 9.5 dB.

17. (a) Higher frequency is greater by a factor of 2; (b) 4.

19. (a) $5.0 \times 10^{-13} \, \text{W}$; (b) $6.3 \times 10^4 \, \text{yr}.$

21. 87 dB.

23. (a) $5.10 \times 10^{-5} \, \text{m}$; (b) 29.8 Pa.

25. (a) $1.5 \times 10^3 \, \text{W}$; (b) $3.4 \times 10^2 \, \text{m}.$

27. (b) 190 dB.

29. (a) 570 Hz; (b) 860 Hz.

31. 8.6 mm $< L <$ 8.6 m.

33. (a) 110 Hz, 330 Hz, 550 Hz, 770 Hz; (b) 220 Hz, 440 Hz, 660 Hz, 880 Hz.

35. (a) 0.656 m; (b) 262 Hz, 1.31 m; (c) 1.31 m, 262 Hz.

37. $-2.6\%.$

39. (a) 0.578 m; (b) 869 Hz.

41. 215 m/s.

43. 0.64, 0.20, -2 dB, -7 dB.

45. 28.5 kHz.

47. (a) 130.5 Hz, or 133.5 Hz; (b) $\pm 2.3\%.$

49. (a) 343 Hz; (b) 1030 Hz, 1715 Hz.

53. 346 Hz.

57. (a) 1690 Hz; (b) 1410 Hz.

59. 30,890 Hz.

61. 120 Hz.

63. 91 Hz.

65. 90 beats/min.

67. (a) 570 Hz; (b) 570 Hz; (c) 570 Hz; (d) 570 Hz; (e) 594 Hz; (f) 595 Hz.

71. (a) 120; (b) 0.96°.

73. (a) 37°; (b) 1.7.

75. 0.278 s.

77. 55 m.

79. 410 km/h (255 mi/h).

81. 1, 0.444, 0.198, 0.0878, 0.0389.

83. 18.1 W.

85. 15 W.

87. 2.3 Hz.

89. $\Delta P_M/\Delta P_{M0} = D_M/D_{M0} = 10^6.$

91. 50 dB.

93. 17.5 m/s.

95. 2.3 kHz.

97. 550 Hz.

99. (a) $2.8 \times 10^3 \, \text{Hz}$; (b) 1.80 m; (c) 0.12 m.

101. (a) $2.2 \times 10^{-7} \, \text{m}$; (b) $5.4 \times 10^{-5} \, \text{m}.$

CHAPTER 17

1. 0.548.

3. (a) 20°C; (b) $\approx$3300°F.

5. 104.0°F.

7. -40°F $= -40$°C.

9. $\Delta L_{\text{Invar}} = 2.0 \times 10^{-6} \, \text{m},$
$\Delta L_{\text{steel}} = 1.2 \times 10^{-4} \, \text{m},$
$\Delta L_{\text{marble}} = 2.5 \times 10^{-5} \, \text{m}.$

11. -69°C.

13. 5.1 mL.

15. 0.06 cm^3.

19. -40 min.

21. -2.8×10^{-3} (0.28%).

23. $3.5 \times 10^7 \, \text{N/m}^2.$

25. (a) 27°C; (b) $4.3 \times 10^3 \, \text{N}.$

27. -459.7°F.

29. 1.07 m^3.

31. 1.43 kg/m^3.

33. (a) 14.8 m³; (b) 1.81 atm.
35. 1.80×10^3 atm.
37. 37°C.
39. 3.43 atm.
41. 0.588 kg/m^3, water vapor is not an ideal gas.
43. 2.69×10^{25} molecules/m³.
45. 4.9×10^{22} molecules.
47. 7.7×10^3 N.
49. (a) 71.2 torr; (b) 157°C.
51. (a) 0.19 K; (b) 0.051%.
53. (a) Low; (b) 0.017%.
55. 1/6.
57. 5.1×10^{27} molecules, 8.4×10^3 mol.
59. 11 L, not advisable.
61. (a) 9.3×10^2 kg; (b) 1.0×10^2 kg.
63. 1.1×10^{44} molecules.
65. 3.3×10^{-7} cm.
67. 1.1×10^3 m.
69. 15 h.
71. 0.66×10^3 kg/m³.
73. ±0.11 C°.
77. 3.6 m.

CHAPTER 18

1. (a) 5.65×10^{-21} J; (b) 7.3×10^3 J.
3. 1.17.
5. (a) 4.5; (b) 5.2.
7. $\sqrt{2}$.
11. (a) 461 m/s; (b) 19 s^{-1}.
13. 1.00429.
17. Vapor.
19. (a) Gas, liquid, vapor;
    (b) gas, liquid, solid, vapor.
21. 0.69 atm.
23. 11°C.
25. 1.96 atm.
27. 120°C.
29. (a) 5.3×10^6 Pa; (b) 5.7×10^6 Pa.
31. (b) $b = 4.28 \times 10^{-5}$ m³/mol,
    $a = 0.365 \text{ N} \cdot \text{m}^4/\text{mol}^2$.
33. (a) 10^{-7} atm; (b) 300 atm.
35. (a) 6.3 cm; (b) 0.58 cm.
37. 2×10^{-7} m.
39. (b) $4.7 \times 10^7 \text{ s}^{-1}$.
43. 7.8 h.
45. (b) 4×10^{-11} mol/s; (c) 0.7 s.
47. 2.6×10^2 m/s,
    $4 \times 10^{-17} \text{ N/m}^2 \approx 4 \times 10^{-22}$ atm.
49. (a) 2.9×10^2 m/s; (b) 12 m/s.
51. Reasonable, 70 cm.

53. $mgh = 4.3 \times 10^{-5}(\frac{1}{2}mv_{rms}^2)$,
    reasonable.
55. $P_2/P_1 = 1.43$, $T_2/T_1 = 1.20$.
57. 1.4×10^5 K.
59. (a) 1.7×10^3 Pa; (b) 7.0×10^2 Pa.
61. 2×10^{13} m.

CHAPTER 19

1. 1.0×10^7 J.
3. 1.8×10^2 J.
5. 2.1×10^2 kg/h.
7. 83 kcal.
9. 4.7×10^6 J.
11. 40 C°.
13. 186°C.
15. 7.1 min.
17. (b) $mc_0[(T_2 - T_1) + a(T_2^2 - T_1^2)/2]$;
    (c) $c_{mean} = c_0[1 + \frac{1}{2}a(T_2 + T_1)]$.
19. 0.334 kg (0.334 L).
21. $\frac{2}{3}m$ steam and $\frac{4}{3}m$ water at 100°C.
23. 9.4 g.
25. 4.7×10^3 kcal.
27. 1.22×10^4 J/kg.
29. 360 m/s.
31. (a) 0; (b) 5.00×10^3 J.

33.
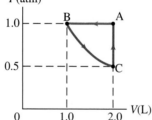

35. (a) 0; (b) −1300 kJ.
37. (a) 1.6×10^2 J; (b) $+1.6 \times 10^2$ J.
39. $W = 3.46 \times 10^3$ J, $\Delta U = 0$,
    $Q = +3.46 \times 10^3$ J (into the gas).
41. +129 J.
45. (a) +25 J; (b) +63 J; (c) −95 J;
    (d) −120 J; (e) −15 J.
47. $W = RT \ln\left(\dfrac{V_2 - b}{V_1 - b}\right) + a\left(\dfrac{1}{V_2} - \dfrac{1}{V_1}\right)$.
49. 22°C/h.
51. 4.98 cal/mol·K, 2.49 kcal/kg·K;
    6.97 cal/mol·K, 3.48 kcal/kg·K.
53. 83.7 g/mol, krypton.
55. 46 C°.
57. (a) 2.08×10^3 J; (b) 8.32×10^2 J;
    (c) 2.91×10^3 J.
59. 0.379 atm, −51°C.
61. 1.33×10^3 J.

63. (a) $T_1 = 317$ K, $T_2 = 153$ K;
    (b) -1.59×10^4 J; (c) -1.59×10^4 J;
    (d) $Q = 0$.
65. (a)
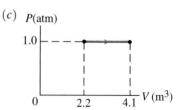
    (b) 231 K;
    (c) $Q_{1\to2} = 0$,
        $\Delta U_{1\to2} = -2.01 \times 10^3$ J,
        $W_{1\to2} = +2.01 \times 10^3$ J;
        $W_{2\to3} = -1.31 \times 10^3$ J,
        $\Delta U_{2\to3} = -1.97 \times 10^3$ J,
        $Q_{2\to3} = -3.28 \times 10^3$ J;
        $W_{3\to1} = 0$,
        $\Delta U_{3\to1} = +3.98 \times 10^3$ J,
        $Q_{3\to1} = +3.98 \times 10^3$ J;
    (d) $W_{cycle} = +0.70 \times 10^3$ J,
        $Q_{cycle} = +0.70 \times 10^3$ J,
        $\Delta U_{cycle} = 0$.
67. (a) 64 W; (b) 22 W.
69. 4.8×10^2 W.
71. 31 h.
73. (a) 1.7×10^{17} W; (b) 278K (5°C).
75. (b) $\Delta Q/\Delta t = A(T_2 - T_1)/\Sigma(\ell_i/k_i)$.
77. 22%.
79. 4×10^{15} J.
81. 2.8 kcal/kg.
83. 30 C°.
85. 682 J.
87. 2.8 C°.
89. 2.58 cm, rod vaporizes.
91. 4.3 kg.
95. (a) 2.3 C°/s; (b) 84°C;
    (c) convection, conduction, evaporation.
97. (a) $\rho = m/V = (mP/nR)/T$;
    (b) $\rho = (m/nRT)P$.
99. (a) 1.9×10^5 J; (b) -1.4×10^5 J;
    (c)
101. 3.2×10^5 s = 3.7 d.
103. 10 C°.

CHAPTER 20

1. 24%.
3. 816 MW.
5. 18%.
7. 13 km³/day, 63 km².
9. 28.0%.
13. 1.2×10^{13} J/h.
15. 1.4×10^3 m/day.
17. 660°C.
19. 3.7×10^8 kg/h.
21. (a) $P_a = 5.15 \times 10^5$ Pa,
    $P_b = 2.06 \times 10^5$ Pa;
    (b) $V_c = 30.0$ L, $V_d = 12.0$ L;
    (c) 2.83×10^3 J; (d) -2.14×10^3 J;
    (e) 0.69×10^3 J; (f) 24%.
23. 5.7.
25. −21°C.
27. 2.9.
29. (a) 3.9×10^4 J; (b) 3.0 min.
31. 76 L.
33. 0.15 J/K.
35. +11 kcal/K.
37. +0.0104 cal/K·s.
39. 1.7×10^2 J/K.
43. (a) 0.312 kcal/K;
    (b) > -0.312 kcal/K.
45. (a) Adiabatic process;
    (b) $\Delta S_i = -nR \ln 2, \Delta S_a = 0$;
    (c) $\Delta S_{surr,i} = nR \ln 2, \Delta S_{surr,a} = 0$.
47. (a) Entropy is a state function.
51. (a) 5/16; (b) 1/64.
53. (b), (a), (c), (d).
55. 69%.
57. 2.6×10^3 J/K.
59. (a) 17; (b) 5.9×10^7 J/h.
61. (a) 5.3 C°; (b) +77 J/kg·K.
63. $\Delta S = K/T$.
65. (a)

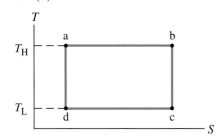

    (b) area $= Q_{net} = W_{net}$.
67. $e_{Stirling} = (T_H - T_L) \ln(V_b/V_a) /$
    $[T_H \ln(V_b/V_a)$
    $+ \frac{3}{2}(T_H - T_L)]$,
    $e_{Stirling} < e_{Carnot}$.

69. 0.091 hp.
71. (a) 1/379; (b) $1/1.59 \times 10^{11}$.

CHAPTER 21

1. 6.3×10^9 N.
3. 2.7×10^{-3} N.
5. 5.5×10^3 N.
7. 8.66 cm.
9. -5.4×10^7 C.
11. 83.8 N away from the center of the triangle.
13. 2.96×10^7 N toward the center of the square.
15. $\mathbf{F}_1 = (kQ^2/\ell^2)[(-2 + 3\sqrt{2}/4)\mathbf{i}$
    $+ (4 - 3\sqrt{2}/4)\mathbf{j}], \mathbf{F}_2 = (kQ^2/\ell^2)$
    $[(2 + 2\sqrt{2})\mathbf{i} + (-6 + 2\sqrt{2})\mathbf{j}]$,
    $\mathbf{F}_3 = (kQ^2/\ell^2)[(-12 -3\sqrt{2}/4)\mathbf{i}$
    $+ (6 + 3\sqrt{2}/4)\mathbf{j}], \mathbf{F}_4 = (kQ^2/\ell^2)$
    $[(12 - 2\sqrt{2})\mathbf{i} + (-4 -2\sqrt{2})\mathbf{j}]$.
17. (a) $Q_1 = Q_2 = \frac{1}{2}Q_T$;
    (b) Q_1 (or Q_2) $= 0$.
19. $0.402Q_0, 0.366\ell$ from Q_0.
21. 60.2×10^{-6} C, 29.8×10^{-6} C;
    -16.8×10^{-6} C, 106.8×10^{-6} C.
23. $\mathbf{F} = -(1.90kQ^2/\ell^2)(\mathbf{i} + \mathbf{j} + \mathbf{k})$.
25. 2.18×10^{-16} N (west).
27. 7.43×10^6 N/C (up).
29. $(1.39 \times 10^2$ N/C)$\mathbf{j}$.
33.

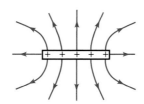

35. 8.26×10^{-10} N/C (south).
37. 4.5×10^6 N/C up, 1.2×10^7 N/C, 56° above the horizontal.
39. 5.61×10^4 N/C away from the opposite corner.
41. $Q_1/Q_2 = \frac{1}{4}$.
43. (a) $2Qy/4\pi\epsilon_0(y^2 + \ell^2)^{3/2}\,\mathbf{j}$.
45. $\dfrac{Q}{4\pi\epsilon_0}\left[\dfrac{x\mathbf{i} - (2a/\pi)\mathbf{j}}{(x^2 + a^2)^{3/2}}\right]$.
49. $\dfrac{-2\lambda \sin \theta_0}{4\pi\epsilon_0 R}\,\mathbf{i}$.
51. (a) $\dfrac{\lambda}{4\pi\epsilon_0 x(x^2 + L^2)^{1/2}}$
    $\{L\mathbf{i} + [x - (x^2 + L^2)^{1/2}]\mathbf{j}\}$.
53. $(\sigma/2\epsilon_0)\mathbf{k}$.
55. (a) $\mathbf{a} = -(3.5 \times 10^{15}$ m/s²$)\,\mathbf{i}$
    $-(1.41 \times 10^{16}$ m/s²$)\,\mathbf{j}$; (b) $\theta = -104°$.
57. $\theta = -28°$.

59. (b) $2\pi(4\pi\epsilon_0 mR^3/qQ)^{1/2}$.
61. (a) 3.4×10^{-20} C; (b) No;
    (c) 8.5×10^{-26} m·N;
    (d) 2.5×10^{-26} J.
63. (a) $\theta \ll 1$;
    (b) $(pE/I)^{1/2}/2\pi$.
65. (b) Direction of the dipole.
67. 6.8×10^3 C.
69. 5.7×10^{13} C.
71. $\mathbf{F}_1 = 0.30$ N, 265° from x-axis,
    $\mathbf{F}_2 = 0.26$ N, 139° from x-axis,
    $\mathbf{F}_3 = 0.26$ N, 30° from x-axis.
73. 4.2×10^5 N/C up.
75. $0.444Q_0, 0.333\ell$ from Q_0.
77. 5.60 m from the positive charge, and 3.60 m from the negative charge.
79. (a) In the direction of the velocity, to the right; (b) 2.1×10^2 N/C.
81. $\theta_0 = 18°$.
83. $(1.08 \times 10^7$ N/C)/
    $\{3.00 - \cos[(12.5\ \text{s}^{-1})t]\}^2$, up.
85. $E_A = 4.2 \times 10^4$ N/C (right),
    $E_B = -1.4 \times 10^4$ N/C (left),
    $E_C = -2.8 \times 10^3$ N/C (left),
    $E_D = -4.2 \times 10^4$ N/C (left).
87. $d(1 + \sqrt{2})$ from the negative charge, and $d(2 + \sqrt{2})$ from the positive charge.

CHAPTER 22

1. (a) 41 N·m²/C; (b) 29 N·m²/C;
    (c) 0.
3. $\Phi_{net} = 0$,
    $\Phi_{x=0} = -(6.50 \times 10^3$ N/C)ℓ^2,
    $\Phi_{x=\ell} = +(6.50 \times 10^3$ N/C)ℓ^2,
    $\Phi_{all\ others} = 0$.
5. 12.8 nC.
7. −1.2 μC.
9. -3.75×10^{-11} C.
11. (a) -1.0×10^4 N/C (toward wire);
    (b) -2.5×10^4 N/C (toward wire).
13.

$E\ (10^7$ N/C)

1.9 ⟋⟍ at peak near 7.5, curve to 15, r (cm)

15. (a) 5.5×10^7 N/C (away from center);
    (b) 0.
17. (a) -8.00 μC;
    (b) $+1.00$ μC
19. (a) 0; (b) σ/ϵ_0;
    (c) unaffected.

21. (a) 0; (b) $\sigma_1 r_1^2/\epsilon_0 r^2$;
    (c) $(\sigma_1 r_1^2 + \sigma_2 r_2^2)/\epsilon_0 r^2$;
    (d) $\sigma_2/\sigma_1 = -(r_1/r_2)^2$; (e) $\sigma_1 = 0$.

23. (a) $q/4\pi\epsilon_0 r^2$;
    (b) $(1/4\pi\epsilon_0)[Q(r^3 - r_1^3)$
      $+ q(r_0^3 - r_1^3)]/(r_0^3 - r_1^3)r^2$;
    (c) $(q + Q)/4\pi\epsilon_0 r^2$.

25. (a) $q/4\pi\epsilon_0 r^2$; (b) $(q + Q)/4\pi\epsilon_0 r^2$;
    (c) $E(r < r_0) = Q/4\pi\epsilon_0 r^2$,
      $E(r > r_0) = 2Q/4\pi\epsilon_0 r^2$;
    (d) $E(r < r_0) = -Q/4\pi\epsilon_0 r^2$,
      $E(r > r_0) = 0$.

27. (a) $\sigma R_0/\epsilon_0 r$; (b) 0; (c) same.

29. (a) 0; (b) $Q/2\pi\epsilon_0 Lr$;
    (c) 0; (d) $eQ/4\pi\epsilon_0 L$.

31. (a) 0;
    (b) -2.3×10^5 N/C (toward the axis);
    (c) -1.8×10^4 N/C (toward the axis).

33. (a) $\rho_E r/2\epsilon_0$; (b) $\rho_E R_1^2/2\epsilon_0 r$;
    (c) $\rho_E(r^2 + R_1^2 - R_2^2)/2\epsilon_0 r$;
    (d) $\rho_E(R_3^2 + R_1^2 - R_2^2)/2\epsilon_0 r$;
    (e)

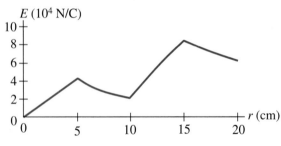

E (10^4 N/C)

35. $Q/\epsilon_0\sqrt{2}$.

37. $\oint \mathbf{g} \cdot d\mathbf{A} = -4\pi GM$.

39. $Q_{enclosed} = \epsilon_0 b\ell^3$.

41. 3.95×10^2 N·m²/C,
    -1.69×10^2 N·m²/C.

43. (a) 0; (b) $Q/25\pi\epsilon_0 r_0^2 \le E \le Q/\pi\epsilon_0 r_0^2$;
    (c) not perpendicular;
    (d) not useful.

45. (a) $0.677e = +1.08 \times 10^{-19}$ C;
    (b) 3.5×10^{11} N/C.

47. (a) $\rho_E r_0/6\epsilon_0$ (right);
    (b) $-17\rho_E r_0/54\epsilon_0$ (left).

49. (a) 0; (b) 5.65×10^5 N/C (right);
    (c) 5.65×10^5 N/C (right);
    (d) -5.00×10^{-6} C/m²;
    (e) $+5.00 \times 10^{-6}$ C/m².

CHAPTER 23

1. -4.2×10^{-5} J (done by the field).

3. 3.4×10^{-15} J.

5. $V_a - V_b = +72.8$ V.

7. 7.04 V.

9. 0.8 μC.

11. (a) $V_{BA} = 0$; (b) $V_{CB} = -2100$ V;
    (c) $V_{CA} = -2100$ V.

13. (a) -9.6×10^8 V;
    (b) $V(\infty) = +9.6 \times 10^8$ V.

15. (a) The same;
    (b) $Q_2 = r_2 Q/(r_1 + r_2)$.

17. (a) $Q/4\pi\epsilon_0 r$;
    (b) $(Q/8\pi\epsilon_0 r_0)[3 - (r^2/r_0^2)]$;
    (c)

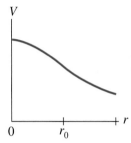

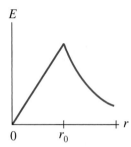

19. (a) $V_0 + (\sigma R_0/\epsilon_0) \ln(R_0/r)$;
    (b) $V = V_0$; (c) $V \ne 0$.

21. (a) 29 V;
    (b) -29 eV $(-4.6 \times 10^{-18}$ J).

23. $+0.19$ J.

25. 4.2 MV.

27. 2.33×10^7 m/s.

29. $V_{BA} = (1/2\pi\epsilon_0)q(2b - d)/b(d - b)$.

31. $\dfrac{\sigma}{2\epsilon_0}[(x^2 + R_2^2)^{1/2} - (x^2 + R_1^2)^{1/2}]$.

33. $\dfrac{Q}{8\pi\epsilon_0 L} \ln\left(\dfrac{x + L}{x - L}\right)$, $x > L$.

35. $\dfrac{a}{6\epsilon_0}[(x^2 + R^2)^{1/2}(R^2 - 2x^2) + 2x^3]$.

37. 3.2 mm.

39. (a) 8.5×10^{-30} C·m; (b) zero.

41. (a) -0.088 V; (b) 1%.

43. (a) 5.2×10^{-20} C.

47. $\mathbf{E} = 2y(2z - 1)\mathbf{i} - 2(y + x - 2xz)\mathbf{j}$
    $+ (4xy)\mathbf{k}$.

49. (a) 9.6×10^4 eV; (b) 1.9×10^5 eV.

51. -2.4×10^4 V.

53. (a) $U = (1/4\pi\epsilon_0)(Q_1 Q_2/r_{12}$
    $+ Q_1 Q_3/r_{13} + Q_1 Q_4/r_{14} + Q_2 Q_3/r_{23}$
    $+ Q_2 Q_4/r_{24} + Q_3 Q_4/r_{34})$.
    (b) $U = (1/4\pi\epsilon_0)(Q_1 Q_2/r_{12}$
    $+ Q_1 Q_3/r_{13} + Q_1 Q_4/r_{14} + Q_1 Q_5/r_{15}$
    $+ Q_2 Q_3/r_{23} + Q_2 Q_4/r_{24} + Q_2 Q_5/r_{25}$
    $+ Q_3 Q_4/r_{34} + Q_3 Q_5/r_{35} + Q_4 Q_5/r_{45})$.

55. (a) 2.0 keV; (b) 42.8.

57. (a) $(-4 + \sqrt{2})Q^2/4\pi\epsilon_0 b$; (b) 0.

59. $3Q^2/20\pi\epsilon_0 r$.

61. 5.4×10^5 V/m.

63. 9×10^2 V.

65. (a) 1.1 MV; (b) 13 kg.

67. 7.2 MV.

69. 1.58×10^{12} electrons.

71. 1.7×10^{-12} V.

73. 1.03×10^6 m/s.

75. $V_a = -3.5\, Q/4\pi\epsilon_0 L$,
    $V_b = -5.2\, Q/4\pi\epsilon_0 L$,
    $V_c = -6.8\, Q/4\pi\epsilon_0 L$.

77. (a) 5.8×10^5 V; (b) 9.2×10^{-14} J.

79. $V_a - V_b = (\lambda/2\pi\epsilon_0) \ln(R_b/R_a)$.

83. (a) $\rho_E(r_2^3 - r_1^3)/3\epsilon_0 r$;
    (b) $(\rho_E/6\epsilon_0)[3r_2^2 - r^2 - (2r_1^3/r)]$;
    (c) $(\rho_E/2\epsilon_0)(r_2^2 - r_1^2)$, potential is
    continuous at r_1 and r_2.

CHAPTER 24

1. 2.6 μF.

3. 6.3 pF.

5. 0.80 μF.

7. 2.0 C.

9. 1.8×10^2 m².

11. 7.1×10^{-4} F.

13. 23 nC.

17. 4.5×10^4 V/m.

19. (a) $\epsilon_0 A/(d - \ell)$; (b) 3.

21. 2880 pF, yes.

23. (a) $(C_1 C_2 + C_1 C_3 + C_2 C_3)/$
           $(C_2 + C_3)$;
    (b) $Q_1 = 350$ μC, $Q_2 = 117$ μC.

25. (a) 3.71 μF; (b) $V_{ab} = V_1 = 26.0$ V,
    $V_2 = 14.9$ V, $V_3 = 11.1$ V.

27. 18 nF (parallel), 1.6 nF (series).

29. (a) $3C/5$; (b) $Q_1 = Q_2 = CV/5$,
    $Q_3 = 2CV/5$, $Q_4 = 3CV/5$,
    $V_1 = V_2 = V/5$, $V_3 = 2V/5$,
    $V_4 = 3V/5$.

31. $Q_1' = C_1 C_2 V_0/(C_1 + C_2)$,
    $Q_2' = C_2^2 V_0/(C_1 + C_2)$.

33. (a) $Q_1 = Q_3 = 30\ \mu$C;
$Q_2 = Q_4 = 60\ \mu$C;
(b) $V_1 = V_2 = V_3 = V_4 = 3.75$ V;
(c) 7.5 V.

35. 3.0 μF.

37. $C \approx (\epsilon_0 A/d)[1 - \tfrac{1}{2}(\theta\sqrt{A}/d)]$.

39. 2.0×10^{-3} J.

41. 2.3×10^3 J.

43. 1.65×10^{-7} J.

45. (a) 2.5×10^{-5} J; (b) 6.2×10^{-6} J;
(c) $Q_{par} = 4.2\ \mu$C, $Q_{ser} = 1.0\ \mu$C.

47. (a) 2.2×10^{-4} J; (b) 8.1×10^{-5} J;
(c) -1.4×10^{-4} J; (d) stored
potential energy is not conserved.

51. 1.5×10^{-10} F.

53. 0.46 μC.

55. 3.3×10^2 J.

57. $C = 2\epsilon_0 A K_1 K_2/d(K_1 + K_2)$.

59. (a) $0.40 Q_0$, $1.60 Q_0$; (b) $0.40 V_0$.

61. (a) 111 pF; (b) 1.66×10^{-8} C;
(c) 1.84×10^{-8} C;
(d) 1.17×10^5 V/m; (e) 3.34×10^4 V/m;
(f) 150 V;
(g) 172 pF; (h) 2.58×10^{-8} C.

63. 22%.

65. $Q = 4.41 \times 10^{-7}$ C,
$Q_{ind} = 3.65 \times 10^{-7}$ C,
$E_{air} = 2.69 \times 10^4$ V/m,
$E_{glass} = 4.64 \times 10^3$ V/m;

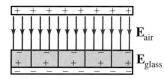

67. 11 μF.

69. (a) 4×; (b) 4×; (c) $\tfrac{1}{2}$×.

71. 10.9 V.

73. (b) 1.5×10^{-10} F/m.

75. $U_2/U_1 = 1/2K$, $E_2/E_1 = 1/K$.

77. (a) 19 J; (b) 0.19 MW.

79. (a) 0.10 MV; (b) voltage will
decrease exponentially.

81. 660 pF in parallel.

83. $Q_1 = 11\ \mu$C, $Q_2 = Q_3 = 13\ \mu$C,
$V_1 = 11$ V, $V_2 = 6.5$ V, $V_3 = 4.4$ V.

85. $Q^2 x/2A\epsilon_0$.

87. (a) 7.4 nF, 0.33 μC, 1.5×10^4 V/m,
7.5×10^{-6} J; (b) 27 nF, 1.2 μC,
1.5×10^4 V/m, 2.7×10^{-5} J.

89. (a) 66 pF; (b) 30 μC; (c) 7.0 mm; (d)
450 V.

CHAPTER 25

1. 9.38×10^{18} electron/s.

3. 2.1×10^{-11} A.

5. 7.5×10^2 V.

7. 2.1×10^{21} electron/min.

9. (a) 16 Ω; (b) 6.8×10^3 C.

11. 0.57 mm.

13. $R_{Al} = 1.2 R_{Cu}$.

15. 1/6 the length, 8.3 Ω, 1.7 Ω.

17. 58.3°C.

19. 1.8×10^3 °C.

21. $R_2 = \tfrac{1}{4} R_1$.

25. $R = (r_2 - r_1)/4\pi\sigma r_1 r_2$.

27. 3.2 W.

29. 37 V.

31. (a) 240 Ω, 0.50 A; (b) 96 Ω, 1.25 A.

33. 0.092 kWh, 22¢/month.

35. 1.1 kWh.

37. 3.

39. 5.3 kW.

41. 0.128 kg/s.

43. 0.094 A.

45. (a) Infinite; (b) 1.9×10^2 Ω.

47. 636 V, 5.66 A.

49. 1.5 kW, 3.0 kW, 0.

51. (a) 7.8×10^{-10} m/s; (b) 10.5 A/m^2
along the wire; (c) 1.8×10^{-7} V/m.

53. 2.7 A/m^2 north.

55. 12 h.

57. 6.67×10^{-2} S.

59. (a) 8.6 Ω, 1.1 W; (b) 4×.

61. (a) \$44; (b) 1.8×10^3 kg/yr.

63. (a) -19.5%; (b) percentage decrease
in the power output would be less.

65. (a) 1.44×10^3 W; (b) 17 W; (c) 11 W;
(d) 0.8¢/day.

67. (a) 1.5 kW; (b) 12.5 A.

69. 2.

71. 0.303 mm, 28.0 m.

73. (a) 1.2 kW; (b) 100 W.

75. 1.4×10^{12} protons.

77. $j_a = 2.8 \times 10^5$ A/m^2,
$j_b = 1.6 \times 10^5$ A/m^2.

CHAPTER 26

1. (a) 8.39 V; (b) 8.49 V.

3. 0.060 Ω.

5. 360 Ω, 23 Ω.

7. 25 Ω, 70 Ω, 95 Ω, 18 Ω.

9. Series connection.

11. 4.6 kΩ.

13. 310 Ω, 3.7%.

15. 960 Ω in parallel.

17. 105 Ω.

19. (a) V_1 and V_2 increase; V_3 and V_4
decrease; (b) I_1 (= I) and I_2
increase; I_3 and I_4 decrease;
(c) increases; (d) $I = I_1 = 0.300$ A,
$I_2 = 0, I_3 = I_4 = 0.150$ A; $I = 0.338$ A,
$I_2 = I_3 = I_4 = 0.113$ A.

21. 0.4 Ω.

23. 0.41 A.

25. $I_1 = 0.68$ A, $I_2 = -0.40$ A.

27. $I_1 = 0.18$ A right, $I_2 = 0.32$ A left,
$I_3 = 0.14$ A up.

29. $I_1 = 0.274$ A, $I_2 = 0.222$ A,
$I_3 = 0.266$ A, $I_4 = 0.229$ A,
$I_5 = 0.007$ A, $I = 0.496$ A.

31. 52 V, –28 V.
The negative value means the
battery is facing the other direction.

33. 70 V.

35. $I_1 = 0.783$ A.

39. (a) $R(3R + 5R')/8(R + R')$; (b) $R/2$.

41. (a) 3.7 nF; (b) 22 μs.

43. $t = 1.23\tau$.

45. (a) $\tau = R_1 R_2 C/(R_1 + R_2)$;
(b) $Q_{max} = \mathscr{E}R_2 C/(R_1 + R_2)$.

47. 2.1 μs.

49. 50 μA.

51. (a) 2.9×10^{-4} Ω in parallel;
(b) 35 kΩ in series.

53. 22 V, 17 V, 14% low.

55. 0.85mA, 4.3V.

57. 9.6 V.

61. 3.6×10^{-2} C°.

63. Two resistors in series.

65. 2.2 V, 116 V.

67. 0.19 MΩ.

69. (a) 0.10 A; (b) 0.10 A; (c) 53 mA.

71. 2.5 V.

73. 46.1 V, 0.71 Ω.

75. (a) 72.0 W; (b) 14.2 W; (c) 3.76 W.

77. (a) 40 kΩ; (b) between b and c.

79. 375 cells, 3.8 m × 0.090 m.

81. (a) 0.50 A; (b) 0.17 A; (c) 3.3 μC;
(d) 32 μs.

83. (a) + 6.8 V, 10.2 μC; (b) 28 μs.

CHAPTER 27

1. (a) 6.7 N/m; (b) 4.7 N/m.

3. 2.68 A.

5. 0.243 T.

7. (a) South pole; (b) 3.5×10^2 A;
(c) 5.22 N.

9. 5.5×10^3 A.

13. 1.05×10^{-13} N north.

15. (a) Down; (b) in; (c) right.

21. 1.6 m.

23. (a) 2.7 cm; (b) 3.8×10^{-7} s.

25. 1.034×10^8 m/s (west), gravity can be ignored.

27. $(6.4\mathbf{i} - 10.3\mathbf{j} - 0.24\mathbf{k})] \times 10^{-16}$ N.

29. 5.3×10^{-5} m, 3.3×10^{-4} m.

31. (a) 45°; (b) 3.5×10^{-3} m.

33. (a) $2\mu B$; (b) 0.

35. (a) 4.33×10^{-5} m · N; (b) north edge.

39. 29 μA.

41. 1.2×10^5 C/kg.

43. (a) 2.2×10^{-4} V/m; (b) 2.7×10^{-4} m/s; (c) 6.4×10^{28} electrons/m³.

45. (a)Determine polarity of the emf; (b) 0.43 m/s.

47. 1.53 mm, 0.76 mm.

51. 3.0 T up.

53. 1.1×10^{-6} m/s west.

55. 0.17 N, 68° above the horizontal toward the north.

57. 0.20 T, 26.6° from the vertical.

59. (c) 48 MeV.

61. Slower protons will deflect more, and faster protons will deflect less, $\theta = 12°$.

63. 2.0 A, down.

65. 7.3×10^{-3} T.

67. -2.1×10^{-20} J.

CHAPTER 28

1. 1.7×10^{-4} T, 3.1×.

3. 0.18 N attraction.

5.

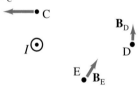

7. 8.9×10^{-5} T, 70° above horizontal.

9. 4.0×10^{-5} T, 15° below the horizontal.

11. (a) $(2.0 \times 10^{-5}$ T/A$)(15$ A $- I)$ up; (b) $(2.0 \times 10^{-5}$ T/A$)(15$ A $+ I)$ down.

13. 21 A down.

15. $[(\mu_0/4\pi)2I(d - 2x)/x(d - x)]\mathbf{j}$.

17. 4.12×10^{-5} T.

19. (b) $(\mu_0/4\pi)(2I/y)$.

21. 0.123 A.

23. (a) 6.4×10^{-3} T; (b) 3.8×10^{-3} T; (c) 2.1×10^{-3} T.

25. (a) 51 cm; (b) 1.3×10^{-2} T.

27. (a) $(\mu_0 I_0/2\pi R_1^2)r$ circular CCW; (b) $\mu_0 I_0/2\pi r$ circular CCW; (c) $(\mu_0 I_0/2\pi r)(R_3^2 - r^2)/(R_3^2 - R_2^2)$ circular CCW; (d) 0.

29. 3.6×10^{-6} T.

31. $\mu_0 I/8R$ out of the page.

33. (a) $\mu_0 I(R_1 + R_2)/4R_1 R_2$ into the page; (b) $\frac{1}{2}\pi I(R_1^2 + R_2^2)$ into the page.

35. (a) $\dfrac{Q\omega R^2}{4}\mathbf{i}$;

(b) $\dfrac{\mu_0 Q\omega}{2\pi R^2}\left[\dfrac{R^2 + 2x^2 - 2x\sqrt{R^2 + x^2}}{\sqrt{R^2 + x^2}}\right]\mathbf{i}$;

(c) yes.

37. (b) $B = \mu_0 IL/4\pi y(L^2 + y^2)^{1/2}$ circular.

39. (a) $(\mu_0 I_0/2\pi R)n \tan (\pi/n)$ into the page.

41. $B = \dfrac{\mu_0 I}{4\pi}\left\{\dfrac{(x^2 + y^2)^{1/2}}{xy}\right.$

$+ \dfrac{[(b - x)^2 + y^2]^{1/2}}{y(b - x)}$

$+ \dfrac{[(a - y)^2 + (b - x)^2]^{1/2}}{(a - y)(b - x)}$

$+ \left.\dfrac{[x^2 + (a - y)^2]^{1/2}}{x(a - y)}\right\}$,

out of the page.

43. (a) 26 A · m²; (b) 31 m · N.

45. 30 T.

47. $F_M/L = 5.84 \times 10^{-5}$ up, $F_N/L = 3.37 \times 10^{-5}$ N/m 60° below the line toward P, $F_P/L = 3.37 \times 10^{-5}$ N/m 60° below the line toward N.

49. 0.27 mm, 1.4 cm.

51. $B = \mu_0 jt/2$ parallel to the sheet, perpendicular to the current (opposite directions on the two sides).

53. Between long, thin and short, fat.

55. 3×10^9 A.

57. B will decrease.

59. 2.1×10^{-6} g.

61. $2\mu_0 I/L\pi$ (left).

63. 4×10^{-6} T, about 10% of the Earth's field.

CHAPTER 29

1. -3.8×10^2 V.

3. Counterclockwise.

5. 0.026 V.

7. (a) Counterclockwise; (b) clockwise; (c) zero; (d) counterclockwise.

9. Counterclockwise.

11. (a) Clockwise; (b) 43 mV; (c) 17 mA.

13. 1.1×10^{-5} J.

15. 4.21 C.

17. (a) 5.2×10^{-2} A; (b) 0.32 mW.

19. 1.7×10^{-2} V.

21. $(\mu_0 Ia/2\pi) \ln 2$.

23. (a) 0.15 V; (b) 5.4×10^{-3} A; (c) 4.5×10^{-4} N.

25. (a) Will move at constant speed; (b) $v = v_0 e^{-B^2 \ell^2 t/mR}$.

27. (a) $\dfrac{\mu_0 Iv}{2\pi} \ln\left(\dfrac{a + b}{b}\right)$ toward long wire;

(b) $\dfrac{\mu_0 Iv}{2\pi} \ln\left(\dfrac{a + b}{b}\right)$ away from long wire.

31. 0.33 kV, 120 rev/s.

33. 100 V.

35. 13 A.

37. 3.54×10^4 turns.

39. 0.18.

41. (a) 5.2 V; (b) step-down transformer.

43. 549 V, 68.6 A.

45. 56.8 kW.

47. 0.188 V/m.

49. (b) Clockwise; (c) $dB/dt > 0$.

51. (a) IR/ℓ (constant); (b) $\dfrac{\mathcal{E}_0}{\ell}e^{-B^2\ell^2 t/mR}$.

53. 31 turns.

55. $v = 0.76$ m/s.

57. 184 kV.

59. 1.5×10^{17}.

61. (a) 23 A; (b) 90 V; (c) 6.9×10^2 W; (d) 75%.

63. (a) 0.85 A; (b) 8.2.

65. $\frac{1}{2}B\omega L^2$ toward the center.

71. $B\omega r$, radially out from the axis.

CHAPTER 30

1. $M = \mu_0 N_1 N_2 A/\ell$.

3. $M/\ell = \mu_0 n_1 n_2 \pi r_2^2$.

5. $M = (\mu_0 w/2\pi) \ln(\ell_2/\ell_1)$.

7. 1.2 H.

9. 2.5×10^{-6} H.

11. $r_1 \geq 2.5$ mm.

13. 3.

15. (a) $L_1 + L_2$; (b) $= L_1 L_2/(L_1 + L_2)$.

17. 15.9 J.

19. (a) $u_E = 4.4 \times 10^{-4}$ J/m^3,
$u_B = 1.6 \times 10^6$ J/m^3, $u_B \gg u_E$;
(b) $E = 6.0 \times 10^8$ V/m.

21. 4.4 J/m^3, 1.6×10^{-14} J/m^3.

23. $(\mu_0 I^2/4\pi) \ln(r_2/r_1)$.

25. $t/\tau = 4.6$.

27. $(dI/dt)_0 = V_0/L$.

29. (a) $(LV^2/2R^2)(1 - 2e^{-t/\tau} + e^{-2t/\tau})$;
(b) $t/\tau = 5.3$.

31. (a) 213 pF; (b) 46.5 μH.

35. (a) $Q = Q_0/\sqrt{2}$; (b) $T/8$.

37. $R = 2.30$ Ω.

41. Decrease, 1.15 kΩ.

43. 20 mH, 95 turns.

45. (a) 21 mH; (b) 45 mA;
(c) 2.2×10^{-5} J.

47. 3.0×10^3 turns, 95 turns.

51. (b) Positioning one coil
perpendicular to the other;
(c) $L_1 L_2/(L_1 + L_2)$;
$(L_1 L_2 - M^2)/(L_1 + L_2 \mp 2M)$.

55. (a) $\frac{1}{2}(Q_0^2/C)e^{-Rt/L}$.

57.

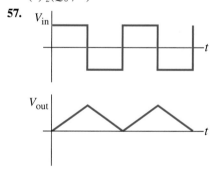

CHAPTER 31

1. (a) 3.7×10^2 Ω; (b) 2.2×10^{-2} Ω.

3. 9.90 Hz.

5.

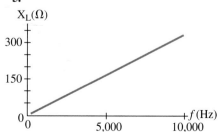

7. 0.13 H.

11. (a) 5.0%; (b) 98%.

13. (a) 9.0 kΩ; (b) 10.2 kΩ.

15. (a) 18 mA; (b) $-29°$; (c) 1.8 W;
(d) $V_R = 105$ V, $V_C = 58$ V.

17. (a) 0.38 A; (b) $-89°$; (c) 0.29 W.

19. 332 Ω.

21. (a) 0; (b) $\frac{2}{\pi}V_0$, $\overline{V}_{1/2} = \frac{2\sqrt{2}}{\pi}V_{rms}$.

23. 8.78 kΩ, $-7.66°$, 91.1 mA.

25. 265 Hz, 324 W.

27. 52.5 mA.

29. (b) $\omega'^2 = [(1/LC) - (R^2/2L^2)]$;
(c) $k \leftrightarrow 1/C, m \leftrightarrow L, b \leftrightarrow R$.

31. (a) $V_0^2 R/[2R^2 + 2(\omega L - 1/\omega C)^2]$;
(b) $\omega'^2 = 1/LC$; (c) $\Delta\omega = R/L$.

33. 4 Ω.

35. 9.76 nF.

37. 27.9 mH.

39. 1.6 kHz.

41. 14 Ω, 75 mH.

43. 2.2×10^3 Hz, 1.1×10^4 Hz.

45. (a) 23.6 kΩ, 10.8°; (b) 1.88×10^{-5} W;
(c) 2.8×10^{-5} A, 0.66 V, 4.7×10^{-4} V,
0.126 V.

49. $\{(R_1 + R_2)^2$
$+ [\omega(L_1 + L_2) -$
$(C_1 + C_2)/\omega C_1 C_2]^2\}^{1/2}$.

51. 19 Ω, 62 mH.

53. $I = \left(\dfrac{V_0}{Z}\right)\sin(\omega t + \phi)$,

$I_C = \left(\dfrac{V_0}{X_C}\right)\left[-\left(\dfrac{R}{Z}\right)\cos(\omega t + \phi) + \cos(\omega t)\right]$,

$I_L = \left(\dfrac{V_0}{X_L}\right)\left[\left(\dfrac{R}{Z}\right)\cos(\omega t + \phi) - \cos(\omega t)\right]$,

$Z = \sqrt{R^2 + \left(\dfrac{X_C X_L}{X_L - X_C}\right)^2}$,

$\tan\phi = \dfrac{X_C X_L}{(X_L - X_C)R}$.

CHAPTER 32

1. 9.2×10^4 V/m $\cdot$ s.

3. 7.9×10^{14} V/m $\cdot$ s.

7. $\oint \mathbf{B} \cdot d\mathbf{A} = \mu_0 Q_m$,
$\oint \mathbf{E} \cdot d\boldsymbol{\ell} = \mu_0\, dQ_m/dt - d\Phi_B/dt$.

9. 1.4×10^{-13} T.

11. (a) $B_0 = E_0/c$, $-y$-direction;
(b) $-z$-direction.

13. (a) 1.08 cm; (b) 3.0×10^{18} Hz.

15. 314 nm, ultraviolet.

17. (a) 4.3 min; (b) 71 min.

19. 1.77×10^{-6} W/m^2.

21. 7.82×10^{-7} J/h.

23. 4.50×10^{-6} J.

25. 3.8×10^{26} W.

29. $r < 3 \times 10^{-7}$ m.

31. 302 pF.

33. 2.59 nH $\leq L \leq$ 3.89 nH.

35. (a) 441 m; (b) 2.979 m.

37. 5.56 m, 0.372 m.

39. (a) 1.28 s; (b) 4.3 min.

43. 469 V/m.

45. Person at the radio hears the voice
0.14 s sooner.

47. (a) 0.40 W; (b) 12 V/m; (c) 12 V.

49. 1.5×10^{11} W.

51. (a) Parallel;
(b) 8.9 pF $\leq C \leq$ 80 pF;
(c) 1.05 mH $\leq L \leq$ 1.12 mH.

CHAPTER 33

1. (a) 2.21×10^8 m/s;
(b) 1.99×10^8 m/s.

3. 8.33 min.

5. 3 m.

7. 3.4×10^3 rad/s.

9. I_3 is the desired image:

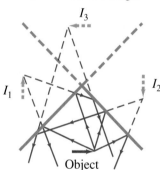

Depending on where you put your eye,
two other images may also be visible.

11. 5°.

15. 36.4 cm.

19. 4.5 m.

21. Convex, -20 m.

23. (a) Center of curvature; (b) real;
(c) inverted; (d) -1.

29. (a) Convex mirror; (b) 22 cm behind
surface; (c) -98 cm; (d) -196 cm.

33. 45.6°.

35. 24.9°.

37. 4.6 m.

43. 3.0%.

45. 0.22°.

47. 61.7°, lucite.

49. 93.5 cm.

51. $n_{liquid} \geq 1.5$.

55. 17.0 cm below the surface of the glass.

59. (a) 3.0 m, 4.0 m, 7.0 m; (b) toward you, away from you, toward you.

61. −3.80 m.

63. Chose different signs for the magnification; 13.3 cm, 26.7 cm.

65. $\geq 56.1°$.

69. 81 cm (inside the glass).

CHAPTER 34

1. (a)

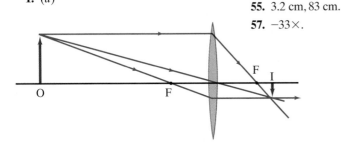

    (b) 24.9 cm.

3. (a) 3.64 D, converging;
    (b) −16.0 cm, diverging.

5. (a) −0.26 mm; (b) −0.47 mm;
    (c) −1.9 mm.

7. (a) 81 mm; (b) 82 mm; (c) 87 mm;
    (d) 24 cm.

9. (a) Virtual; (b) converging lens;
    (c) 7.5 D.

11. (a) −10.5 cm (diverging), virtual;
    (b) +203 cm (converging).

13. 22.9 cm, 53.1 cm.

15. Real and upright.

17. Real, 21.3 cm beyond second lens, −0.708 (inverted).

19. (a) +7.14 cm; (b) −0.357 (inverted);
    (c)

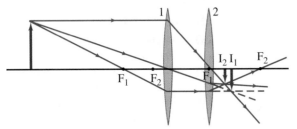

23. 1.54.

25. 8.1 cm.

27. −1.87 m (concave).

29. +1.15 D.

31. $f/2.3$.

35. 41 mm.

37. +2.3 D.

39. Glasses would be better.

41. (a) −1.33 D; (b) 38 cm.

43. −26.8 cm.

45. 17 cm, 100 cm.

47. 8.3 cm.

49. 6.3 cm from the lens, 3.9×.

51. (a) −49.4 cm; (b) 4.7×.

53. (a) 7.2×; (b) 2.2×.

55. 3.2 cm, 83 cm.

57. −33×.

59. 12×.

61. $f_o = 4.0$ m, $r = 8.0$ m.

63. 7.5×.

65. 1.7 cm.

67. (a) 0.85 cm; (b) 230×.

69. (a) 14.4 cm; (b) 137×.

71. (a) 15.9 cm; (b) 14.3 cm; (c) 1.6 cm;
    (d) $r = 0.46$ cm.

73. 6.87 m $\leq d_o \leq \infty$.

75. 100 mm, 200 mm.

77. 79.4 cm, 75.5 cm.

79. 0.101 m, −2.7 m.

81. $0 < -d_o < -f$.

83. (c) $\Delta d = \sqrt{d_T^2 - 4d_T f}$,
$$m = \left(\frac{d_T + \sqrt{d_T^2 - 4d_T f}}{d_T - \sqrt{d_T^2 - 4d_T f}} \right)^2.$$

85. $1/f' = [(n/n') - 1][(1/R_1) + (1/R_2)]$;
    $(1/d_o) + (1/d_i) = 1/f'$, where
    $1/f' = [(n/n') - 1]/f (n - 1)$;
    $m = -d_i/d_o$.

87. +3.6 D.

89. 2.9×, 4.1×.

91. (a) −2.5×; (b) 5.0 D.

93. −20×.

CHAPTER 35

3. 5.9 μm.

5. 3.9 cm.

7. 0.21 mm.

9. 613 nm.

11. 533 nm.

15. (a) $I_\theta/I_0 = (1 + 4 \cos \delta + 4 \cos^2 \delta)/9$;
    (b) $\sin \theta_{max} = m\lambda/d$, $m = 0, \pm1, \pm2, \ldots$;
    $\sin \theta_{min} = (m + \frac{1}{3}k)\lambda/d; k = 1, 2$;
    $m = 0, \pm1, \pm2, \ldots$.

17. Orange-red.

19. 179 nm.

21. 9.1 μm.

23. 120 nm, 240 nm.

25. 1.26.

29. 0.47, 0.23.

31. 0.221 mm.

33. 0.289 mm.

35. (a) 17 lm/W; (b) 156.

37. (a) Constructive; (b) destructive.

39. 464 nm.

41. 646 nm.

43. (a) 81.5 nm; (b) 127 nm.

45. 0.5 cm.

47. $\theta = 63.3°$.

49. $\sin \theta_{max} = (m + \frac{1}{2})\lambda/2S, m = 0, 1, 2, \ldots$,
    $\sin \theta_{min} = m\lambda/2S, m = 0, 1, 2, \ldots$.

51. $I/I_0 = \cos^2 (2\pi x/\lambda)$.

CHAPTER 36

1. 2.26°.

3. 2.4 m.

5. 5.8 cm.

7. 4.28 cm.

9. (b) Average intensity is 2×.

11. 10.7°.

13. $d = 4a$.

15. (a) 1.8 cm; (b) 11.0 cm.

17.

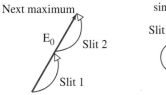

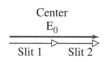

19. 2.4×10^{-7} rad $= (1.4 \times 10^{-5})°$
$= 0.050''$.

21. 820 lines/mm, 102 lines/mm.

23. 5.61°.

25. Two full orders.

27. 497 nm, 612 nm, 637 nm, 754 nm.

29. 600 nm to 750 nm of second order overlaps with 400 nm to 500 nm of third order.

31. 621 nm, 909 nm.

35. (a) Two orders; (b) 6.44×10^{-5} rad $= 13.3''$, 7.36×10^{-5} rad $= 15.2''$, 2.52×10^{-4} rad $= 52.0''$.

37. (a) 1.60×10^4, 3.20×10^4; (b) 0.026 nm, 0.013 nm.

39. $\Delta f = f/mN$.

41. (a) 62.0°; (b) 0.21 nm.

43. 0.033.

45. 57.3°.

47. (a) 35°; (b) 63°.

49. 36.9°, 53.1°.

51. $I_0/32$.

55. 31° on either side of the normal.

57. 12,500 lines/cm.

59. $\sin\theta = \sin 20° - (m\lambda/a)$,
$m = \pm 1, \ \pm 2, \ldots$.

61. Two orders.

63. 11.7°.

65. (a) 16 km; (b) 0.42'.

67. (a) 0; (b) $0.094I_0$; (c) no light gets transmitted.

69. (a) 30°; (b) 18°; (c) 5.7°.

73. 0.245 nm.

CHAPTER 37

1. (a) 1.00; (b) 0.99995; (c) 0.995; (d) 0.436; (e) 0.141; (f) 0.0447.

3. 2.07×10^{-6} s.

5. $0.773c$.

7. $0.141c$.

9. (a) 99.0 yr; (b) 27.7 yr; (c) 26.6 ly; (d) $0.960c$.

11. $0.89c = 2.7 \times 10^8$ m/s.

13. (a) (470 m, 20 m, 0); (b) (1820 m, 20 m, 0).

15. (a) $0.80c$; (b) $-0.80c$.

17. 60 m/s, 24°.

19. 2.7×10^8 m/s, 43°.

21. (a) $L_0\sqrt{1 - (v/c)^2\cos^2\theta}$; (b) $\tan\theta' = \tan\theta/\sqrt{1 - (v/c)^2}$.

23. Not possible in the boy's frame.

25. $0.866c$.

27. (a) 0.5%; (b) 13%.

29. 5.36×10^{-13} kg.

31. 8.20×10^{-14} J, 0.511 MeV.

33. 9×10^2 kg.

37. (a) 11.2 GeV (1.79×10^{-9} J); (b) 6.45×10^{-18} kg·m/s.

39. 7.49×10^{-19} kg·m/s.

41. $0.941c$.

43. $M = 2m/\sqrt{1 - (v^2/c^2)}$, $K_{\text{loss}} = 2mc^2\{[1/\sqrt{1 - (v^2/c^2)}] - 1\}$.

45. $0.866c$, 4.73×10^{-22} kg·m/s.

47. 39 MeV (6.3×10^{-12} J), 1.5×10^{-19} kg·m/s, -6%, -4%.

49. $0.804c$.

51. 3.0 T.

57. (a) $0.80c$; (b) $2.0c$.

61. 3.8×10^{-5} s.

63. $\rho = \rho_0/[1 - (v^2/c^2)]$.

65. (a) 1.5 m/s less than c; (b) 30 cm.

67. 1.02 MeV (1.64×10^{-13} J).

69. 2.2 mm.

71. 0.78 MeV.

73. Electron.

75. $5.19 \times 10^{-13}\%$.

81. (a) $\alpha = \tan^{-1}[(c^2/v^2) - 1]^{1/2}$; (c) $\tan\theta = c/v$, $u = \sqrt{v^2 + c^2}$.

CHAPTER 38

1. 6.59×10^3 K.

3. 5.4×10^{-20} J, 0.34 eV.

5. (a) 114 J; (b) 228 J; (c) 342 J; (d) $114n$ J; (e) -456 J.

7. (b) $h = 6.63 \times 10^{-34}$ J·s.

9. 3.67×10^{-7} eV.

11. 2.4×10^{13} Hz, 1.2×10^{-5} m.

13. 400 nm.

15. (a) 2.18 eV; (b) 0.92 V.

17. 3.46 eV.

19. 1.88 eV, 43.3 kcal/mol.

21. (a) 2.43×10^{-12} m; (b) 1.32×10^{-15} m.

23. (a) 5.90×10^{-3}, 1.98×10^{-2}, 3.89×10^{-2}; (b) 60.8 eV, 204 eV, 401 eV.

27. 1.82 MeV.

29. 212 MeV, 5.85×10^{-15} m.

31. 3.2×10^{-32} m.

33. 19 V.

35. (a) 0.39 nm; (b) 0.12 nm; (c) 0.039 nm.

37. 1.84×10^3.

39. 3.3×10^{-38} m/s, no diffraction.

41. 0.026 nm.

43. 3.4 eV.

45. 122 eV.

49. 52.5 nm.

51.

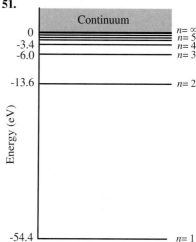

53. $U = -27.2$ eV, $K = +13.6$ eV.

55. Justified.

61. 3.28×10^{15} Hz.

65. 2.78×10^{21} photons/s·m².

67. 8.3×10^6 photons/s.

69. $\theta = 89.4°$.

71. 4.7×10^{-14} m.

73. 10.2 eV.

75. 4.4×10^{-40}, yes.

77. 653 nm, 102 nm, 122 nm.

79. 0.64 V.

83. 5×10^{-12} m.

85. 3.

CHAPTER 39

1. 1.8×10^{-7} m.

3. $\pm 1.3 \times 10^{-11}$ m.

5. 7.2×10^3 m/s.

7. 2.4×10^{-3} m, 1.4×10^{-32} m.

9. (a) 6.6×10^{-8} eV; (b) 6.5×10^{-9}; (c) 7.9×10^{-7} nm.

13. (a) $\psi = A \sin (3.5 \times 10^9 \text{ m}^{-1})x + B \cos (3.5 \times 10^9 \text{ m}^{-1})x$; (b) $\psi = A \sin (6.3 \times 10^{12} \text{ m}^{-1})x + B \cos (6.3 \times 10^{12} \text{ m}^{-1})x$.

17. 3.6×10^6 m/s.

19. (a) 52 nm; (b) 0.22 nm.

23. $E_1 = 0.094$ eV, $\psi_1 = (1.00 \text{ nm}^{-1/2}) \sin (1.57 \text{ nm}^{-1} x)$; $E_2 = 0.38$ eV, $\psi_2 = (1.00 \text{ nm}^{-1/2}) \sin (3.14 \text{ nm}^{-1} x)$; $E_3 = 0.85$ eV, $\psi_3 = (1.00 \text{ nm}^{-1/2}) \sin (4.71 \text{ nm}^{-1} x)$; $E_4 = 1.51$ eV, $\psi_4 = (1.00 \text{ nm}^{-1/2}) \sin (6.28 \text{ nm}^{-1} x)$.

25. (a) 4 GeV; (b) 2 MeV; (c) 2 MeV.

27. (a) 0.18; (b) 0.50; (c) 0.50.

29.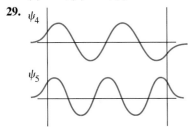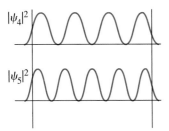

31. 0.03 nm.

33. 9.2 eV.

35. (a) Decreases by 8%; (b) decreases by 5%.

37. (a) 32 MeV; (b) 56 fm; (c) 8.8×10^{20} s^{-1}, 10^{10} yr.

39. 21 MeV.

41. 3.00×10^{-10} eV/c^2.

43. r_1, the Bohr radius.

45. 0.5 MeV, 5×10^6 m/s.

47. (b) 4 s.

49. 14% decrease.

CHAPTER 40

1. $\ell = 0, 1, 2, 3, 4, 5.$

3. 32 states, $(4, 0, 0, -\frac{1}{2}), (4, 0, 0, +\frac{1}{2}),$ $(4, 1, -1, -\frac{1}{2}), (4, 1, -1, +\frac{1}{2}), (4, 1, 0, -\frac{1}{2}),$ $(4, 1, 0, +\frac{1}{2}), (4, 1, 1, -\frac{1}{2}), (4, 1, 1, +\frac{1}{2}),$ $(4, 2, -2, -\frac{1}{2}), (4, 2, -2, +\frac{1}{2}),$ $(4, 2, -1, -\frac{1}{2}), (4, 2, -1, +\frac{1}{2}),$ $(4, 2, 0, -\frac{1}{2}), (4, 2, 0, +\frac{1}{2}), (4, 2, 1, -\frac{1}{2}),$ $(4, 2, 1, +\frac{1}{2}), (4, 2, 2, -\frac{1}{2}), (4, 2, 2, +\frac{1}{2}),$ $(4, 3, -3, -\frac{1}{2}), (4, 3, -3, +\frac{1}{2}),$ $(4, 3, -2, -\frac{1}{2}), (4, 3, -2, +\frac{1}{2}),$ $(4, 3, -1, -\frac{1}{2}), (4, 3, -1, +\frac{1}{2}), (4, 3, 0, -\frac{1}{2}),$ $(4, 3, 0, +\frac{1}{2}), (4, 3, 1, -\frac{1}{2}), (4, 3, 1, +\frac{1}{2}),$ $(4, 3, 2, -\frac{1}{2}), (4, 3, 2, +\frac{1}{2}), (4, 3, 3, -\frac{1}{2}),$ $(4, 3, 3, +\frac{1}{2}).$

5. $n \geq 5, m_\ell = -4, -3, -2, -1, 0, 1, 2, 3, 4,$ $m_s = -\frac{1}{2}, +\frac{1}{2}.$

7. (a) 6; (b) -0.378 eV; (c) $\ell = 4$, $\sqrt{20}\hbar = 4{,}72 \times 10^{-34}$ kg·m^2/s; (d) $m_\ell = -4, -3, -2, -1, 0, 1, 2, 3, 4.$

13. (a) $-[3/(32\pi r_0^3)^{1/2}] e^{-5/2}$; (b) $(9/32\pi r_0^3) e^{-5}$; (c) $(225/8 r_0) e^{-5}.$

15. 1.85.

17. 17.3%.

19. (a) $1.34 r_0$; (b) $2.7 r_0$; (c) $4.2 r_0$.

21. $\dfrac{r^4}{24 r_0^5} e^{-r/r_0}.$

27. (a) $\dfrac{4r^2}{27 r_0^3}\left(1 - \dfrac{2r}{3r_0} + \dfrac{2r^2}{27 r_0^2}\right)^2 e^{-2r/3r_0}$;

(b)

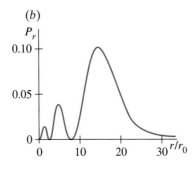

(c) $r = 13 r_0.$

29. (a) $(1, 0, 0, -\frac{1}{2}), (1, 0, 0, +\frac{1}{2}),$ $(2, 0, 0, -\frac{1}{2}), (2, 0, 0, +\frac{1}{2}), (2, 1, -1, -\frac{1}{2}),$ $(2, 1, -1, +\frac{1}{2})$; (b) $(1, 0, 0, -\frac{1}{2}),$ $(1, 0, 0, +\frac{1}{2}), (2, 0, 0, -\frac{1}{2}), (2, 0, 0, +\frac{1}{2}),$ $(2, 1, -1, -\frac{1}{2}), (2, 1, -1, +\frac{1}{2}), (2, 1, 0, -\frac{1}{2}),$ $(2, 1, 0, +\frac{1}{2}), (2, 1, -1, -\frac{1}{2}), (2, 1, -1, +\frac{1}{2}),$ $(3, 0, 0, -\frac{1}{2}), (3, 0, 0, +\frac{1}{2}).$

31. (a) $1s^2 2s^2 2p^6 3s^2 3p^6 3d^{10} 4s^2 4p^4$; (b) $1s^2$ $2s^2 2p^6 3s^2 3p^6 3d^{10} 4s^2 4p^6 4d^{10} 4f^{14} 5s^2 5p^6$ $5d^{10} 6s^1$; (c) $1s^2 2s^2 2p^6 3s^2 3p^6 3d^{10} 4s^2 4p^6$ $4d^{10} 4f^{14} 5s^2 5p^6 5d^{10} 6s^2 6p^6 5f^3 6d^1 7s^2.$

33. 5.8×10^{-13} m, 0.115 MeV.

37. 0.041 nm, 1 nm.

41. 0.19 nm.

43. Chromium.

47. (a) 0.25 mm; (b) 0.13 mm.

49. (a) $\frac{1}{2}, \frac{3}{2}, \frac{1}{2}\sqrt{3}\ \hbar, \frac{1}{2}\sqrt{15}\hbar$; (b) $\frac{5}{2}, \frac{7}{2}, \frac{1}{2}\sqrt{35}\hbar, \frac{1}{2}\sqrt{63}\hbar$; (c) $\frac{3}{2}, \frac{5}{2}, \frac{1}{2}\sqrt{15}\hbar, \frac{1}{2}\sqrt{35}\hbar.$

51. (a) 0.4 T; (b) 0.4 T.

53. 5.6×10^{-4} rad, 1.7×10^2 m.

55. 3.7×10^4 K.

57. (a) 1.56; (b) 1.4×10^{-10} m.

59. (a) $1s^2 2s^2 2p^6 3s^2 3p^6 3d^7 4s^2$; (b) $1s^2 2s^2 2p^6 3s^2 3p^6 3d^{10} 4s^2 4p^6$; (c) $1s^2 2s^2 2p^6 3s^2 3p^6 3d^{10} 4s^2 4p^6 5s^2.$

61. (a) 2.5×10^{74}; (b) $5.0 \times 10^{74}.$

63. $r = 5.24 r_0.$

65. (a) ϕ is unknown; (b) L_x and L_y are unknown.

67. (a) 1.2×10^{-4} eV; (b) 1.1 cm; (c) no difference.

69. (a) $3 \times 10^{-171}, 7 \times 10^{-203}$; (b) $1.1 \times 10^{-8}, 6.3 \times 10^{-10}$; (c) $6.6 \times 10^{15}, 3.8 \times 10^{14}$; (d) 7×10^{23} photons/s, 4×10^{22} photons/s.

71. 182.

73. 2.25.

CHAPTER 41

1. 5.1 eV.

3. 0.7 eV.

7. (a) 13.941 u; (b) 7.0034 u; (c) 0.9801 u.

9. (a) 6.86 u; (b) 1.85×10^3 N/m.

11. (a) 1.79×10^{-4} eV; (b) 7.16×10^{-4} eV, 1.73 mm.

13. 2.36×10^{-10} m.

15. -7.9 eV.

17. 0.283 nm.

19. (b) -6.9 eV; (c) -10.8 eV; (d) 3%.

21. $1.8 \times 10^{21}.$

23. (a) 6.9 eV; (b) 6.8 eV.

25. 6.3%.

27. 3.2 eV, 1.05×10^6 m/s.

29. (a) 1.79×10^{29} m^{-3}; (b) 3.

33. (a) 0.021, reasonable; (b) 0.979; (c) 0.021.

37. A large energy is required to create a conduction electron by raising an electron from the valence band to the conduction band.

39. 1.1 μm.

41. $5 \times 10^6.$

43. 1.91 eV.

45. 13 mA.

47. (a) 2.1 mA; (b) 4.3 mA.

49. (a) 9.4 mA (smooth);
(b) 6.7 mA (rippled).

51. 4.0 kΩ.

53. 0.43 mA.

55. (a) 3.5×10^4 K; (b) 1.2×10^3 K.

57. (a) 0.9801 u; (b) 482 N/m, 88% of the constant for H_2.

59. States with higher values of L are less likely to be occupied, so less likely to absorb a photon; I will depend on L.

61. 2.8×10^4 J/kg.

63. 1.24 eV.

65. 1.09 μm, could be used.

67. (a) 0.094 eV; (b) 0.63 nm.

69. (a) $145 \text{ V} \leq V \leq 343$ V;
(b) $3.34 \text{ k}\Omega \leq R_{\text{load}} < \infty$.

CHAPTER 42

1. 3729 MeV/c^2.

3. 1.9×10^{-15} m.

5. (a) 2.3×10^{17} kg/m^3; (b) 184 m;
(c) 2.6×10^{-10} m.

7. 28 MeV.

9. $^{31}_{15}$P.

11. (a) 1.8×10^3 MeV;
(b) 7.3×10^2 MeV.

13. 7.48 MeV.

15. (a) 32.0 MeV, 5.33 MeV;
(b) 1636 MeV, 7.87 MeV.

17. 12.4 MeV, 7.0 MeV, neutron is more closely bound in ^{23}Na.

19. (b) Stable.

21. 0.782 MeV.

23. β^+ emitter, 1.82 MeV.

25. $^{228}_{90}$Th, 228.02883 u.

27. (a) $^{32}_{16}$S; (b) 31.97207 u.

29. 0.862 MeV.

31. 5.31 MeV.

33. (b) 0.961 MeV, 0.961 MeV to 0.

35. 3.0 h.

37. 1.2×10^9 decays/s.

39. 1.78×10^{20} nuclei.

41. 7 α particles, 4 β^- particles.

43. (a) 4.8×10^{16} nuclei;
(b) 3.2×10^{15} nuclei;
(c) 7.2×10^{13} decays/s; (d) 26 min.

45. 1.68×10^{-10} g.

47. 2.6 min.

49. 3.4 mg.

51. 8.6×10^{-7}.

53. $^{211}_{82}$Pb.

55. $N_D = N_0(1 - e^{-\lambda t})$.

57. (a) 0.99946; (b) 1.2×10^{-14};
(c) 2.3×10^{17} kg/m^3, $10^{14}\times$.

59. 28.6 eV.

61. (a) 7.2×10^{55}; (b) 1.2×10^{29} kg;
(c) 3.2×10^{11} m/s^2.

63. 6×10^3 yr.

65. 6.64 $T_{1/2}$.

69. Calcium, stored by body in bones,
193 yr, $^{90}_{38}$SR $\rightarrow$ $^{90}_{39}$Y + $^{0}_{-1}$e + $\bar{\nu}$,
$^{90}_{39}$Y is radioactive,
$^{90}_{39}$Y $\rightarrow$ $^{90}_{40}$Zr + $^{0}_{-1}$e + $\bar{\nu}$,
$^{90}_{40}$Zr is stable.

73. (a) 0.002603 u, 2.425 MeV/c^2; (b) 0;
(c) -0.094909 u, -88.41 MeV/c^2;
(d) 0.043924 u, 40.92 MeV/c^2;
(e) $\Delta \geq 0$ for $0 \leq Z \leq 8$
and $Z \geq 85$, $\Delta < 0$ for $9 \leq Z \leq 84$.

75. 0.083%.

77. $^{228}_{88}$RA, $^{228}_{89}$Ac, $^{228}_{90}$Th, $^{224}_{88}$Ra, $^{220}_{87}$Rn,
$^{231}_{90}$Th, $^{231}_{91}$Pa, $^{227}_{89}$Ac, $^{227}_{90}$Th, $^{223}_{88}$Ra.

79. (b) $\approx 10^{17}$ yr; (c) ≈ 60 yr; (d) 0.4.

CHAPTER 43

1. $^{28}_{13}$Al, β^- emitter, $^{28}_{14}$Si.

3. Possible.

5. 5.701 MeV is released.

7. (a) Can occur; (b) 19.85 MeV.

9. $+4.730$ MeV.

11. (a) $^{7}_{3}$Li; (b) neutron is stripped from the deuteron; (c) $+5.025$ MeV, exothermic.

13. (a) $^{31}_{15}$P(p, γ)$^{32}_{16}$S; (b) $+8.864$ MeV.

15. $\sigma = \pi(R_1 + R_2)^2$.

17. Rate at which incident particles pass through target without scattering.

19. (a) 0.7 μm; (b) 1 mm.

21. 173.2 MeV.

23. 0.116 g.

25. 25.

27. 0.11.

29. 1.3 keV.

33. 6.1×10^{23} MeV/g, 4.9×10^{23} MeV/g,
2.1×10^{24} MeV/g,
5.1×10^{23} MeV/g.

35. Not independent.

37. 3.23×10^9 J, $65\times$.

39. (a) $4.9\times$; (b) 1.5×10^9 K.

41. 400 rad.

43. 167 rad.

45. 1.7×10^2 counts/s.

49. 0.225 μg.

51. (a) 0.03 mrem $\approx 0.006\%$ of allowed dose; (b) 0.3 mrem $\approx 0.06\%$ of allowed dose.

53. (a) 1; (b) $1 \leq m \leq 2.7$.

55. (a) $^{12}_{6}$C; (b) $+5.70$ MeV.

57. 1.004.

59. 51 mrem/yr.

61. 4.6 m.

63. 18.000953 u.

65. 6.31×10^{14} J/kg, $\approx 10^7\times$ the heat of combustion of coal.

67. 1×10^{24} neutrinos/yr.

69. (a) 6.8 bn; (b) 3×10^{-14} m.

71. (a) 3.7×10^3 decays/s;
(b) 5.2×10^{-4} Sv/yr ≈ 0.15 background.

CHAPTER 44

1. 7.29 GeV.

3. 1.8 T.

5. 13 MHz.

7. $\lambda_\alpha = 2.6 \times 10^{-15}$ m $\approx$ size of nucleon, $\lambda_p = 5.2 \times 10^{-15}$ m $\approx$ 2(size of nucleon), α particle is better.

9. 1.4×10^{-18} m.

11. 2.2×10^6 km, 7.5 s.

15. 33.9 MeV.

17. 1.879 GeV.

19. 67.5 MeV.

21. 2.3×10^{-18} m.

23. (b) Uncertainty principle allows energy to not be conserved.

25. 69.3 MeV.

27. 8.6 MeV, 57.4 MeV.

29. 52.3 MeV.

31. 7.5×10^{-21} s.

33. (a) 1.3 keV; (b) 8.9 keV.

35. (a) n = d d u; (b) $\bar{n} = \bar{d}\,\bar{d}\,\bar{u}$;
(c) $\Lambda^0 = $ u d s; (d) $\Sigma^0 = \bar{u}\,\bar{d}\,\bar{s}$.

37. $D^0 = c\bar{u}$.

39. (a)

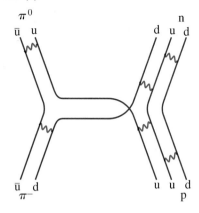

(b)

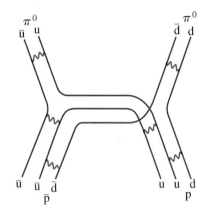

41. 26 GeV, 4.8×10^{-17} m.

43. 5.5 T.

45. (a) Possible, through the strong interaction; (b) possible, through the strong interaction; (c) possible, through the strong interaction; (d) forbidden, charge is not conserved; (e) possible, through the weak interaction; (f) forbidden, charge is not conserved; (g) possible, through the strong interaction; (h) forbidden, strangeness is not conserved; (i) possible, through the weak interaction; (j) possible, through the weak interaction.

49. −135.0 MeV, −140.9 MeV.

51. 64.

53. (b) 10^{27} K.

55. 6.58×10^{-5} m.

57. $\bar{u}\,\bar{u}\,\bar{d} + u\,d\,d \rightarrow \bar{u}\,d + d\,\bar{d}$.

CHAPTER 45

3. 4.8 ly.

5. 0.059″, 17 pc.

7. Less, $\phi_1/\phi_2 = \frac{1}{2}$.

9. 48 W/m².

11. 1.4×10^{-4} kg/m³.

13. 1.8×10^9 kg/m³, 3.3×10^5.

15. −92.2 keV, 7.366 MeV.

17. (a) 9.594 MeV is released; (b) 7.6 MeV; (c) 6×10^{10} K.

19. $d_2/d_1 = 6.5$.

23. 540°.

25. 1.4×10^8 ly.

27. $0.86c$.

29. $0.328c$, 4.6×10^9 ly.

31. 1.1 mm.

33. (a) 10^{-3}; (b) 10^{-10}; (c) 10^{-13}; (d) 10^{-27}.

35. (a) Temperature increases, luminosity is constant, size decreases; (b) temperature is constant, luminosity decreases, size decreases; (c) temperature decreases, luminosity increases, size increases.

37. 8×10^3 rev/s.

39. 7×10^{24} W.

41. (a) 46 eV; (b) 18 eV.

43. $\approx (2 \times 10^{-5})°$.

45. 1.4×10^{16} K, hadron era.

47. Venus is brighter, $\ell_V/\ell_S = 16$.

49. $R \geq GM/c^2$.

51. $R = \dfrac{h^2}{16 m_n^{8/3} G M^{1/3}} \left(\dfrac{18}{\pi^2} \right)^{2/3}$.

Index

Bimetallic-strip thermometer, 448
Binary system, 1146, 1155
Binding energy:
  in atoms, 968
  in molecules, 1031, 1033, 1036
  in nuclei, 1064–66
  in solids, 1044
  total, 1064
Binding energy per nucleon, 1065
Binoculars, 827, 856
Binomial expansion, A-1
Biological damage by radiation,
    1101–2
Biot, Jean Baptist, 719
Biot-Savart law, 719–21
Blackbody, 950
Blackbody radiation, 950–51, 1160
Black dwarf, 1149
Black hole, 149, 1144, 1151, 1155,
    1158, 1166, 1168
Blood flow, 344–45
Blueshift, 1156
Blue sky, 911
Bohr, Niels, 958, 965, 969, 985
Bohr magneton, 1016, 1064
Bohr model of the atom, 965–71,
    977, 1003–5
Bohr radius, 590 pr, 966, 1004
Boiling, 475–76, 490–93 (see also
    Phase, changes of)
Boiling point, 475–76
Boltzmann, Ludwig, 535–36
Boltzmann's constant, 459
Boltzmann distribution, 1020
Boltzmann factor, 1020, 1046
Bomb calorimeter, 490
Bond:
  covalent, 1030–33, 1044
  dipole, 1036–37
  hydrogen, 1036
  ionic, 1031–33, 1044
  metallic, 1044
  molecular, 1030–33
  partially covalent/ionic, 1032–33,
      1044
  strong (defn), 1036
  van der Waals, 1036–37
  weak, 1036–37, 1045
Bond energy, 1031, 1036
Bonding:
  in molecules, 1030–33
  in solids, 1044–45
Bone density, 957
Bose, S. N., 1011 fn

Bose-Einstein statistics, 1046 fn
Bosons, 1011 fn, 1046 fn, 1126–27,
    1132–34
Bottomness and bottom quark,
    1130–31
Boundary conditions, 990, 995
Bound charge, 625
Bound state, 994
Bow wave, 436
Boyle, Robert, 455
Boyle's law, 455–56, 467
Bragg, W. H., 906
Bragg, W. L., 906
Bragg equation, 906
Brahe, Tycho, 143
Brake, hydraulic, 338
Braking distance of car, 29, 167
Brayton cycle, 544 pr
Breakdown voltage, 596
Break-even (fusion), 1099
Breaking point, 309
Breeder reactor, 1094
Bremsstrahlung, 1014–15
Brewster, D., 910
Brewster's angle and law, 910
Bridges, 315–18
Brightness, apparent, 1145
British system of units, 8
Broglie, Louis de (see de Broglie)
Bronchoscope, 827
Brown, Robert, 446
Brown dwarfs, 1166
Brownian movement, 446
Brushes, 698, 740
BSCCO, 650
Btu, 486
Bubble chamber, 1120, 1124
Bulk modulus, 310, 312
Buoyancy, 340–43
Buoyant force, 340–43
Burglar alarm, 954

C

Calculator errors, 5
Caloric theory, 486
Calorie (unit), 486–87
  related to the joule, 486
Calorimeter, 489–90
Calorimetry, 489–90
Camera, 848–50
  autofocusing, 419
  gamma, 1104

Camera flash unit, 621
Cancer, 1101, 1104
Candela (unit), 882
Cantilever, 304–5
Capacitance, 614–21
Capacitance bridge, 630 pr
Capacitive reactance, 775
Capacitor discharge, 672
Capacitor microphone, 677
Capacitors, 613–21
  charging of, 789–90
  in circuits, 669–73, 764–65, 774 ff
  as filter, 776
  reactance of, 775
  in series and parallel, 617–20
  uses of, 776
Capacity (see Capacitance)
Capillaries, 352
Capillarity, 351–52
Car, stopping distance, 29, 167
Carbon cycle, 1097, 1112 pr, 1148
Carbon dating, 1061, 1078–79
Carburetor of car, 348
Carnot, S., 520
Carnot cycle, 520–25
Carnot efficiency, 522–23
Carnot engine, 520–25
Carnot's theorem, 522–23
Carrier frequency, 804
Carrier of forces, 1121 ff
Cassegrainian focus, 856
Cathedrals, 319
Cathode, 605
Cathode rays, 605, 699 (see also
    Electron)
Cathode ray tube (CRT), 605–6, 700,
    805
CAT scan, 1105–6
Causality, 146
Cavendish, Henry, 135, 137, 148
CD player, 1022, 1054
Cell, electric, 635
Cellular telephone, 806
Celsius temperature scale, 448–49
Center of buoyancy, 357 pr
Center of gravity (CG), 223
Center of mass (CM), 221–24
  for human body, 221–22
  translational motion and, 225–27,
      262–65
Centigrade (see Celsius temperature
    scale)
Centipoise (unit), 350
Centrifugal (pseudo) force, 115, 291

Equipotential lines, 600–601
Equipotential surface, 600–601
Equivalence, principle of, 148, 1151–52
Erg (unit), 156
Escape velocity, 192–94
Estimated uncertainty, 4
Estimating, 9–12
Ether, 919–22
Euclidean space, 1153
Evaporation, 474–75, 493
Event horizon, 1155
Everest, Mt., 138
Evolution and entropy, 534
Exact differential, 494 *fn*
Exchange coupling, 723
Excited state, of atom, 957 *ff*
Exclusion principle, 1010–12, 1046, 1132–33, 1150
Exoergic reaction (*defn*), 1086
Exothermic reaction, 1086
Expansion, thermal (*see* Thermal expansion)
Expansion of universe, 1156–59
Expansions, in waves, 391
Expansions, mathematical, A-1
Exponential curves, 1074–75
Exponential decay, 1074–75
Extragalactic (*defn*), 1143
Eye:
    accommodation, 851
    defects of, 851
    far and near points of, 851
    normal (*defn*), 851
    resolution of, 899
    structure and function of, 850–52
Eyeglass lenses, 851–52
Eyepiece, 854

F

Fahrenheit temperature scale, 448–49
Falling bodies, 31–36
Fallout, radioactive, 1095
Farad (unit of capacitance), 614
Faraday, Michael, 147, 554, 622, 734, 736, 787
Faraday's law, 736–38, 747–48, 756, 792
Far point of eye, 851
Farsighted eye, 851–52
Fermat's principle, 835 *pr*

Fermi, E., 11, 958, 978, 1011 *fn*, 1046 *fn*, 1071, 1088, 1090–95, 1128
Fermi-Dirac statistics, 1046–48
Fermi energy, 1046–48
Fermi factor, 1046–47
Fermi gas, 1046
Fermilab, 1114, 1117
Fermi level, 1046–48
Fermions, 1011 *fn*, 1046, 1132–33
Fermi speed, 1047
Fermi temperature, 1060 *pr*
Ferris wheel, 124 *pr*
Ferromagnetism and ferromagnetic materials, 687, 722–23
Feynman, R., 1121
Feynman diagram, 1121–22
Fiber optics, 826–27
Fick's law of diffusion, 480
Field, 147 (*see also* Electric field, Gravitational field, Magnetic field)
    conservative and nonconservative, 748–49
    in elementary particles, 1121
Film badge, 1103
Filter circuit, 776, 785 *pr*
Fine structure, 977, 1003, 1006, 1018
Fine structure constant, 1018
First law of motion, 78 *ff*
First law of thermodynamics, 493–98
Fission, nuclear, 1090–95
Fission bomb, 1095
Fission fragments, 1090–95
Fitzgerald, G. F., 922
Flashlight battery, 636
Flashlight bulb, 639
Flavor (of elementary particles), 1132–33
Floating, 342
Flow of fluids, 343–45
    laminar, 343–45
    streamline, 343–45
    in tubes, 351
    turbulent, 343
Flow rate, 343–45
Fluids, 332–61 (*see also* Gases, Liquids)
Fluorescence, 1019
Fluorescent lightbulbs, 1019
Flux:
    electric, 575–78, 789
    magnetic, 736, 791
Flying buttress, 319
FM radio, 804

Focal length:
    of lens, 838, 840–41
    of spherical mirror, 817, 821
Focal plane, 838
Focal point, 817, 821, 837–38
Focus, 817
Focusing, of camera, 849
Foot-candle (*defn*), 882 *fn*
Foot-pound (unit), 156, 167
Forbidden transition, 1007, 1020 *fn*, 1041 *fn*
Force, in general, 77–85, 147–48, 177–78 (*see also* Electric force, Gravitational force, Magnetic force)
    buoyant, 340–43
    centrifugal, 115
    centripetal, 114
    color, 1133–34
    conservative and nonconservative, 177–78
    contact, 86, 148
    Coriolis, 292–94
    damping, 374–77 (*see also* Drag force)
    dissipative, 189–92
    drag, 122–23
    elastic, 162
    electromagnetic, 1121–22, 1126–27, 1134–36
    electroweak, 148
    in equilibrium, 300–303
    exerted by inanimate object, 83–84
    fictitious, 291
    of friction, 78, 106–14, 118–20, 178, 190–92
    of gravity, 85–86, 133 *ff*, 1123, 1150–55
    impulsive, 212
    inertial, 291–92
    long and short range, 1066
    measurement of, 78
    in muscles and joints, 306
    net, 80, 88
    in Newton's laws, 78–85, 207–9
    nonconservative, 178
    normal, 85–87, 107
    nuclear, 1064–66, 1072, 1121–36
    pseudoforce, 291
    relation of momentum to, 206–8
    resistive, 122–23
    restoring, 162, 363
    strong, 1064–66, 1121–36

types of, in nature, 147–48, 545 *fn*, 1123, 1136
units of, 81
van der Waals, 1036–37
velocity-dependent, 122–23
viscous, 344, 350–51
weak, 1066, 1072, 1121–36
work done by, 156 *ff*
Forced convection, 505
Force diagram (*see* Free-body diagram)
Forced vibrations, 378–79
Force pump, 353
Fossil-fuel power plants, 741
Foucault, J., 868
Four-dimensional space-time, 932
Fourier integral, 401
Fourier's theorem, 401
Fovea, 850, 899
Fractal behavior, 446
Fracture, 312–15
Frames of reference, 17, 79, 291
Franklin, Benjamin, 546, 586
Fraunhofer diffraction, 888 *fn*
Free-body diagram, 88–96, 303
Free charge, 625
Free electrons, 548
Free-electron theory of metals, 1045–48
Free expansion, 498, 531, 537
Free fall, 31–33
Free particle, and Schrödinger equation, 989–94
Freezing (*see* Phase, changes of)
Frequency, 65, 242
angular, 366
of audible sound, 422
beat, 429–32
carrier, 804
of circular motion, 65
cyclotron, 694
fundamental, 406, 424–26
infrasonic, 419
of light, 799, 869
natural, 366, 378, 405
resonant, 378–79, 405, 780
of rotation, 242–43
ultrasonic, 418, 437
of vibration, 363, 366
of wave, 390
Frequency modulation (FM), 804
Fresnel, A., 887
Fresnel diffraction, 888 *fn*
Friction, 78, 106–14, 118–20, 178, 184

coefficients of, 107, 112
kinetic, 106 *ff*
in rolling, 106, 263–64, 267
static, 107 *ff*
Fringes, interference, 871
Frisch, O., 1090
f-stop, 848
Full-wave rectifier, 1053
Fundamental frequency, 406, 424–26
Fuse, 644–45
Fusion, heat of, 490–93
Fusion, nuclear, 1095–1100
in stars, 1148–49, 1163
Fusion reactor, 1097

G

g-factor, 1017
Galaxies, 1141–45, 1155 *ff*
Galaxy clusters, 1143
Galilean-Newton relativity, 917–19, 932–36
Galilean telescope, 856
Galilean transformation, 932–36
Galileo, 2–3, 16, 31–32, 56–57, 78–79, 146, 339, 448, 811, 854–56, 854 *fn*, 917–18, 932, 1141, 1150
Galvani, Luigi, 635
Galvanometer, 674, 697, 755 *pr*
Gamma camera, 1104
Gamma decay, 1067, 1072–73
Gamma rays, 957, 1067, 1072–73
Gamow, G., 1160
Gas constant, 457
Gases, 332, 445–46, 454–81
adiabatic expansion of, 502–3
change of phase, 473–76, 490–93
heat capacities of, 498–99
ideal, 456–60, 466–72
real, 473–74
work done by, 496–98
Gas laws, 454–59
Gauge bosons, 1026–27, 1132, 1134
Gauge pressure, 337
Gauges, pressure, 338–40
Gauge theory, 1134
Gauss, C. F., 1154
Gauss (unit), 691
Gauss's law, 575–86, 713
for magnetism, 791
Gay-Lussac, Joseph, 456
Gay-Lussac's law, 456

Geiger counter, 612 *pr*, 1080
Gell-Mann, M., 1130
General motion, 262–68, 283–84
General theory of relativity, 148, 149, 169, 922, 1140–41, 1151–55
Generator, electric, 740–42
Genetic damage, 1101
Geodesic, 1153–54
Geological dating, 1079
Geometric optics, 811 *ff*
Geophone, 759–60
Geosynchronous satellite, 140, 145
Germer, L. H., 960
Glaser, D., 1120
Glashow, S., 1134
Glasses, eye, 851–52
Glueballs, 1133 *fn*
Gluino, 1136
Gluons, 1123, 1133
Gradient:
concentration, 480
electric potential, 602
pressure, 351
velocity, 350
Gradient operator, 603 *fn*
Gram (unit), 7
Grand unified era, 1162
Grand unified theories, 148, 1134–36
Grating, diffraction (*see* Diffraction grating)
Gravitation, universal law of, 133 *ff*
Gravitational collapse, 1155
Gravitational constant (*G*), 135
Gravitational field, 146–47, 562, 1151–55
Gravitational force, 85–86, 133 *ff*
due to spherical mass distribution, A-6–A-8
Gravitational mass, 148, 1152
Gravitational potential energy, 178–80, 192–94
Gravitino, 1164 *fn*
Graviton, 1122–23
Gravity, 32–33, 85, 133 *ff*, 1123, 1150–55
acceleration of, 32–33, 85, 137–39
center of, 223
free fall under, 31–36
specific, 333
Gravity anomalies, 138
Gray (unit), 1102
Greek alphabet, inside front cover
Grimaldi, F., 867, 872
Grounding, electrical, 651–52

Induction, electromagnetic, 734 *ff*
Induction, Faraday's law, 736–38, 792
Inductive reactance, 774
Inductor, 758–59
  in circuits, 758–63, 772–83
  energy stored in, 760–61
  reactance of, 774
Inelastic collision, 214, 217–18
Inelastic scattering, 1089
Inertia, 79
  law of, 78–79
  moment of, 249–56, 374
  rotational (*defn*), 249–50
Inertial confinement, 1098–99
Inertial forces, 291–92
Inertial mass, 148, 1152
Inertial reference frame, 79, 291, 917 *ff*
  equivalence of all, 918–19, 922
  transformations, 932–36
Inflationary universe, 1164
Infrared radiation (IR), 506, 799, 825
Infrasonic waves, 419
Initial conditions, 365
Instantaneous axis, 246
Instruments, wind and stringed, 424–29
Insulators:
  electrical, 547–48, 640, 1049–50
  thermal, 504, 1049–50
Integrals and integration, 36–37, 161 *ff*, A-4–A-5, B-5
Integrated circuits, 1054–55
Intensity:
  of coherent and incoherent sources, 877
  in interference and diffraction patterns, 874–77, 890–93
  of light, 877, 882
  of sound, 420–23
  of waves, 395–96, 420–23
Intensity level (*see* Sound level)
Interference, 404–5
  constructive, 404, 430, 871, 1031
  destructive, 404, 430, 871, 1031
  as distinguished from diffraction, 895
  of light waves, 870–81, 894–96, 900
  of sound waves, 429–32
  by thin films, 877–81
  of water waves, 404–5
  of waves on a string, 404
Interference fringes, 871

Interference pattern:
  double slit, 870–73, 978–79
  double slit, including diffraction, 893–95
  multiple slit, 900–905
Interferometers, 881
Internal combustion engine, 518–19
Internal conversion, 1073
Internal energy, 189–90, 487–88
Internal reflection, total, 415 *pr*, 826–27
Internal resistance, 659
Intrinsic semiconductor (*defn*), 1049
Inverted population, 1020
Ion, 547
Ionic bonds, 1031–33, 1044
Ionic cohesive energy, 1044
Ionization energy, 968, 970
Ionizing radiation (*defn*), 1100
IR radiation, 506, 799, 825
Irreversible process, 520–25
Isobaric process (*defn*), 496–97
Isochoric process (*defn*), 496–97
Isolated system, 209, 493
Isomer, 1073
Isotherm, 495
Isothermal process (*defn*), 495
Isotopes, 702, 1062
  table of, A-9–A-13
Isotropic material, 867

J

J (total angular momentum), 1017–18
J/ψ particle, 1131
Jeweler's loupe, 854
Joint, 315
Joule, James Prescott, 486
Joule (unit), 156, 167, 248 *fn*
  relation to calorie, 486
Jump start, 668–69
Junction diode, 1054–55
Junction transistor, 1054–55
Jupiter, moons of, 152, 854

K

Kant, Immanuel, 1143
Kaon, 1127
K-capture, 1072
K lines, 1013

Kelvin (unit), 455
Kelvin-Planck statement of second law of thermodynamics, 520, 523, 533
Kelvin temperature scale, 455, 460, 538
Kepler, Johannes, 143, 854 *fn*, 1150
Keplerian telescope, 854
Kepler's laws, 143–46, 288–89
Keyboard, computer, 616
Kilocalorie (unit), 486–87
Kilogram (unit), 7, 79
Kilowatt-hour (unit), 643
Kinematics, 16–67, 240–44 (*see also* Motion)
  for rotational motion, 240–44
  for translational motion, 16–67
  for uniform circular motion, 63–65
  vector kinematics, 53–55
Kinetic energy, 164–69, 182 *ff*, 260–65
  in collisions, 214–17, 219
  molecular, relation to temperature, 469
  relativistic, 938–42
  rotational, 260–65
  total, 262–65
  translational, 165–69
Kinetic friction, 106 *ff*
Kinetic theory, 445, 466 *ff*
Kirchhoff, G. R., 665
Kirchhoff's rules, 664–69

L

Ladder, 306–7
Laminar flow, 343–45
Land, Edwin, 907
Large Magellanic Cloud, 1150
Laser light, 873
Lasers, 873, 887, 1019–22, 1100
Latent heats, 490–93
Lateral magnification, 819, 841
Lattice structure, 446–47, 1044, 1051–52
Laue, Max von, 905
Lawrence, E. O., 1116
Laws, 3–4 (*see also* specific name of law)
Lawson, J. D., 1099
Lawson criterion, 1099–1100
LC circuit, 764–65
LCD, 909

Polarization of light, 907–10
  plane, 907–10
  by reflection, 910, 911
  of skylight, 911
Polarized light (*see* Polarization of
    light)
Polarizer, 908
Polarizing angle, 910
Polar molecules, 547, 565, 624–25,
    1032–33
Polaroid, 907–9
Poles, magnetic, 686–87
  of Earth, 688
Pole vault, 176, 185–86
Pollution, 525
  thermal, 525
Position vector, 53, 55, 240, 243–44
Positive electric charge, 546
Positive lens, 841
Positron, 1072, 1106, 1121
Positron emission tomography, 1106–7
Post-and-beam construction, 319
Potential difference, electric, 591–94
    (*see also* Electric potential)
Potential energy, 178–82, 192–93,
    197–98
  diagrams, 197–98
  elastic, 181, 369
  electric, 591–94, 603–5
  gravitational, 178–80, 192–94
  for molecular bonds, 1033–36, 1041
  for square well and barriers, 994–99
Potential well, 994–96, 1009 *fn*
Potentiometer, 685 *pr*
Pound (unit), 81
Powell, C.F., 1122
Power, 195–97, 642–44, 778 (*see also*
    Electric power)
Power factor (ac circuit), 778
Power generation, 741
Power of a lens, 838
Power plants, 741, 1094
Power reactor, 1093
Powers of ten, 5
Power transmission, 746–47, 782
Poynting, J. H., 801 *fn*
Poynting vector, 800–802
Precession, 290
Presbyopia, 851
Pressure, 333–40
  absolute, 337
  atmospheric, 336–40
  in fluids, 333–36
  in a gas, 337, 454–56, 468

gauge, 337
  hydraulic, 338
  measurement of, 338–40
  partial, 476
  radiation, 802–3
  units for and conversions, 333, 337
  vapor, 475–76
Pressure amplitude, 420
Pressure cooker, 476
Pressure gradient, 351
Pressure head, 335
Pressure transducers, 677
Pressure waves, 417 *ff*
Prestressed concrete, 313
Principal axis, 817
Principle of complementarity, 958–59
Principle of correspondence, 943,
    971, 976 *pr*
Principle of equivalence, 148,
    1151–52
Principle of superposition, 400–402
Principle quantum number, 1005–6
Principles (*see* names of)
Prism, 825
Prism binoculars, 856
Probability, in quantum mechanics,
    980, 984–85
Probability and entropy, 536–37
Probability density, 980, 989, 992, 995,
    1004, 1008–10, 1031–32
Problem solving techniques, 28, 89,
    96, 159
Process, reversible and irreversible
    (*defn*), 520–25
Projectile, horizontal range of, 60–61
Projectile motion, 55–63
Proper length, 930
Proper time, 928, 1139 *pr*
Proportional limit, 309
Proton, 1062–63, 1127
  decay (?) of, 1135
  electric charge on, 547
Proton-proton collision, 220
Proton-proton cycle, 999–1100, 1148
Protostar, 1148
Proxima Centauri, 1141
Pseudoforce, 291
Pseudovector, 246 *fn*
psi (unit), 333
PT diagram, 474
Pulley, 93
Pulsar, 1150
Pulsating universe, 1168
Pulsed laser, 1021–22

Pulse-echo technique, 437
Pumps, 353
  heart as, 353
  heat, 527–28
Pupil, 850, 899
PV diagram, 473
P waves, 394
Pythagorean theorem, A-2

Q

Q-value, 380, 771 *pr*, 785 *pr*, 1068–69,
    1086
QCD, 1123, 1133–35
QED, 1121
Quadratic formula, 35, A-1
Quadrupole, 573 *pr*
Quality factor, 1102
Quality of a sound, 429
Quality value (*Q*-value) of a
    resonant system, 380, 771 *pr*, 785
    pr, 1068–69, 1086
Quantities, base and derived, 8
Quantization:
  of angular momentum, 965
  of electric charge, 550
  of energy, 951, 965–71, 991–92
Quantum chromodynamics, 1123,
    1133–35
Quantum condition, Bohr's, 965, 972
Quantum electrodynamics, 1121
Quantum hypothesis, Planck's,
    949–51
Quantum mechanics, 972, 977–1060
Quantum numbers, 965 *ff*, 992,
    1004–7
  in H atom, 1004–7
  for complex atoms, 1010–11
  for molecules, 1038–43
Quantum theory, 916, 949–72
  of atoms, 965–71
  of blackbody radiation, 951
  of light, 949–58
Quarks, 550 *fn*, 1121, 1123, 1130–32,
    1134, 1162
Quasars, 1144, 1157–58
Quasistatic process (*defn*), 495

R

Rad (unit), 1102
Radar, 435, 876–77

Radial acceleration, 64–65, 69, 114 *ff*, 240
Radial probability density, 1008–10
Radian measure for angles, 240
Radiant flux, 882
Radiation:
  blackbody, 950–51, 1160
  Čerenkov, 886 *pr*
  cosmic, 1102, 1114
  detection of, 1080, 1102
  from hot objects, 506–8, 950–51
  infrared (IR), 506, 799, 825
  ionizing (*defn*), 1100
  measurement of, 1101–3
  microwave, 798
  nuclear, 1067–80, 1100–1103
  synchrotron, 1117
  thermal, 506–8
  ultraviolet (UV), 799, 825
  X (*see* X-rays)
Radiation damage, 1100–1101
Radiation-dominated universe, 1161, 1163
Radiation dose, 1101–3
Radiation era, 1162–63
Radiation field, 793
Radiation film badge, 1103
Radiation pressure, 802–3
Radiation sickness, 1103–4
Radiation therapy, 1104
Radio, 632, 745, 780, 803–6
Radioactive dating, 1078–79
Radioactive decay (*see* Radioactivity)
Radioactive decay constant, 1073–76
Radioactive decay law, 1074
Radioactive decay series, 1077–78
Radioactive fallout, 1095
Radioactive tracers, 1104
Radioactive waste, 1093–94
Radioactivity, 998, 1067–80
  alpha, 1067–70, 1073
  artificial, 1067
  beta, 1067, 1070–73
  dosage of, 1101–3
  gamma, 1037, 1072–73
  natural, 1067, 1102
  probabilistic nature of, 1073
  rate of decay, 1073–78
Radionuclide (*defn*), 1101
Radio waves, 787, 798
Radius of gyration, 273 *pr*
Radius of nuclei, 1062
Radon, 1102

Rainbow, 825
Random access memory (RAM), 613
Range of projectile, 60–61
Rapid estimating, 9–12
Rarefaction, in wave, 391
Ray, 403, 811
  paraxial (*defn*), 817
Ray diagramming, 837 *ff*
Rayleigh, Lord, 897, 951
Rayleigh criterion, 897
Rayleigh-Jeans theory, 951
Ray model of light, 811 *ff*
RBE, 1102
RC circuit, 669–73
Reactance, 759, 774–77 (*see also* Impedance)
Reaction energy, 1086
Reactions, nuclear, 1085–88
Reactors, nuclear, 1090–95
Real image, 814, 818, 839
Rearview mirror, 822
Receivers, radio and television, 805
Recombination, 1170 *pr*
Rectifiers, 1053–54
Red giant star, 1144, 1146, 1148
Redshift, 435, 943, 1145, 1156 *ff*
Reduced mass, 1039
Reference frames, 17, 917 *ff*
  accelerating, 79, 148, 291
  inertial, 79, 291, 917 *ff*
  noninertial, 79, 291
  rotating, 291
  transformations between, 932–36
Reflecting telescope, 856
Reflection:
  angle of, 403, 813
  diffuse, 813
  law of, 403, 813
  of light, 813–22
  phase changes during, 877–81
  polarization by, 910
  specular, 813
  from thin films, 877–81
  total internal, 415 *pr*, 826–27
  of water waves, 402–3
  of waves on a string, 402
Reflection coefficient, 997
Reflection grating, 900
Refracting telescope, 854
Refraction, 408–9, 822–24
  angle of, 408–9, 822
  index of, 811–12
  law of, 409, 823–24, 867–69
  of light, 822–24, 867–69

  and Snell's law, 822–24, 868
  at spherical surface, 828–30
  by thin lenses, 837–40
  of water waves, 408–9
Refrigerator, 525–26
Reinforced concrete, 313
Relative biological effectiveness (RBE), 1102
Relative humidity, 476
Relative motion, 66–68, 916–41
Relative permeability, 725
Relative velocity, 66–68
Relativistic mass, 936–38
Relativistic momentum, 936–38, 941
Relativity, Galilean-Newtonian, 917–19, 932–36
Relativity, general theory of, 148, 149, 169, 922, 1140–41, 1151–55
Relativity, special theory of, 916–44
  and appearance of object, 931
  constancy of speed of light, 923
  four-dimensional space-time, 932
  and length, 930–31
  and Lorentz transformation, 932–36
  and mass, 936–38
  mass-energy relation in, 938–42
  postulates of, 922–23
  simultaneity in, 924–26
  and time, 926–29, 932
Relativity principle, 917–19, 923 *ff*
Relay, 727
Rem (unit), 1102–3
Research reactor, 1093
Resistance and resistors, 638–39
  in ac circuit, 773, 776–79
  with capacitor, 669–73, 776–79
  color code, 639
  electric currents and, 634–52
  with inductor, 762–63, 766–67, 776–79
  internal, 659
  in series and parallel, 660–64
  shunt, 674
Resistance thermometer, 641, 676
Resistive force, 122–23
Resistivity, 640–42
  temperature coefficient of, 641–42
Resolution:
  of diffraction grating, 903–5
  of electron microscope, 961
  of eye, 899
  of lens, 896–99
  of light microscope, 898–99

Total mechanical energy, 183
Total reaction cross section, 1089
Townsend, J. S., 700
Tracers, 1104
Transducers, 676–77
Transformations:
  Galilean, 932–36
  Lorentz, 932–36
Transformer, 744–47, 758
Transformer equation, 745
Transient ischemic attack (TIA), 349
Transistors, 1054–55
Transition elements, 1013
Transition temperature, 650
Translational motion, 16–238, 262–65
Transmission coefficient, 997
Transmission electron microscope, 961
Transmission grating, 900 *ff*
Transmission lines, 746–47, 782
Transmission of electricity, 746–47, 782
Transmutation of elements, 1068, 1070, 1085–88
Transverse waves, 391 *ff*, 794, 907
Trap, sink, 349
Triangulation, 11, 1144 *fn*, 1153
Trigonometric functions and identities, 49, A-2–A-3, inside back cover
Triple point, 460, 474
Tritium, 1084 *pr*, 1097–99
Trusses, 315–18
Tube, vibrating column of air in, 425–28
Tubes, flow in, 351
Tunnel diode, 998–99
Tunneling through a barrier, 996–99, 1069–70, 1155 *fn*, 1168
Turbine, 741
Turbulent flow, 343
Turning points, 197
Turn signal, automobile, 673
Tweeter, 776
Twin paradox, 926–29
Tycho Brahe, 143

U

UHF, 804–5
Ultimate speed, 938
Ultimate strength, 309, 312–13
Ultrasonic waves, 418, 437

Ultrasound, 437
Ultrasound imaging, 437
Ultraviolet (UV) light, 799, 825
Unavailability of energy, 534–35
Uncertainty (in measurements), 4, 981
Uncertainty principle, 981–84, 996, 1031
Underdamping, 376
Underexposure, 848
Unification scale, 1134
Unified atomic mass unit, 7, 446, 1063
Uniform circular motion, 63–65, 114 *ff*, 371
  dynamics of, 114–17
  kinematics of, 63–65
Uniformly accelerated motion, 26 *ff*, 54–55
Uniformly accelerated rotational motion, 243–44
Unit conversion, 8–9, inside front cover
Units of measurement, 6–8
Unit vectors, 52–53
Universal gas constant, 457
Universal law of gravitation, 133–37 and *ff*
Universe (*see also* Cosmology):
  age of, 1158–59
  Big Bang theory of, 1136, 1159–64
  curvature of, 149, 1151–55, 1165
  expanding, 1156–59
  future of, 1165–69
  inflation scenario of, 1164
  matter-dominater, 1161, 1164
  open or closed, 1151, 1165–69
  pulsating, 1168
  radiation-dominated, 1161, 1163
  standard model of, 1161–64
  steady state model of, 1159
Unpolarized light (*defn*), 907
Unstable equilibrium, 198, 308
Uranium, 1077, 1079, 1090–95
Uranus, 146
Useful magnification, 899, 961
UV light, 799, 825

V

Vacuum pump, 353
Vacuum tube, 803
Valence, 1013
Valence band, 1049–50

Van de Graaff generator and accelerator, 612 *pr*, 1115
van der Waals, J. D., 477
van der Waals bonds and forces, 1036–37
van der Waals equation of state, 477–78
Vapor, 473 (*see also* Gases)
Vaporization, latent heat of, 491
Vapor pressure, 474–76
Variable-mass systems, 227–29
Vector displacement, 46–47, 53
Vector model (atoms) 1028 *pr*
Vector product, 279–80
Vectors, 17, 46–53, 159–61, 279–80, A-3
  addition of, 46–52
  components of, 48–52
  cross product, 279–80
  multiplication of, 48, 159–61, 279–80
  multiplication by a scalar, 48
  position, 53, 55, 240, 243–44
  resolution of, 48–52
  resultant, 47, 50
  scalar (dot) product, 159–61
  subtraction of, 48
  sum, 46–52
  unit, 52–53
  vector (cross) product, 279–80
Velocity, 18–22 and *ff*, 36–37, 45, 54–55
  addition of, 66–68
  angular, 241–44
  average, 18–19, 21, 54
  drift, 647–49, 701
  of EM waves, 794–97
  escape, 192–94
  gradient, 350
  instantaneous, 20–22, 26–27, 54–55
  of light, 6, 797, 811–12, 867–68, 919–23, 938
  molecular, and relation to temperature, 466–72
  phase, 397
  relative, 66–68
  rms, 469
  of sound, 418
  supersonic, 418, 435–36
  terminal, 32 *fn*, 122–23
  of waves, 390–94, 397, 418
Velocity selector, 695
Venturi meter, 348
Venturi tube, 348

done by a gas, 496–98
in first law of thermodynamics,
  494 *ff*
from heat engines, 517 *ff*
relation to energy, 164–69, 177–83,
  190–91, 193, 195, 260–61
units of, 156
Work-energy principle, 165–68, 182,
  190–91, 261
Work function, 953, 1048
Working substance (*defn*), 519
Wright, Thomas, 1141

X

X-ray crystallography, 906
X-ray diffraction, 905–6
X-rays, 799, 905–6, 1013–15, 1073,
  1105–9

and atomic number, 1013–15
in electromagnetic spectrum, 799
spectra, 1013–15

Y

Young, Thomas, 870
Young's double-slit experiment,
  870–73
Young's modulus, 309–10
Yukawa, H., 1121–22

Z

Z^0 particle, 1123, 1126–28, 1133
Zeeman effect, 708 *pr*, 1005, 1016,
  1018, 1107
Zener diode, 1053

Zero, absolute, 455, 538
Zero-point energy, 992, 1041
Zeroth law of thermodynamics,
  449–50
Zoom lens, 850
Zweig, G., 1130

Photo Credits

CO-1 NOAA/Phil Degginger/Color-Pic, Inc. 1-1a Philip H. Coblentz/Tony Stone Images 1-1b Richard Berenholtz/The Stock Market
1-1c Antranig M. Ouzoonian, P.E./Weidlinger Associates, Inc. 1-2 Mary Teresa Giancoli 1-3a Oliver Meckes/E.O.S./MPI-Tubingen/Photo Researchers, Inc. 1-3b Douglas C. Giancoli 1-4 International Bureau of Weights and Measures, Sevres, France
1-5 Douglas C. Giancoli 1-6 Doug Martin/Photo Researchers, Inc. 1-9 David Parker/Science Photo Library/Photo Researchers, Inc.
1-10 The Image Works CO-2 Joe Brake Photography/Crest Communications, Inc. 2-22 Justus Sustermans, painting of Galileo
Galilei/The Granger Collection 2-23 Photograph by Dr. Harold E. Edgerton, © The Harold E. Edgerton 1992 Trust. Courtesy Palm
Press, Inc. CO-3 Michel Hans/Vandystadt Allsport Photography (USA), Inc 3-19 Berenice Abbott/Commerce Graphics Ltd., Inc.
3-21 Richard Megna/Fundamental Photographs 3-30a Don Farrall/PhotoDisc, Inc. 3-30b Robert Frerck/Tony Stone Images
3-30c Richard Megna/Fundamental Photographs CO-4 Mark Wagner/Tony Stone Images 4-1 AP/Wide World Photos 4-3 Central
Scientific Company 4-5 Sir Godfrey Kneller, Sir Isaac Newton, 1702. Oil on canvass. The Granger Collection 4-6 Gerard
Vandystadt/Agence Vandystadt/Photo Researchers, Inc. 4-8 David Jones/Photo Researchers, Inc. 4-11 Tsado/NASA/Tom Stack &
Associates 4-31 Lars Ternblad/The Image Bank 4-34 Kathleen Schiaparelli 4-36 Jeff Greenberg/Photo Researchers, Inc. CO-5a Jess
Stock/Tony Stone Images CO-5b Werner H. Muller/Peter Arnold, Inc. 5-12 Jay Brousseau/The Image Bank 5-17 Guido Alberto
Rossi/The Image Bank 5-33 C. Grzimek/Okapia/Photo Researchers, Inc. 5-34 Photofest 5-38 Cedar Point Photo by Dan Feicht
CO-6 Earth Imaging/Tony Stone Images 6-9 NASA/Johnson Space Center 6-13 I. M. House/Tony Stone Images 6-14L Jon
Feingersh/The Stock Market 6-14M © Johan Elbers 1995 6-14R Peter Grumann/The Image Bank CO-7 Eric Miller/Reuters/Corbis
7-20a Stanford Linear Accelerator Center/Science Photo Library/Photo Researchers, Inc. 7-20b Account Phototake/Phototake NYC
7-24 Official U.S. Navy Photo CO-8 and 8-10 Photograph by Dr. Harold E. Edgerton, © The Harold E. Edgerton 1992 Trust. Courtesy
Palm Press 8-11 David Madison/David Madison Photography 8-16 AP/Wide World Photos 8-23 M. C. Escher's "Waterfall" © 1996
Cordon Art-Baarn-Holland. All rights reserved. CO-9 Richard Megna/Fundamental Photographs 9-9 Photograph by Dr. Harold E.
Edgerton, © The Harold E. Edgerton 1992 Trust. Courtesy Palm Press 9-16 D.J. Johnson 9-20 Courtesy Brookhaven National
Laboratory 9-22 Berenice Abbott/Photo Researchers, Inc. CO-10 Ch. Russeil/Kipa/Sygma Photo News 10-10b Mary Teresa Giancoli
10-14a Richard Megna/Fundamental Photographs 10-14b Photoquest, Inc. 10-47 Jens Hartmann/AP/Wide World Photos 10-48 Focus
on Sports, Inc. 10-49 Karl Weatherly/PhotoDisc, Inc CO-11 AP/Wide World Photos 11-19c NOAA/Phil Degginger/Color-Pic, Inc.
11-19d NASA/TSADO/Tom Stack & Associates 11-34 Michael Kevin Daly/The Stock Market CO-12 Steve Vidler/Leo de Wys, Inc.
12-2 AP/Wide World Photos 12-7 T. Kitchin/Tom Stack & Associates 12-10 The Stock Market 12-21 Douglas C. Giancoli 12-23 Mary
Teresa Giancoli 12-27 Fabricius & Taylor/Liaison Agency, Inc. 12-36 Henryk T. Kaiser/Leo de Wys, Inc. 12-38 Douglas C. Giancoli
12-39 Galen Rowell/Mountain Light Photography, Inc. 12-41 Douglas C. Giancoli 12-43 Giovanni Paolo Panini (Roman, 1691-1765),
Interior of the Pantheon, Rome c. 1734. Oil on canvas. 1.283 × .991 (50 1/2 × 39); framed: 1.441 × 1.143 (56 3/4 × 45). © 1995 Board
of Trustees, National Gallery of Art, Washington. Samuel H. Kress Collection. Photo by Richard Carafelli. 12-44 Robert Holmes/Corbis
12-45 Italian Government Tourist Board CO-13 Steven Frink/Tony Stone Images 13-12 Corbis 13-20 Department of Mechanical and
Aerospace Engineering, Princeton University 13-31 Rod Planck/Thomas Stack & Associates 13-33 Alan Blank/Bruce Coleman Inc.
13-42 Douglas C. Giancoli 13-45 Galen Rowell/Mountain Light Photography, Inc. 13-51 NASA/Goddard Space Flight Center/Science
Source/Photo Researchers, Inc. CO-14 Richard Megna/Fundamental Photographs 14-4 Mark E. Gibson/Visuals Unlimited
14-14 Fundamental Photographs 14-15 Douglas C. Giancoli 14-22 Taylor Devices, Inc. 14-25 Martin Bough/Fundamental Photographs
14-26a AP/Wide World Photos 14-26b Paul X. Scott/Sygma Photo News CO-15 Richard Megna/Fundamental Photographs
15-1 Douglas C. Giancoli 15-24 Douglas C. Giancoli 15-30 Martin G. Miller/Visuals Unlimited 15-32 Richard Megna/Fundamental
Photographs CO-16 Photographic Archives, Teatro alla Scala, Milan, Italy 16-6 Yoav Levy/Phototake NYC 16-9a David Pollack/The
Stock Market 16-9b Andrea Brizzi/The Stock Market 16-10 Bob Daemmrich/The Image Works 16-13 Bildarchiv Foto Marburg/Art
Resource, N.Y. 16-24a Norman Owen Tomalin/Bruce Coleman Inc. 16-25b Sandia National Laboratories, New Mexico
16-28a P. Saada/Eurelios/Science Photo Library/Photo Researchers, Inc. 16-28b Howard Sochurek/Medical Images Inc.
CO-17 Le Matin de Lausanne/Sygma Photo News 17-3 Bob Daemmrich/Stock Boston 17-5 Leonard Lessin/Peter Arnold, Inc.
17-14 Leonard Lessin/Peter Arnold, Inc. 17-17 Brian Yarvin/Photo Researchers, Inc. CO-18 Tom Till/DRK Photo 18-9 Paul
Silverman/Fundamental Photographs 18-14 Mary Teresa Giancoli CO-19 Tom Bean/DRK Photo 19-25 Science Photo Library/Photo
Researchers, Inc. CO-20a David Woodfall/Tony Stone Images CO-20b AP/Wide World Photos 20-7a Sandia National Laboratories,
New Mexico 20-7b Martin Bond/Science Photo Library/Photo Researchers, Inc 20-7c Lionel Delevingne/Stock Boston 20-13 Leonard
Lessin/Peter Arnold, Inc. CO-21 Fundamental Photographs 21-38 Michael J. Lutch/Boston Museum of Science CO-23 Gene
Moore/Phototake NYC 23-21 Jon Feingersh/The Stock Market CO-24 Paul Silverman/Fundamental Photographs CO-25 Mahaux
Photography/The Image Bank 25-1 J.-L. Charmet/Science Photo Library/Photo Researchers, Inc. 25-10 T.J. Florian/Rainbow
25-13 Richard Megna/Fundamental Photographs 25-16 Barbara Filet/Tony Stone Images 25-26 Takeshi Takahara/Photo Researchers,
Inc. 25-33 Liaison Agency, Inc. CO-26a Steve Weinrebe/Stock Boston CO-26b Sony Electronics, Inc. 26-22 Paul Silverman/
Fundamental Photographs 26-26 Paul Silverman/Fundamental Photographs CO-27 Richard Megna/Fundamental Photographs
27-4 Richard Megna/Fundamental Photographs 27-6 Mary Teresa Giancoli 27-8 Richard Megna/Fundamental Photographs
27-19 Pekka Parviainen/Science Photo Library/Photo Researchers, Inc. CO-28 Manfred Kage/Peter Arnold, Inc. 28-21 Richard
Megna/Fundamental Photographs CO-29 Richard Megna/Fundamental Photographs 29-10 Werner H. Muller/Peter Arnold, Inc.

29-15 Tomas D.W. Friedmann/Photo Researchers, Inc. **29-20** Jon Feingersh/Comstock **29-28b** National Earthquake Information Center, U.S.G.S. **CO-30** Adam Hart-Davis/Science Photo Library/Photo Researchers, Inc. **CO-31v1** Albert J. Copley/Visuals Unlimited **CO-31v2** Mary Teresa Giancoli **CO-32** Richard Megna/Fundamental Photographs **32-1** AIP Emilio Segre' Visual Archives **32-14** The Image Works **CO-33** Douglas C. Giancoli **33-6** Douglas C. Giancoli **33-11a** Mary Teresa Giancoli and Suzanne Saylor **33-11b** Mary Teresa Giancoli **33-21** Mary Teresa Giancoli **33-25** David Parker/Science Photo Library/Photo Researchers, Inc. **33-28b** Michael Giannechini/Photo Researchers, Inc. **33-34b** S. Elleringmann/Bilderberg/Aurora & Quanta Productions **33-40** Douglas C. Giancoli **33-42** Mary Teresa Giancoli **CO-34** Mary Teresa Giancoli **34-1c** Douglas C. Giancoli **34-1d** Douglas C. Giancoli **34-2** Douglas C. Giancoli and Howard Shugat **34-4** Douglas C. Giancoli **34-7a** Douglas C. Giancoli **34-7b** Douglas C. Giancoli **34-18** Mary Teresa Giancoli **34-19a** Mary Teresa Giancoli **34-19b** Mary Teresa Giancoli **34-26** Mary Teresa Giancoli **34-30a** Franca Principe/Istituto e Museo di Storia della Scienza, Florence, Italy **34-32** Yerkes Observatory, University of Chicago **34-33c** Palomar Observatory/California Institute of Technology **34-33d** Joe McNally/Joe McNally Photography **34-35b** Olympus America Inc. **CO-35** Larry Mulvehill/Photo Researchers, Inc. **35-4a** John M. Dunay IV/Fundamental Photographs **35-9a** Bausch & Lomb Incorporated **35-16a** Paul Silverman/Fundamental Photographs **35-16b** Richard Megna/Fundamental Photographs **35-16c** Yoav Levy/Phototake NYC **35-18b** Ken Kay/Fundamental Photographs **35-20b** Bausch & Lomb Incorporated **35-20c** Bausch & Lomb Incorporated **35-22** Kristen Brochmann/Fundamental Photographs **CO-36** Richard Megna/Fundamental Photographs **36-2a** Reprinted with permission from P.M. Rinard, American Journal of Physics, Vol. 44, #1, 1976, p. 70. Copyright 1976 American Association of Physics Teachers. **36-2b** Ken Kay/Fundamental Photographs **36-2c** Ken Kay/Fundamental Photographs **36-11a** Richard Megna/Fundamental Photographs **36-11b** Richard Megna/Fundamental Photographs **36-12a and b** Reproduced by permission from M. Cagnet, M. Francon, and J. Thrier, The Atlas of Optical Phenomena. Berlin: Springer-Verlag, 1962. **36-15** Space Telescope Science Institute **36-16** The Arecibo Observatory is part of the National Astronomy and Ionosphere Center which is operated by Cornell University under a cooperative agreement with the National Science Foundation. **36-21** Wabash Instrument Corp./Fundamental Photographs **36-26** Photo by W. Friedrich/Max von Laue. Burndy Library, Dibner Institute for the History of Science and Technology, Cambridge, Massachusetts. **36-29b** Bausch & Lomb Incorporated **36-36** Diane Schiumo/Fundamental Photographs **36-39a** Douglas C. Giancoli **36-39b** Douglas C. Giancoli **CO-37** Image of Albert Einstein licensed by Einstein Archives, Hebrew University, Jerusalem, represented by Roger Richman Agency, Beverly Hills, California **37-1** AIP Emilio Segre Visual Archives **CO-38** Wabash Instrument Corp./Fundamental Photographs **38-10** Photo by S.A. Goudsmit, AIP Emilio Segre' Visual Archives **38-11** Education Development Center, Inc. **38-20b** Richard Megna/Fundamental Photographs **38-21abc** Wabash Instrument Corp./Fundamental Photographs **CO-39** Institut International de Physique/American Institute of Physics/Emilio Segre Visual Archives **39-01** American Institute of Physics/Emilio Segre Visual Archives **39-02** F.D. Rosetti/American Institute of Physics/Emilio Segre Visual Archives **39-04** Advanced Research Laboratory/Hitachi Metals America, Ltd. **39-18** Driscoll, Youngquist, and Baldeschwieler, Caltech/Science Photo Library/Photo Researchers, Inc. **CO-40** Patricia Peticolas/Fundamental Photographs **40-16** Paul Silverman/Fundamental Photographs **40-22** Yoav Levy/Phototake NYC **40-24b** Paul Silverman/Fundamental Photographs **CO-41** Charles O'Rear/Corbis **CO-42** Reuters Newmedia Inc/Corbis **42-03** Chemical Heritage Foundation **42-07** University of Chicago, Courtesy of AIP Emilio Segre Visual Archives **CO-43** AP/Wide World Photos **43-07** Gary Sheahan "Birth of the Atomic Age" Chicago (Illinois); 1957. Chicago Historical Society. **43-10** Liaison Agency, Inc. **43-11** LeRoy N. Sanchez/Los Alamos National Laboratory **43-12** Corbis **43-16a** Lawrence Livermore National Laboratory/Science Source/Photo Researchers, Inc. **43-16b** Gary Stone/Lawrence Livermore National Laboratory **43-22a** Martin M. Rotker/Martin M. Rotker **43-22b** Simon Fraser/Science Photo Library/Photo Researchers, Inc. **43-26b** Southern Illinois University/Peter Arnold, Inc. **43-28** Mehau Kulyk/Science Photo Library/Photo Researchers, Inc. **CO-44** Fermilab Visual Media Services **44-06** CERN/Science Photo Library/Photo Researchers, Inc. **44-08** Fermilab Visual Media Services **CO-45** Jeff Hester and Paul Scowen, Arizona State University, and NASA **45-01** NASA Headquarters **45-02c** NASA/Johnson Space Center **45-03** U.S. Naval Observatory Photo/NASA Headquarters **45-04** National Optical Astronomy Observatories **45-05a** R.J. Dufour, Rice University **45-05b** U.S. Naval Observatory **45-05c** National Optical Astronomy Observatories **45-10** Space Telescope Science Institute/NASA Headquarters **45-11** National Optical Astronomy Observatories **45-15** European Space Agency/NASA Headquarters **45-22** Courtesy of Lucent Technologies/Bell Laboratories

Table of Contents Photos **p. v** (left) NOAA/Phil Degginger/Color-Pic, Inc. **p. v** (right) Mark Wagner/Tony Stone Images **p. vi** (left) Jess Stock/Tony Stone Images **p. vi** (right) Photograph by Dr. Harold E. Edgerton, © The Harold E. Edgerton 1992 Trust. Courtesy Palm Press **p. vii** (left) Ch. Russeil/Kipa/Sygma Photo News **p.vii** (right) Tibor Bognar/The Stock Market **p. viii** (left) Richard Megna/Fundamental Photographs **p. viii** (right) Douglas C. Giancoli **p. ix** (top) S. Feval/Le Matin de Lausanne/Sygma Photo News **p. ix** (bottom) Richard A. Cooke III/Tony Stone Images (right) David Woodfall/Tony Stone Images and AP/ Wide World Photos **p. x** (left) Fundamental Photographs **p. x** (right) Richard Kaylin/Tony Stone Images **p. xi** (left) Mahaux Photography/The Image Bank **p. xi** (right) Manfred Cage/Peter Arnold, Inc. **p. xii** Werner H. Muller/Peter Arnold, Inc. **p. xiii** (left) Mary Teresa Giancoli. **p. xiii** (right) Larry Mulvehill/Photo Researchers, Inc. **p. xiv** Image of Albert Einstein licensed by Einstein Archives, Hebrew University, Jerusalem, represented by Roger Richman Agency, Beverly Hills, California **p. xiv** (right) Donna McWilliam/AP/Wide World Photos **p. xv** (left) Charles O'Rear/Corbis (right) Fermilab **p. xvi** (left) Jeff Hester and Paul Scowen, Arizona State University, and NASA (right) R. J. Dufour, Rice University.

Periodic Table of the Elements[§]

Transition Elements

Key:

Symbol	Cl 17 — Atomic Number
Atomic Mass[§]	35.4527
	$3p^5$ — Electron Configuration (outer shells only)

Group I	Group II												Group III	Group IV	Group V	Group VI	Group VII	Group VIII
H 1 1.00794 $1s^1$																		He 2 4.002602 $1s^2$
Li 3 6.941 $2s^1$	Be 4 9.012182 $2s^2$												B 5 10.811 $2p^1$	C 6 12.0107 $2p^2$	N 7 14.00674 $2p^3$	O 8 15.9994 $2p^4$	F 9 18.9984032 $2p^5$	Ne 10 20.1797 $2p^6$
Na 11 22.989770 $3s^1$	Mg 12 24.3050 $3s^2$												Al 13 26.981538 $3p^1$	Si 14 28.0855 $3p^2$	P 15 30.973761 $3p^3$	S 16 32.066 $3p^4$	Cl 17 35.4527 $3p^5$	Ar 18 39.948 $3p^6$
K 19 39.0983 $4s^1$	Ca 20 40.078 $4s^2$	Sc 21 44.955910 $3d^14s^2$	Ti 22 47.867 $3d^24s^2$	V 23 50.9415 $3d^34s^2$	Cr 24 51.9961 $3d^54s^1$	Mn 25 54.938049 $3d^54s^2$	Fe 26 55.845 $3d^64s^2$	Co 27 58.933200 $3d^74s^2$	Ni 28 58.6934 $3d^84s^2$	Cu 29 63.546 $3d^{10}4s^1$	Zn 30 65.39 $3d^{10}4s^2$		Ga 31 69.723 $4p^1$	Ge 32 72.61 $4p^2$	As 33 74.92160 $4p^3$	Se 34 78.96 $4p^4$	Br 35 79.904 $4p^5$	Kr 36 83.80 $4p^6$
Rb 37 85.4678 $5s^1$	Sr 38 87.62 $5s^2$	Y 39 88.90585 $4d^15s^2$	Zr 40 91.224 $4d^25s^2$	Nb 41 92.90638 $4d^45s^1$	Mo 42 95.94 $4d^55s^1$	Tc 43 (98) $4d^55s^2$	Ru 44 101.07 $4d^75s^1$	Rh 45 102.90550 $4d^85s^1$	Pd 46 106.42 $4d^{10}5s^0$	Ag 47 107.8682 $4d^{10}5s^1$	Cd 48 112.411 $4d^{10}5s^2$		In 49 114.818 $5p^1$	Sn 50 118.710 $5p^2$	Sb 51 121.760 $5p^3$	Te 52 127.60 $5p^4$	I 53 126.90447 $5p^5$	Xe 54 131.29 $5p^6$
Cs 55 132.90545 $6s^1$	Ba 56 137.327 $6s^2$	57–71[†]	Hf 72 178.49 $5d^26s^2$	Ta 73 180.9479 $5d^36s^2$	W 74 183.84 $5d^46s^2$	Re 75 186.207 $5d^56s^2$	Os 76 190.23 $5d^66s^2$	Ir 77 192.217 $5d^76s^2$	Pt 78 195.078 $5d^96s^1$	Au 79 196.96655 $5d^{10}6s^1$	Hg 80 200.59 $5d^{10}6s^2$		Tl 81 204.3833 $6p^1$	Pb 82 207.2 $6p^2$	Bi 83 208.98038 $6p^3$	Po 84 (209) $6p^4$	At 85 (210) $6p^5$	Rn 86 (222) $6p^6$
Fr 87 (223) $7s^1$	Ra 88 (226) $7s^2$	89–103[‡]	Rf 104 (261) $6d^27s^2$	Db 105 (262) $6d^37s^2$	Sg 106 (266) $6d^47s^2$	Bh 107 (264)	Hs 108 (269)	Mt 109 (268)	110 (271)	111 (272)	112 (277)			114 (289)		116 (289)		118 (293)

†Lanthanide Series

La 57 138.9055 $5d^16s^2$	Ce 58 140.115 $4f^15d^16s^2$	Pr 59 140.90765 $4f^35d^06s^2$	Nd 60 144.24 $4f^45d^06s^2$	Pm 61 (145) $4f^55d^06s^2$	Sm 62 150.36 $4f^65d^06s^2$	Eu 63 151.964 $4f^75d^06s^2$	Gd 64 157.25 $4f^75d^16s^2$	Tb 65 158.92534 $4f^95d^06s^2$	Dy 66 162.50 $4f^{10}5d^06s^2$	Ho 67 164.93032 $4f^{11}5d^06s^2$	Er 68 167.26 $4f^{12}5d^06s^2$	Tm 69 168.93421 $4f^{13}5d^06s^2$	Yb 70 173.04 $4f^{14}5d^06s^2$	Lu 71 174.967 $4f^{14}5d^16s^2$

‡Actinide Series

Ac 89 (227.02775) $6d^17s^2$	Th 90 232.0381 $6d^27s^2$	Pa 91 231.03588 $5f^26d^17s^2$	U 92 238.0289 $5f^36d^17s^2$	Np 93 (237) $5f^46d^17s^2$	Pu 94 (244) $5f^66d^07s^2$	Am 95 (243) $5f^76d^07s^2$	Cm 96 (247) $5f^76d^17s^2$	Bk 97 (247) $5f^96d^07s^2$	Cf 98 (251) $5f^{10}6d^07s^2$	Es 99 (252) $5f^{11}6d^07s^2$	Fm 100 (257) $5f^{12}6d^07s^2$	Md 101 (258) $5f^{13}6d^07s^2$	No 102 (259) $5f^{14}6d^07s^2$	Lr 103 (262) $5f^{14}6d^17s^2$

[§] Atomic mass values averaged over isotopes in percentages they occur on Earth's surface. For many unstable elements, mass of the longest-lived known isotope is given in parentheses. 1999 revisions. (See also Appendix D.)

Trigonometric Table

Angle in Degrees	Angle in Radians	Sine	Cosine	Tangent	Angle in Degrees	Angle in Radians	Sine	Cosine	Tangent
0°	0.000	0.000	1.000	0.000					
1°	0.017	0.017	1.000	0.017	46°	0.803	0.719	0.695	1.036
2°	0.035	0.035	0.999	0.035	47°	0.820	0.731	0.682	1.072
3°	0.052	0.052	0.999	0.052	48°	0.838	0.743	0.669	1.111
4°	0.070	0.070	0.998	0.070	49°	0.855	0.755	0.656	1.150
5°	0.087	0.087	0.996	0.087	50°	0.873	0.766	0.643	1.192
6°	0.105	0.105	0.995	0.105	51°	0.890	0.777	0.629	1.235
7°	0.122	0.122	0.993	0.123	52°	0.908	0.788	0.616	1.280
8°	0.140	0.139	0.990	0.141	53°	0.925	0.799	0.602	1.327
9°	0.157	0.156	0.988	0.158	54°	0.942	0.809	0.588	1.376
10°	0.175	0.174	0.985	0.176	55°	0.960	0.819	0.574	1.428
11°	0.192	0.191	0.982	0.194	56°	0.977	0.829	0.559	1.483
12°	0.209	0.208	0.978	0.213	57°	0.995	0.839	0.545	1.540
13°	0.227	0.225	0.974	0.231	58°	1.012	0.848	0.530	1.600
14°	0.244	0.242	0.970	0.249	59°	1.030	0.857	0.515	1.664
15°	0.262	0.259	0.966	0.268	60°	1.047	0.866	0.500	1.732
16°	0.279	0.276	0.961	0.287	61°	1.065	0.875	0.485	1.804
17°	0.297	0.292	0.956	0.306	62°	1.082	0.883	0.469	1.881
18°	0.314	0.309	0.951	0.325	63°	1.100	0.891	0.454	1.963
19°	0.332	0.326	0.946	0.344	64°	1.117	0.899	0.438	2.050
20°	0.349	0.342	0.940	0.364	65°	1.134	0.906	0.423	2.145
21°	0.367	0.358	0.934	0.384	66°	1.152	0.914	0.407	2.246
22°	0.384	0.375	0.927	0.404	67°	1.169	0.921	0.391	2.356
23°	0.401	0.391	0.921	0.424	68°	1.187	0.927	0.375	2.475
24°	0.419	0.407	0.914	0.445	69°	1.204	0.934	0.358	2.605
25°	0.436	0.423	0.906	0.466	70°	1.222	0.940	0.342	2.747
26°	0.454	0.438	0.899	0.488	71°	1.239	0.946	0.326	2.904
27°	0.471	0.454	0.891	0.510	72°	1.257	0.951	0.309	3.078
28°	0.489	0.469	0.883	0.532	73°	1.274	0.956	0.292	3.271
29°	0.506	0.485	0.875	0.554	74°	1.292	0.961	0.276	3.487
30°	0.524	0.500	0.866	0.577	75°	1.309	0.966	0.259	3.732
31°	0.541	0.515	0.857	0.601	76°	1.326	0.970	0.242	4.011
32°	0.559	0.530	0.848	0.625	77°	1.344	0.974	0.225	4.331
33°	0.576	0.545	0.839	0.649	78°	1.361	0.978	0.208	4.705
34°	0.593	0.559	0.829	0.675	79°	1.379	0.982	0.191	5.145
35°	0.611	0.574	0.819	0.700	80°	1.396	0.985	0.174	5.671
36°	0.628	0.588	0.809	0.727	81°	1.414	0.988	0.156	6.314
37°	0.646	0.602	0.799	0.754	82°	1.431	0.990	0.139	7.115
38°	0.663	0.616	0.788	0.781	83°	1.449	0.993	0.122	8.144
39°	0.681	0.629	0.777	0.810	84°	1.466	0.995	0.105	9.514
40°	0.698	0.643	0.766	0.839	85°	1.484	0.996	0.087	11.43
41°	0.716	0.656	0.755	0.869	86°	1.501	0.998	0.070	14.301
42°	0.733	0.669	0.743	0.900	87°	1.518	0.999	0.052	19.081
43°	0.750	0.682	0.731	0.933	88°	1.536	0.999	0.035	28.636
44°	0.768	0.695	0.719	0.966	89°	1.553	1.000	0.017	57.290
45°	0.785	0.707	0.707	1.000	90°	1.571	1.000	0.000	∞